SCHAUM'S OUTLINE OF

THEORY AND PROBLEMS

OF

SET THEORY
and Related Topics

•

BY

SEYMOUR LIPSCHUTZ, Ph.D.
Professor of Mathematics
Temple University

•

SCHAUM'S OUTLINE SERIES
McGRAW-HILL BOOK COMPANY
New York, St. Louis, San Francisco, Toronto, Sydney

ISBN 07-037986-6

16 17 18 19 20 21 22 23 24 25 26 27 28 29 30 SH SH 8 7 6 5 4

Preface

The theory of sets lies at the foundations of mathematics. Concepts in set theory, such as functions and relations, appear explicitly or implicitly in every branch of mathematics. This text is an informal, non-axiomatic treatment of the theory of sets.

The material is divided into three Parts, since the logical development is thereby not disturbed while the usefulness as a text and reference book on any of several levels is increased. Part I contains an introduction to the elementary operations of sets and a detailed discussion of the concept of a function and of a relation. Part II develops the theory of cardinal and ordinal numbers in the classical approach of Cantor. It also considers partially ordered sets and the axiom of choice and its equivalents, including Zorn's lemma. Part III treats of those topics which are usually associated with elementary set theory. Of course, the particular presentation of certain topics is influenced by the author's preferences. For example, functions are introduced before relations and are not initially defined as sets of ordered pairs.

Each chapter begins with clear statements of pertinent definitions, principles and theorems together with illustrative and other descriptive material. This is followed by graded sets of solved and supplementary problems. The solved problems serve to illustrate and amplify the theory, bring into sharp focus those fine points without which the student continually feels himself on unsafe ground, and provide the repetition of basic principles so vital to effective learning. Numerous proofs of theorems and derivations of basic results are included among the solved problems. The supplementary problems serve as a complete review of the material of each chapter.

Considerably more material has been included here than can be covered in most first courses. This has been done to make the book more flexible, to provide a more useful book of reference, and to stimulate further interest in the topics.

The following texts are suggested references. Those by Halmos and Kamke are especially recommended as auxiliary reading for Part II.

Bourbaki, N., *Theorie des Ensembles,* Hermann, Paris, 1958
Halmos, P. R., *Naive Set Theory,* Van Nostrand, 1960
Hausdorff, F., *Set Theory,* Chelsea, 1957
Kamke, E., *Theory of Sets,* Dover, 1950
Kuratowski, C., *Introduction to Set Theory and Topology,* Addison-Wesley, 1962
Natanson, I. P., *Theory of Functions of a Real Variable,* Chap. 1, 2, 14, Ungar, 1955

I wish to take this opportunity to thank many of my friends and colleagues for invaluable suggestions and critical review of the manuscript. Particular thanks are extended to the staff of the Schaum Publishing Company for their excellent cooperation.

SEYMOUR LIPSCHUTZ

Polytechnic Institute of Brooklyn
January, 1964

CONTENTS

Part I

Elementary Theory of Sets

Part II

Cardinals, Ordinals, and Transfinite Induction

Part III

Related Topics

Part I – Elementary Theory of Sets

Chapter 1

Sets and Subsets

SETS

A fundamental concept in all branches of mathematics is that of a *set*. Intuitively, a set is any well-defined list, collection, or class of objects. The objects in sets, as we shall see from our examples, can be anything: numbers, people, letters, rivers, etc. These objects are called the *elements* or *members* of the set.

Although we shall study sets as abstract entities, we now list ten particular examples of sets.

Example 1.1: The numbers 1, 3, 7, and 10.

Example 1.2: The solutions of the equation $x^2 - 3x - 2 = 0$.

Example 1.3: The vowels of the alphabet: a, e, i, o and u.

Example 1.4: The people living on the earth.

Example 1.5: The students Tom, Dick and Harry.

Example 1.6: The students who are absent from school.

Example 1.7: The countries England, France and Denmark.

Example 1.8: The capital cities of Europe.

Example 1.9: The numbers 2, 4, 6, 8, ...

Example 1.10: The rivers in the United States.

Notice that the sets in the odd numbered examples are *defined,* that is, presented, by actually listing its members; and the sets in the even numbered examples are defined by stating properties, that is, rules, which decide whether or not a particular object is a member of the set.

NOTATION

Sets will usually be denoted by capital letters

$$A,\ B,\ X,\ Y,\ \ldots$$

The elements in our sets will usually be represented by lower case letters

$$a,\ b,\ x,\ y,\ \ldots$$

If we define a particular set by actually listing its members, for example, let A consist of the numbers 1, 3, 7 and 10, then we write

$$A \;=\; \{1, 3, 7, 10\}$$

that is, the elements are separated by commas and enclosed in brackets { }. We call this the *tabular form* of a set. But if we define a particular set by stating properties which its elements must satisfy, for example, let B be the set of all even numbers, then we use a letter, usually x, to represent an arbitrary element and we write

$$B \;=\; \{x \mid x \text{ is even}\}$$

which reads "B is the set of numbers x such that x is even". We call this the *set-builder form* of a set. Notice that the vertical line "|" is read "such that".

In order to illustrate the use of the above notation, we rewrite the sets in Examples 1.1-1.10. We denote the sets by $A_1, A_2, \ldots, A_{10}$ respectively.

Example 2.1: $A_1 = \{1, 3, 7, 10\}$.

Example 2.2: $A_2 = \{x \mid x^2 - 3x - 2 = 0\}$.

Example 2.3: $A_3 = \{\text{a, e, i, o, u}\}$.

Example 2.4: $A_4 = \{x \mid x$ is a person living on the earth$\}$.

Example 2.5: $A_5 = \{$Tom, Dick, Harry$\}$.

Example 2.6: $A_6 = \{x \mid x$ is a student and x is absent from school$\}$.

Example 2.7: $A_7 = \{$England, France, Denmark$\}$.

Example 2.8: $A_8 = \{x \mid x$ is a capital city and x is in Europe$\}$.

Example 2.9: $A_9 = \{2, 4, 6, 8, \ldots\}$.

Example 2.10: $A_{10} = \{x \mid x$ is a river and x is in the United States$\}$.

If an object x is a member of a set A, i.e., A contains x as one of its elements, then we write

$$x \in A$$

which can also be read "x *belongs* to A" or "x is in A". If, on the other hand, an object x is not a member of a set A, i.e. A does not contain x as one of its elements, then we write

$$x \notin A$$

It is common custom in mathematics to put a vertical line "|" or "/" through a symbol to indicate the opposite or negative meaning of the symbol.

Example 3.1: Let $A = \{\text{a, e, i, o, u}\}$. Then $a \in A$, $b \notin A$, $e \in A$, $f \notin A$.

Example 3.2: Let $B = \{x \mid x \text{ is even}\}$. Then $3 \notin B$, $6 \in B$, $11 \notin B$, $14 \in B$.

FINITE AND INFINITE SETS

Sets can be *finite* or *infinite*. Intuitively, a set is finite if it consists of a specific number of different elements, i.e., if in counting the different members of the set the counting process can come to an end. Otherwise a set is infinite. In a later chapter we give a precise definition of infinite and finite sets.

Example 4.1: Let M be the set of the days of the week. Then M is finite.

Example 4.2: Let $N = \{2, 4, 6, 8, \ldots\}$. Then N is infinite.

Example 4.3: Let $P = \{x \mid x$ is a river on the earth$\}$. Although it may be difficult to count the number of rivers in the world, P is still a finite set.

EQUALITY OF SETS

Set A is *equal* to set B if they both have the same members, i.e. if every element which belongs to A also belongs to B and if every element which belongs to B also belongs to A. We denote the equality of sets A and B by

$$A = B$$

Example 5.1: Let $A = \{1, 2, 3, 4\}$ and $B = \{3, 1, 4, 2\}$. Then $A = B$, that is, $\{1, 2, 3, 4\} = \{3, 1, 4, 2\}$, since each of the elements 1, 2, 3 and 4 of A belongs to B and each of the elements 3, 1, 4 and 2 of B belongs to A. Note therefore that a set does not change if its elements are rearranged.

Example 5.2: Let $C = \{5, 6, 5, 7\}$ and $D = \{7, 5, 7, 6\}$. Then $C = D$, that is, $\{5, 6, 5, 7\} = \{7, 5, 7, 6\}$, since each element of C belongs to D and each element of D belongs to C. Note that a set does not change if its elements are repeated. Also, the set $\{5, 6, 7\}$ equals C and D.

Example 5.3: Let $E = \{x \mid x^2 - 3x = -2\}$, $F = \{2, 1\}$ and $G = \{1, 2, 2, 1\}$. Then $E = F = G$.

NULL SET

It is convenient to introduce the concept of the *empty set*, that is, a set which contains no elements. This set is sometimes called the *null set*. We say that such a set is *void* or empty, and we denote it by the symbol $\emptyset$.

Example 6.1: Let A be the set of people in the world who are older than 200 years. According to known statistics A is the null set.

Example 6.2: Let $B = \{x \mid x^2 = 4, x \text{ is odd}\}$. Then B is the empty set.

SUBSETS

If every element in a set A is also a member of a set B, then A is called a *subset* of B. More specifically, A is a subset of B if $x \varepsilon A$ implies $x \varepsilon B$. We denote this relationship by writing

$$A \subset B$$

which can also be read "A is contained in B".

Example 7.1: The set $C = \{1, 3, 5\}$ is a subset of $D = \{5, 4, 3, 2, 1\}$, since each number 1, 3 and 5 belonging to C also belongs to D.

Example 7.2: The set $E = \{2, 4, 6\}$ is a subset of $F = \{6, 2, 4\}$, since each number 2, 4 and 6 belonging to E also belongs to F. Note, in particular, that $E = F$. In a similar manner it can be shown that every set is a subset of itself.

Example 7.3: Let $G = \{x \mid x \text{ is even}\}$, i.e. $G = \{2, 4, 6, \ldots\}$, and let $F = \{x \mid x \text{ is a positive power of } 2\}$, i.e. let $F = \{2, 4, 8, 16, \ldots\}$. Then $F \subset G$, i.e. F is contained in G.

With the above definition of a subset, we are able to restate the definition of the equality of two sets:

Definition 1.1: Two sets A and B are equal, i.e. $A = B$, if and only if $A \subset B$ and $B \subset A$.

If A is a subset of B, then we can also write

$$B \supset A$$

which reads "B is a superset of A" or "B contains A". Furthermore, we write

$$A \not\subset B \quad \text{or} \quad B \not\supset A$$

if A is not a subset of B.

In conclusion, we state:

Remark 1.1: The null set $\emptyset$ is considered to be a subset of every set.

Remark 1.2: If A is not a subset of B, that is, if $A \not\subset B$, then there is at least one element in A that is not a member of B.

PROPER SUBSET

Since every set A is a subset of itself, we call B a *proper subset* of A if, first, B is a subset of A and, secondly, if B is not equal to A. More briefly, B is a proper subset of A if

$$B \subset A \quad \text{and} \quad B \neq A$$

In some books "B is a subset of A" is denoted by

$$B \subseteq A$$

and "B is a proper subset of A" is denoted by

$$B \subset A$$

We will continue to use the previous notation in which we do not distinguish between a subset and a proper subset.

COMPARABILITY

Two sets A and B are said to be *comparable* if

$$A \subset B \quad \text{or} \quad B \subset A$$

that is, if one of the sets is a subset of the other set. Moreover, two sets A and B are said to be *not comparable* if

$$A \not\subset B \quad \text{and} \quad B \not\subset A$$

Note that if A is not comparable to B then there is an element in A which is not in B and, also, there is an element in B which is not in A.

Example 8.1: Let $A = \{a, b\}$ and $B = \{a, b, c\}$. Then A is comparable to B, since A is a subset of B.

Example 8.2: Let $R = \{a, b\}$ and $S = \{b, c, d\}$. Then R and S are not comparable, since $a \in R$ and $a \notin S$ and $c \in S$ and $c \notin R$.

THEOREM AND PROOF

In mathematics, many statements can be proven to be true by the use of previous assumptions and definitions. In fact, the essence of mathematics consists of theorems and their proofs. We now prove our first

Theorem 1.1: If A is a subset of B and B is a subset of C then A is a subset of C, that is,

$$A \subset B \text{ and } B \subset C \quad \text{implies} \quad A \subset C$$

Proof. (Notice that we must show that any element in A is also an element in C.) Let x be an element of A, that is, let $x \in A$. Since A is a subset of B, x also belongs to B, that is, $x \in B$. But, by hypothesis, $B \subset C$; hence every element of B, which includes x, is a member of C. We have shown that $x \in A$ implies $x \in C$. Accordingly, by definition, $A \subset C$.

SETS OF SETS

It sometimes will happen that the objects of a set are sets themselves; for example, the set of all subsets of A. In order to avoid saying "set of sets", it is common practice to say "family of sets" or "class of sets". Under these circumstances, and in order to avoid confusion, we sometimes will let script letters

$$\mathcal{A}, \mathcal{B}, \ldots$$

denote families, or classes, of sets since capital letters already denote their elements.

Example 9.1: In geometry we usually say "a family of lines" or "a family of curves" since lines and curves are themselves sets of points.

Example 9.2: The set $\{\{2, 3\}, \{2\}, \{5, 6\}\}$ is a family of sets. Its members are the sets $\{2, 3\}$, $\{2\}$ and $\{5, 6\}$.

Theoretically, it is possible that a set has some members which are sets themselves and some members which are not sets, although in any application of the theory of sets this case arises infrequently.

Example 9.3: Let $A = \{2, \{1, 3\}, 4, \{2, 5\}\}$. Then A is not a family of sets; here some elements of A are sets and some are not.

UNIVERSAL SET

In any application of the theory of sets, all the sets under investigation will likely be subsets of a fixed set. We call this set the *universal set* or *universe of discourse*. We denote this set by U.

Example 10.1: In plane geometry, the universal set consists of all the points in the plane.

Example 10.2: In human population studies, the universal set consists of all the people in the world.

POWER SET

The family of all the subsets of any set S is called the *power set* of S. We denote the power set of S by

$$2^S$$

Example 11.1: Let $M = \{a, b\}$. Then

$$2^M = \{\{a, b\}, \{a\}, \{b\}, \emptyset\}$$

Example 11.2: Let $T = \{4, 7, 8\}$. Then

$$2^T = \{T, \{4, 7\}, \{4, 8\}, \{7, 8\}, \{4\}, \{7\}, \{8\}, \emptyset\}$$

If a set S is finite, say S has n elements, then the power set of S can be shown to have 2^n elements. This is one reason why the class of subsets of S is called the power set of S and is denoted by 2^S.

DISJOINT SETS

If sets A and B have no elements in common, i.e. if no element of A is in B and no element of B is in A, then we say that A and B are *disjoint*.

Example 12.1: Let $A = \{1, 3, 7, 8\}$ and $B = \{2, 4, 7, 9\}$. Then A and B are not disjoint since 7 is in both sets, i.e. $7 \in A$ and $7 \in B$.

Example 12.2: Let A be the positive numbers and let B be the negative numbers. Then A and B are disjoint since no number is both positive and negative.

Example 12.3: Let $E = \{x, y, z\}$ and $F = \{r, s, t\}$. Then E and F are disjoint.

VENN-EULER DIAGRAMS

A simple and instructive way of illustrating the relationships between sets is in the use of the so-called Venn-Euler diagrams or, simply, Venn diagrams. Here we represent a set by a simple plane area, usually bounded by a circle.

Example 13.1: Suppose $A \subset B$ and, say, $A \neq B$. Then A and B can be described by either diagram:

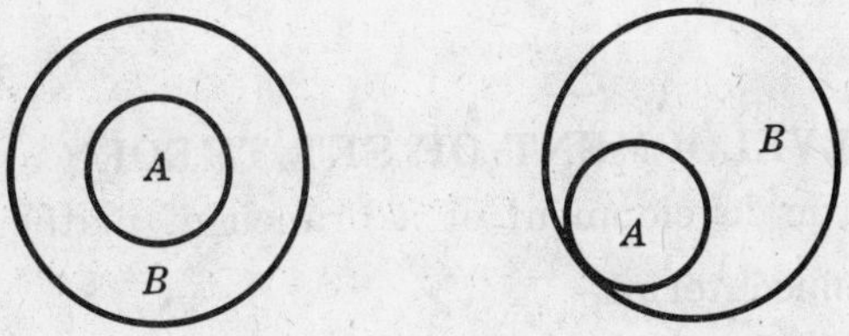

Example 13.2: Suppose A and B are not comparable. Then A and B can be represented by the diagram on the right if they are disjoint, or the diagram on the left if they are not disjoint.

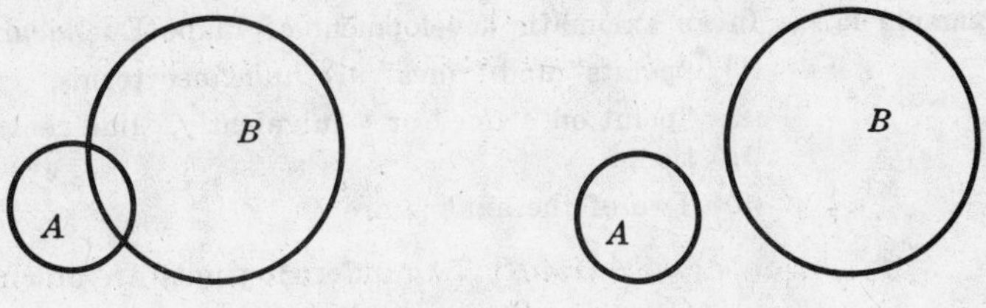

Example 13.3: Let $A = \{a, b, c, d\}$ and $B = \{c, d, e, f\}$. Then we illustrate these sets with a Venn diagram of the form

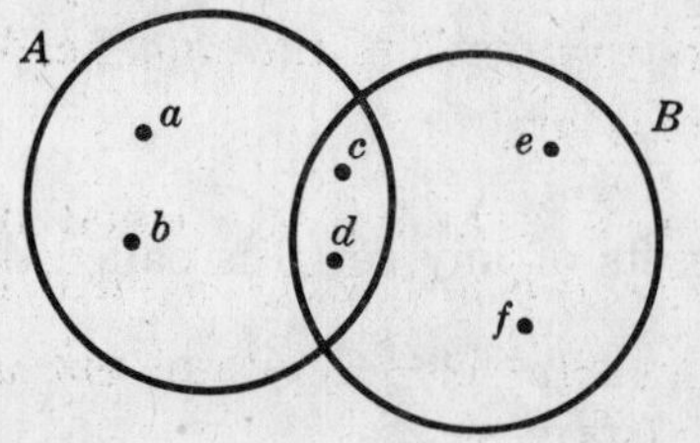

LINE DIAGRAMS

Another useful and instructive way of illustrating the relationships between sets is by the use of the so-called line diagrams. If $A \subset B$, then we write B on a higher level than A and connect them by a line:

If $A \subset B$ and $B \subset C$, we write

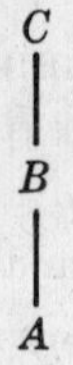

Example 14.1: Let $A = \{a\}$, $B = \{b\}$ and $C = \{a, b\}$. Then the line diagram of A, B and C is

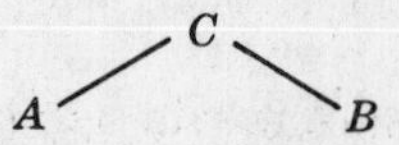

Example 14.2: Let $X = \{x\}$, $Y = \{x, y\}$, $Z = \{x, y, z\}$ and $W = \{x, y, w\}$. Then the line diagram of X, Y, Z and W is

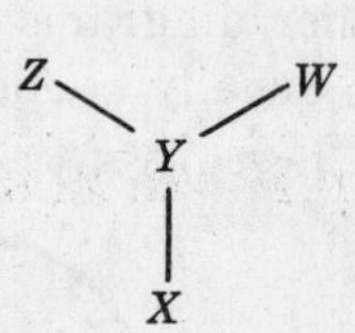

AXIOMATIC DEVELOPMENT OF SET THEORY

In an axiomatic development of a branch of mathematics, one begins with:

(1) undefined terms

(2) undefined relations

(3) axioms relating the undefined terms and undefined relations.

Then, one develops theorems based upon the axioms and definitions.

Example 15.1: In an axiomatic development of Plane Euclidean Geometry:

(1) "points" and "lines" are undefined terms,

(2) "point on a line" or, equivalently, "line contains a point" is an undefined relation,

(3) Two of the axioms are:

Axiom 1: Two different points are on one and only one line.

Axiom 2: Two different lines cannot contain more than one point in common.

In an axiomatic development of set theory:

(1) "element" and "set" are undefined terms.

(2) "element belongs to a set" is the undefined relation.

(3) Two of the axioms are:

Axiom of Extension: Two sets A and B are equal if and only if every element in A belongs to B and every element in B belongs to A.

Axiom of Specification: Let $P(x)$ be any statement and let A be any set. Then there exists a set

$$B = \{a \mid a \in A, P(a) \text{ is true}\}$$

Here, $P(x)$ is a sentence in one variable for which $P(a)$ is true or false for any $a \in A$. For example, $P(x)$ could be the sentence "$x^2 = 4$" or "x is a member of the United Nations".

There are other axioms which are not listed since the axioms concern concepts which are discussed later. Furthermore, as our treatment of set theory is mainly intuitive, especially Part I, we will refrain from any further discussion of the axiomatic development of set theory.

Solved Problems

NOTATION

1. Rewrite the following statements using set notation:

(1) x does not belong to A.
(2) R is a superset of S.
(3) d is a member of E.
(4) F is not a subset of G.
(5) H does not include D.

Solution:

(1) $x \notin A$ (2) $R \supset S$ (3) $d \in E$ (4) $F \not\subset G$ (5) $H \not\supset D$.

2. Let $A = \{x \mid 2x = 6\}$ and let $b = 3$. Does $b = A$?

Solution:

A is a set which consists of the single element 3, that is, $A = \{3\}$. The number 3 belongs to A; it does not equal A. There is a basic difference between an element x and the set $\{x\}$.

3. Let $M = \{r, s, t\}$. In other words, M consists of the elements r, s and t. State whether each of the four statements is correct or incorrect. If a statement is incorrect, tell why.

(a) $r \in M$ (b) $r \subset M$ (c) $\{r\} \in M$ (d) $\{r\} \subset M$

Solution:

(a) Correct.

(b) Incorrect. The symbol $\subset$ must connect two sets; it indicates that one set is a subset of the other. Accordingly, $r \subset M$ is incorrect since r is a member of M, not a subset.

(c) Incorrect. The symbol $\in$ should connect an object to a set; it indicates that the object is a member of the set. Therefore $\{r\} \in M$ is incorrect since $\{r\}$ is a subset of M, not a member of M.

(d) Correct.

4. State in words and then write in tabular form:

(1) $A = \{x \mid x^2 = 4\}$
(2) $B = \{x \mid x - 2 = 5\}$
(3) $C = \{x \mid x$ is positive, x is negative$\}$
(4) $D = \{x \mid x$ is a letter in the word "correct"$\}$.

Solution:

(1) It reads "A is the set of x such that x squared equals four". The only numbers which when squared give four are 2 and -2; hence $A = \{2, -2\}$.

(2) It reads "B is the set of x such that x minus 2 equals 5". The only solution is 7; hence $B = \{7\}$.

(3) It reads "C is the set of x such that x is positive and x is negative". There is no number which is both positive and negative; hence C is empty, that is, $C = \emptyset$.

(4) It reads "D is the set of x such that x is a letter in the word 'correct' ". The indicated letters are c, o, r, e and t; thus $D = \{c, o, r, e, t\}$.

5. Write these sets in a set-builder form:

(1) Let A consist of the letters a, b, c, d and e.
(2) Let $B = \{2, 4, 6, 8, \ldots\}$.
(3) Let C consist of the countries in the United Nations.
(4) Let $D = \{3\}$.
(5) Let E be the Presidents Truman, Eisenhower and Kennedy.

Solution:

Notice first that a set-description, i.e. set-builder, form of a set need not be unique. It is only necessary that any description define the same set. We give a few of the many possible answers to this problem.

(1) $A = \{x \mid x$ appears before f in the alphabet$\}$
$= \{x \mid x$ is one of the first five letters in the alphabet$\}$
(2) $B = \{x \mid x$ is even and positive$\}$
(3) $C = \{x \mid x$ is a country, x is in the United Nations$\}$
(4) $D = \{x \mid x - 2 = 1\} = \{x \mid 2x = 6\}$
(5) $E = \{x \mid x$ was President after Franklin D. Roosevelt$\}$

FINITE AND INFINITE SETS

6. Which sets are finite?

(1) The months of the year.
(2) $\{1, 2, 3, \ldots, 99, 100\}$.
(3) The people living on the earth.
(4) $\{x \mid x$ is even$\}$
(5) $\{1, 2, 3, \ldots\}$.

Solution:

The first three sets are finite. Although physically it might be impossible to count the number of people on the earth, the set is still finite. The last two sets are infinite. If we ever try to count the even numbers we would never come to the end.

EQUALITY OF SETS

7. Which of these sets are equal: $\{r, t, s\}$, $\{s, t, r, s\}$, $\{t, s, t, r\}$, $\{s, r, s, t\}$?

Solution:

They are all equal to each other. Note that order and repetition do not change a set.

8. Which of these sets are equal?

(1) $\{x \mid x$ is a letter in the word "follow"$\}$.

(2) The letters which appear in the word "wolf".

(3) $\{x \mid x$ is a letter in the word "flow"$\}$.

(4) The letters f, l, o and w.

Solution:

If sets are written in tabular form then it is easy to decide whether or not they are equal. After writing the four sets in tabular form, we see that they are all equal to the set {f, l, o, w}.

NULL SET

9. Which word is different from the others, and why: (1) empty, (2) void, (3) zero, (4) null?

Solution:

The first, second and fourth words refer to the set which contains no elements. The word zero refers to a specific number. Hence zero is different.

10. Which of the following sets are different: $\emptyset$, $\{0\}$, $\{\emptyset\}$?

Solution:

Each is different from the others. The set $\{0\}$ contains one element, the number zero. The set $\emptyset$ contains no elements; it is the null set. The set $\{\emptyset\}$ also contains one element, the null set; it is a set of sets.

11. Which of these sets is the null set?

(1) $A = \{x \mid x$ is a letter before a in the alphabet$\}$.

(2) $B = \{x \mid x^2 = 9$ and $2x = 4\}$.

(3) $C = \{x \mid x \neq x\}$.

(4) $D = \{x \mid x + 8 = 8\}$.

Solution:

(1) Since a is the first letter of the alphabet, set A contains no elements; hence $A = \emptyset$.

(2) There is no number which satisfies both equations $x^2 = 9$ and $2x = 4$; hence B is also the null set.

(3) We assume that every object is itself, so C is empty. In fact, some books define the null set in this way, that is,

$$\emptyset = \{x \mid x \neq x\}$$

(4) The number zero satisfies the equation $x + 8 = 8$, so D consists of the element zero. Accordingly D is not the empty set.

SUBSETS

12. Let $A = \{x, y, z\}$. How many subsets does A contain, and what are they?

Solution:

We list all the possible subsets of A. They are: $\{x, y, z\}$, $\{y, z\}$, $\{x, z\}$, $\{x, y\}$, $\{x\}$, $\{y\}$, $\{z\}$, and the null set $\emptyset$. There are eight subsets of A.

13. Define the following sets of figures in the Euclidean plane:

$$Q = \{x \mid x \text{ is a quadrilateral}\} \qquad H = \{x \mid x \text{ is a rhombus}\}$$
$$R = \{x \mid x \text{ is a rectangle}\} \qquad S = \{x \mid x \text{ is a square}\}.$$

Decide which sets are proper subsets of the others.

Solution:

Since a square has 4 right angles it is a rectangle, since it has 4 equal sides it is a rhombus, and since it has 4 sides it is a quadrilateral. Accordingly, $S \subset Q$, $S \subset R$, $S \subset H$, that is, S is a subset of the other three. Also, since there are examples of rectangles, rhombuses and quadrilaterals which are not squares, S is a proper subset of the other three. In a similar manner we see that R is a proper subset of Q, and H is a proper subset of Q. There are no other relations among the sets.

14. Does every set have a proper subset?

Solution:

The null set $\emptyset$ does not have a proper subset. Every other set does have $\emptyset$ as a proper subset. Some books do not call the null set a proper subset; in such case, sets which contain only one element would not contain a proper subset.

15. Prove: If A is a subset of the null set $\emptyset$, then $A = \emptyset$.

Solution:

The null set $\emptyset$ is a subset of every set; in particular $\emptyset \subset A$. By hypothesis, $A \subset \emptyset$. Hence by Definition 1.1, $A = \emptyset$.

16. How does one prove that a set A is not a subset of a set B? Prove that $A = \{2, 3, 4, 5\}$ is not a subset of $B = \{x \mid x \text{ is even}\}$.

Solution:

It is necessary to show that there is at least one element in A which is not in B. Since $3 \in A$ and $3 \notin B$, we see that A is not a subset of B, that is, $A \not\subset B$. Notice that it is not necessary to know whether or not there are other elements in A which are not in B.

17. Let $V = \{d\}$, $W = \{c, d\}$, $X = \{a, b, c\}$, $Y = \{a, b\}$ and $Z = \{a, b, d\}$. Determine whether each of the following statements is true or false.

(1) $Y \subset X$ (3) $W \neq Z$ (5) $V \not\subset Y$ (7) $V \subset X$ (9) $X = W$

(2) $W \not\supset V$ (4) $Z \supset V$ (6) $Z \not\supset X$ (8) $Y \not\subset Z$ (10) $W \subset Y$

Solution:

(1) Since each element in Y is a member of X, we conclude that $Y \subset X$ is true.

(2) The only element in V is d, and d is also in W; thus W is a superset of V and hence $W \not\supset V$ is false.

(3) Since $a \in Z$ and $a \notin W$, $W \neq Z$ is true.

(4) Z is a superset of V since the only element in V is a member of Z; hence $Z \supset V$ is true.

(5) Since $d \in V$ and $d \notin Y$, $V \not\subset Y$ is true.

(6) Since $c \in X$ and $c \notin Z$, then Z is not a superset of X, i.e. $Z \not\supset X$ is true.

(7) V is not a subset of X since $d \in V$ and $d \notin X$; hence $V \subset X$ is false.

(8) Each element in Y is a member of Z; hence $Y \not\subset Z$ is false.

(9) Since $a \in X$ and $a \notin W$, $X = W$ is false.

(10) Since $c \in W$ and $c \notin Y$, W is not a subset of Y and hence $W \subset Y$ is false.

18. Let $A = \{r, s, t, u, v, w\}$, $B = \{u, v, w, x, y, z\}$, $C = \{s, u, y, z\}$, $D = \{u, v\}$, $E = \{s, u\}$ and $F = \{s\}$. Let X be an unknown set. Determine which sets A, B, C, D, E or F can equal X if we are given the following information:

(1) $X \subset A$ and $X \subset B$ (3) $X \not\subset A$ and $X \not\subset C$

(2) $X \not\subset B$ and $X \subset C$ (4) $X \subset B$ and $X \not\subset C$

Solution:

(1) The only set which is a subset of both A and B is D. Notice that C, E and F are not subsets of B since $s \in C, E, F$ and $s \notin B$.

(2) Set X can equal C, E or F since these are subsets of C and, as was noted previously, they are not subsets of B.

(3) Only B is not a subset of either A or C. D and A are subsets of A; and C, E and F are subsets of C. Thus $X = B$.

(4) Both B and D are subsets of B, and are not subsets of C. All other sets violate at least one of the conditions. Hence $X = B$ or $X = D$.

19. Let A be a subset of B and let B be a subset of C, that is, let $A \subset B$ and $B \subset C$. Suppose $a \in A$, $b \in B$, $c \in C$, and suppose $d \notin A$, $e \notin B$, $f \notin C$. Which statements must be true?

$$(1)\ a \in C, \quad (2)\ b \in A, \quad (3)\ c \notin A, \quad (4)\ d \in B, \quad (5)\ e \notin A, \quad (6)\ f \notin A$$

Solution:

(1) By Theorem 1.1, A is a subset of C. Then $a \in A$ implies $a \in C$, and the statement is always true.

(2) Since the element $b \in B$ need not be an element in A, the statement can be false.

(3) The element $c \in C$ could be an element in A; hence $c \notin A$ need not be true.

(4) The element d, which is not in A, need not be in B; hence the statement might not be true.

(5) Since $e \notin B$ and $A \subset B$, $e \notin A$ is always true.

(6) Since $f \notin C$ and $A \subset C$, $f \notin A$ is always true.

LINE DIAGRAMS

20. Construct a line diagram for the sets $A = \{a, b, c\}$, $B = \{a, b\}$ and $C = \{a, c\}$.

Solution:

Since $A \supset B$, $A \supset C$ and B and C are not comparable, we write

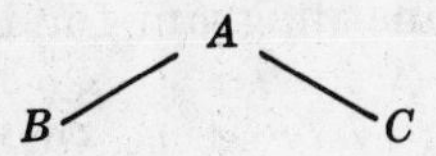

21. Construct a line diagram of the sets $X = \{a, b, c\}$, $Y = \{a, b\}$ and $Z = \{b\}$.

Solution:

Here $Z \subset Y$ and $Y \subset X$. So we write

We do not write

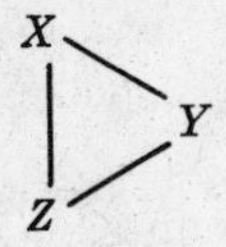

for the line from Z to X is redundant since $Z \subset Y$ and $Y \subset X$ already implies $Z \subset X$.

22. Construct a line diagram of the sets $R = \{r, s, t\}$, $S = \{s\}$ and $T = \{s, t, u\}$.

Solution:

Here $S \subset R$ and $S \subset T$. Also, R and T are not comparable. Accordingly we write

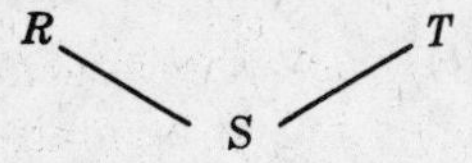

23. Let Q, R, H and S be the sets in Problem 13. Construct a line diagram for these sets.

Solution:

Since $Q \supset R$ and $Q \supset H$, we first write

We now add S to the diagram. Since $S \subset R$ and $S \subset H$, we complete the diagram as follows:

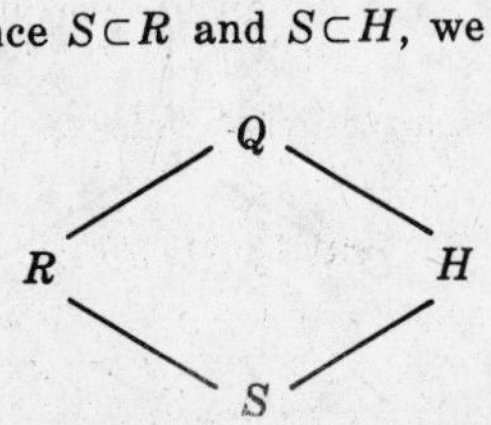

24. Construct a line diagram for the sets V, W, X, Y and Z in Problem 17.

Solution:

Since $V \subset W$ and $V \subset Z$, we write

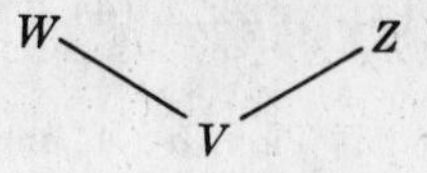

Since $Y \subset Z$, we add Y to the diagram:

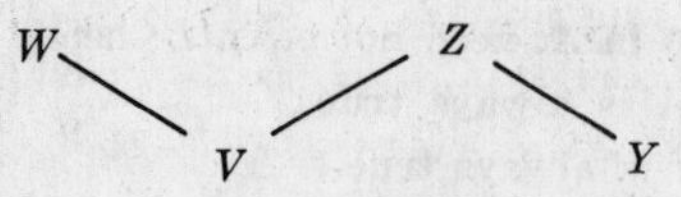

Finally, since $Y \subset X$ we complete the diagram as follows:

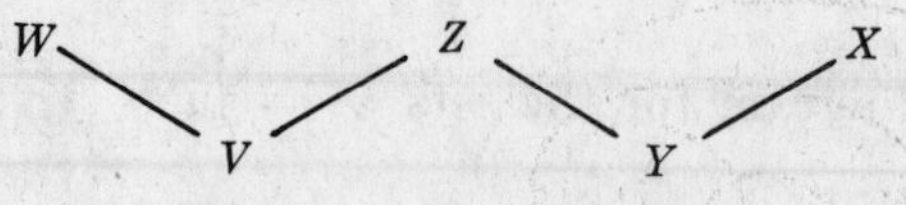

25. Let S be any set. Construct a line diagram for the sets $\emptyset$, S and the universal set U.

Solution:

Since the null set $\emptyset$ is a subset of every set, i.e. $\emptyset \subset S$, we write

Furthermore, since the universal set U is a superset of every set including S, we complete the diagram as follows:

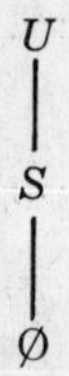

MISCELLANEOUS PROBLEMS

26. Consider the following five statements: (1) $A \subset B$, (2) $A \supset B$, (3) $A = B$, (4) A and B are disjoint, (5) A and B are not comparable. Which statement best describes each Venn diagram?

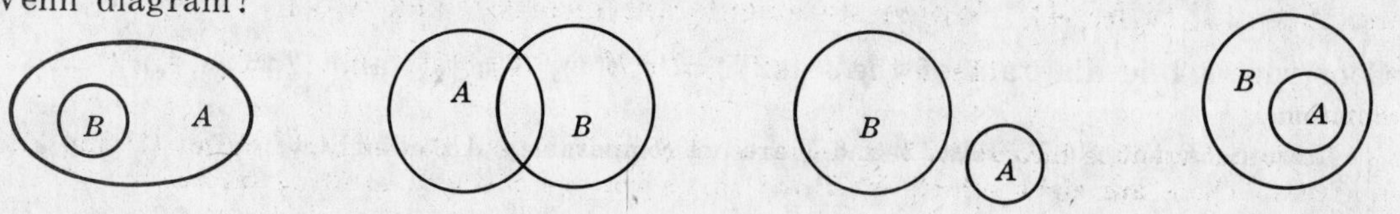

(a) (b) (c) (d)

Solution:

(a) The area of B is part of the area of A; hence $A \supset B$.

(b) There are points in A which are not in B, and points in B which are not in A; hence A and B are not comparable. The sets are not disjoint since there are points which belong to both sets.

(c) Here the sets are disjoint since no point lies in both sets. The sets are also not comparable.

(d) The area of A is part of the area of B; hence $A \subset B$.

27. Consider the following line diagram of sets A, B, C and D.

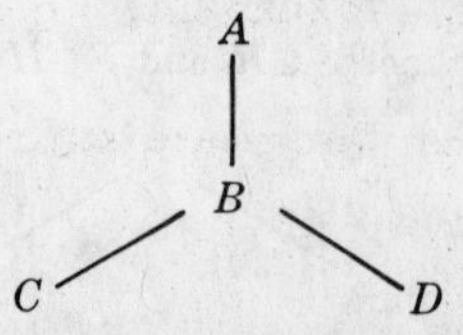

Write a statement that relates each pair of sets in the diagram. There should be six statements.

Solution:

We see first that $C \subset B$, $D \subset B$ and $B \subset A$, since these sets are connected by lines. By Theorem 1.1, we conclude that $C \subset A$ and $D \subset A$. Finally, the sets C and D are not comparable since they are not connected by an increasing path.

28. Construct possible Venn diagrams of sets A, B, C and D which have a line diagram in Problem 27.

Solution:

We make two possible diagrams:

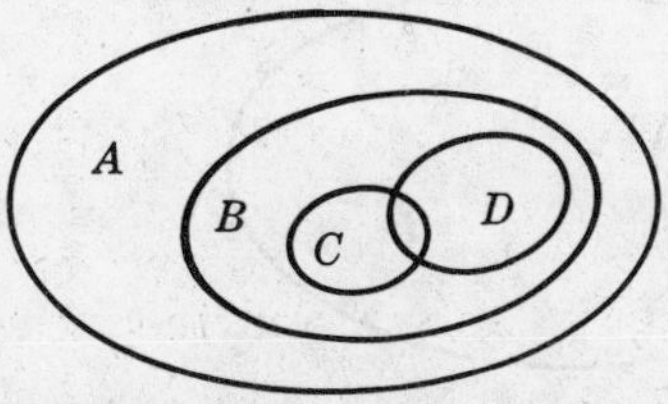

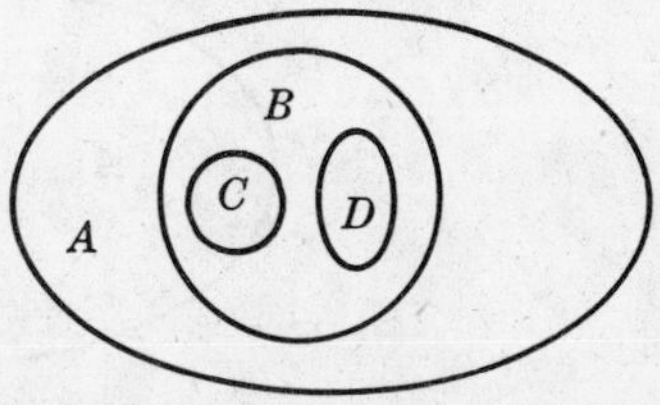

The main difference in the diagrams is that the sets C and D are disjoint in the second diagram. But both have the same line diagram.

29. What is meant by the symbol $\{\{2, 3\}\}$?

Solution:

We have a set which contains one element, the set consisting of the elements 2 and 3. Notice that $\{2, 3\}$ belongs to $\{\{2, 3\}\}$; it is not a subset of $\{\{2, 3\}\}$. Also, we can say that $\{\{2, 3\}\}$ is a set of sets.

30. Let $A = \{2, \{4, 5\}, 4\}$. Which statements are incorrect and why?

(1) $\{4, 5\} \subset A$ (2) $\{4, 5\} \in A$ (3) $\{\{4, 5\}\} \subset A$

Solution:

The elements of A are 2, 4 and the set $\{4, 5\}$. Therefore (2) is correct and (1) is an incorrect statement. (3) is a correct statement since the set consisting of the single element, $\{4, 5\}$, is a subset of A.

31. Let $E = \{2, \{4, 5\}, 4\}$. Which statements are incorrect and why?

(1) $5 \in E$ (2) $\{5\} \in E$ (3) $\{5\} \subset E$

Solution:

Each statement is incorrect. The elements of E are 2, 4 and the set $\{4, 5\}$; hence (1) and (2) are incorrect. There are eight subsets of E and $\{5\}$ is not one of them; so (3) is incorrect.

32. Find the power set 2^S of the set $S = \{3, \{1, 4\}\}$.

Solution:

Notice first that S contains two elements, 3 and the set $\{1, 4\}$. Therefore 2^S contains $2^2 = 4$ elements: S itself, the null set, $\{3\}$ and the set which contains $\{1, 4\}$ alone, i.e. $\{\{1, 4\}\}$. More briefly,

$$2^S = \{S, \{3\}, \{\{1, 4\}\}, \emptyset\}$$

33. Which of the following are undefined in an axiomatic development of set theory: (1) set, (2) subset of, (3) disjoint, (4) element, (5) equals, (6) belongs to, (7) superset of.

Solution:

The only undefined concepts in set theory are set, element, and the relation "belongs to", i.e. (1), (4) and (6).

34. Prove: Let A and B be non-empty, i.e. $A \neq \emptyset$ and $B \neq \emptyset$. If A and B are disjoint, then A and B are not comparable.

Solution:

Since A and B are non-empty, there are elements $a \in A$ and $b \in B$. Furthermore, since A and B are disjoint, $a \notin B$ and $b \notin A$. Hence $A \not\subset B$ and $B \not\subset A$, i.e. A and B are not comparable.

35. Let A and B be not comparable. Must A and B be disjoint?

Solution:

No. The sets in the following Venn diagram are not comparable; these sets are also not disjoint.

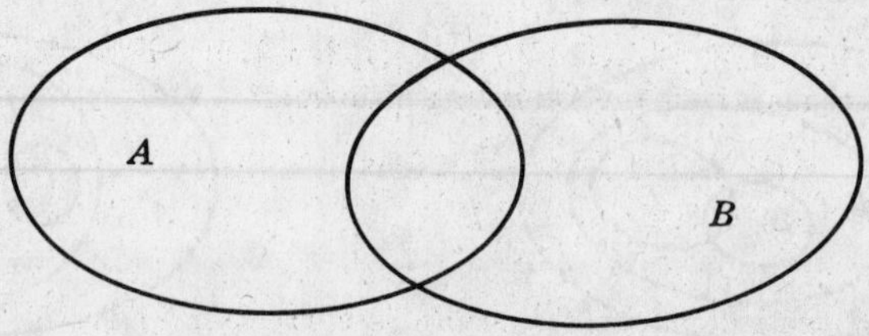

Supplementary Problems

NOTATION

36. Write in set notation:

(1) R is a superset of T
(2) x is a member of Y
(3) M is not a subset of S
(4) the power set of W
(5) z does not belong to A
(6) B is included in F
(7) the empty set
(8) R belongs to $\mathcal{A}$

37. State in words:

(1) $A = \{x \mid x \text{ lives in Paris}\}$
(2) $B = \{x \mid x \text{ speaks Danish}\}$
(3) $C = \{x \mid x \text{ is older than 21 years}\}$
(4) $D = \{x \mid x \text{ is a citizen of France}\}$.

38. Write in tabular form:

(1) $P = \{x \mid x^2 - x - 2 = 0\}$
(2) $Q = \{x \mid x \text{ is a letter in the word "follow"}\}$
(3) $R = \{x \mid x^2 = 9,\ x - 3 = 5\}$
(4) $S = \{x \mid x \text{ is a vowel}\}$
(5) $T = \{x \mid x \text{ is a digit in the number 2324}\}$.

39. Let $E = \{1, 0\}$. State whether each of the following statements is correct or incorrect.

(1) $\{0\} \in E$ (2) $\emptyset \in E$ (3) $\{0\} \subset E$ (4) $0 \in E$ (5) $0 \subset E$

40. In an axiomatic development of set theory, state which of these symbols represents an undefined relation: (1) $\not\subset$, (2) $\in$, (3) $\supset$.

SUBSETS

41. Let $B = \{0, 1, 2\}$. Find all the subsets of B.

42. Let $F = \{0, \{1, 2\}\}$. Find all the subsets of F.

43. Let

$$A = \{2, 3, 4\} \qquad C = \{x \mid x^2 - 6x + 8 = 0\}$$
$$B = \{x \mid x^2 = 4, \ x \text{ is positive}\} \qquad D = \{x \mid x \text{ is even}\}.$$

Complete the following statements by inserting $\subset$, $\supset$ or "nc" (not comparable) between each pair of sets:
(1) $A \ldots B$, (2) $A \ldots C$, (3) $B \ldots C$, (4) $A \ldots D$, (5) $B \ldots D$, (6) $C \ldots D$.

44. Let $A = \{1, 2, \ldots, 8, 9\}$, $B = \{2, 4, 6, 8\}$, $C = \{1, 3, 5, 7, 9\}$, $D = \{3, 4, 5\}$, and $E = \{3, 5\}$. Which sets can equal X if we are given the following information?

(1) X and B are disjoint
(2) $X \subset D$ and $X \not\subset B$
(3) $X \subset A$ and $X \not\subset C$
(4) $X \subset C$ and $X \not\subset A$

45. State whether each of the following statements is correct or incorrect.

(1) Every subset of a finite set is finite.
(2) Every subset of an infinite set is infinite.

MISCELLANEOUS PROBLEMS

46. Draw a line diagram for the sets A, B, C and D of Problem 43.

47. Draw a line diagram for the sets A, B, C, D and E of Problem 44.

48. State whether each of these statements is correct or incorrect:

(1) $\{1, 4, 3\} = \{3, 4, 1\}$
(2) $\{1, 3, 1, 2, 3, 2\} \subset \{1, 2, 3\}$
(3) $\{4\} \in \{\{4\}\}$
(4) $\{4\} \subset \{\{4\}\}$
(5) $\emptyset \subset \{\{4\}\}$

49. State whether each of the following sets is finite or infinite:
(1) The set of lines which are parallel to the x-axis.
(2) The set of letters in the English alphabet.
(3) The set of numbers which are multiples of 5.
(4) The set of animals living on earth.
(5) The set of numbers which are roots of the equation $x^{38} + 42x^{23} - 17x^{18} - 2x^5 + 19 = 0$.
(6) The set of circles through the origin $(0, 0)$.

50. State whether each of the following statements is correct or incorrect. Here S is any set that is not empty.
(1) $S \in 2^S$ (2) $S \subset 2^S$ (3) $\{S\} \in 2^S$ (4) $\{S\} \subset 2^S$

51. Draw a line diagram for the sets in the following Venn diagram.

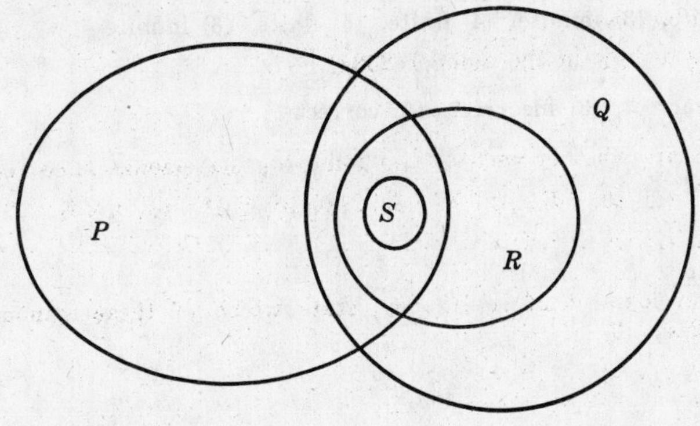

Answers to Supplementary Problems

36. (1) $R \supset T$, (2) $x \,\varepsilon\, Y$, (3) $M \not\subset S$, (4) 2^W, (5) $z \notin A$, (6) $B \subset F$, (7) $\emptyset$, (8) $R \,\varepsilon\, \mathcal{A}$.

37. (1) A is the set of x such that x lives in Paris.
(2) B is the set of x such that x speaks Danish.
(3) C is the set of x such that x is older than 21 years.
(4) D is the set of x such that x is a citizen of France.

38. (1) $P = \{2, -1\}$, (2) $Q = \{\text{f, o, l, w}\}$, (3) $R = \emptyset$, (4) $S = \{\text{a, e, i, o, u}\}$ (5) $T = \{2, 3, 4\}$.

39. (1) incorrect, (2) incorrect, (3) correct, (4) correct, (5) incorrect

40. The symbol ε represents an undefined relation.

41. There are eight subsets: B, $\{0,1\}$, $\{0,2\}$, $\{1,2\}$, $\{0\}$, $\{1\}$, $\{2\}$, $\emptyset$.

42. There are four subsets: F, $\{0\}$, $\{\{1,2\}\}$, $\emptyset$.

43. (1) $\supset$, (2) $\supset$, (3) $\subset$, (4) nc, (5) $\subset$, (6) $\subset$.

44. (1) C, E (2) D, E (3) A, B, D (4) None

45. (1) correct, (2) incorrect

46.
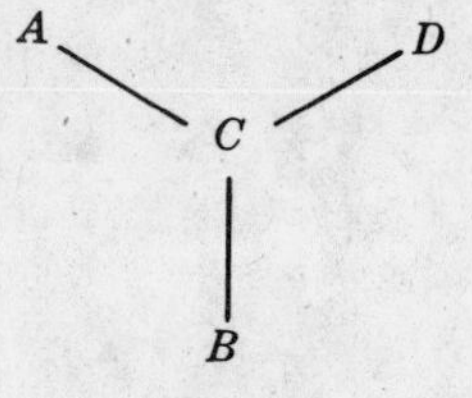

47.
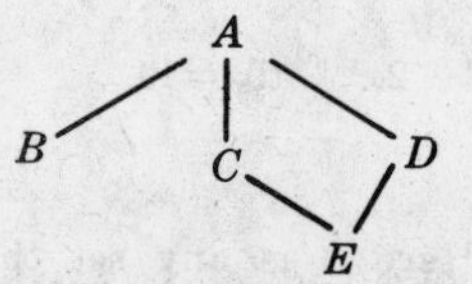

48. (1) correct, (2) correct, (3) correct, (4) incorrect, (5) correct

49. (1) infinite, (2) finite, (3) infinite, (4) finite, (5) finite, (6) infinite

50. (1) correct, (2) incorrect, (3) incorrect, (4) correct

51.
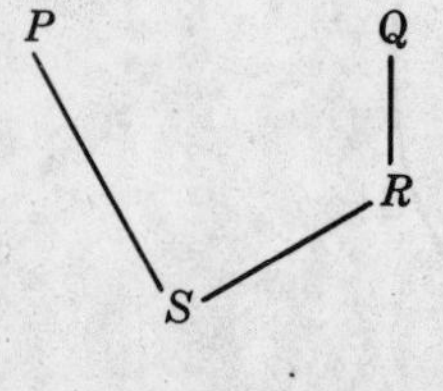

Chapter 2

Basic Set Operations

SET OPERATIONS

In arithmetic we learn to add, subtract and multiply, that is, we assign to each pair of numbers x and y a number $x+y$ called the sum of x and y, a number $x-y$ called the difference of x and y, and a number xy called the product of x and y. These assignments are called the operations of addition, subtraction and multiplication of numbers. In this chapter we define the operations of *union, intersection* and *difference* of sets, that is, we will assign new sets to pairs of sets A and B. In a later chapter we will see that these set operations behave in a manner somewhat similar to the above operations on numbers.

UNION

The union of sets A and B is the set of all elements which belong to A or to B or to both. We denote the union of A and B by

$$A \cup B$$

which is usually read "A union B".

Example 1.1: In the Venn diagram in Fig. 2-1, we have shaded $A \cup B$, i.e. the area of A and the area of B.

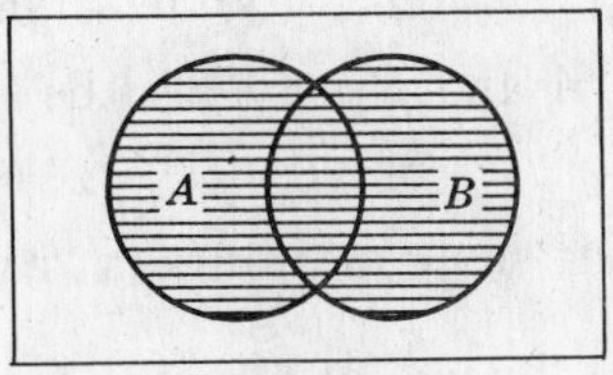

$A \cup B$ is shaded

Fig. 2-1

Example 1.2: Let $S = \{a, b, c, d\}$ and $T = \{f, b, d, g\}$. Then

$$S \cup T = \{a, b, c, d, f, g\}$$

Example 1.3: Let P be the set of positive real numbers and let Q be the set of negative real numbers. Then $P \cup Q$, the union of P and Q, consists of all the real numbers except zero.

The union of A and B may also be defined concisely by

$$A \cup B = \{x \mid x \in A \text{ or } x \in B\}$$

Remark 2.1: It follows directly from the definition of the union of two sets that $A \cup B$ and $B \cup A$ are the same set, i.e.,

$$A \cup B = B \cup A$$

Remark 2.2: Both A and B are always subsets of $A \cup B$, that is,

$$A \subset (A \cup B) \quad \text{and} \quad B \subset (A \cup B)$$

In some books the union of A and B is denoted by $A + B$ and is called the set-theoretic sum of A and B or, simply, A plus B.

INTERSECTION

The *intersection* of sets A and B is the set of elements which are common to A and B, that is, those elements which belong to A and which also belong to B. We denote the intersection of A and B by

$$A \cap B$$

which is read "A intersection B".

Example 2.1: In the Venn diagram in Fig. 2-2, we have shaded $A \cap B$, the area that is common to both A and B.

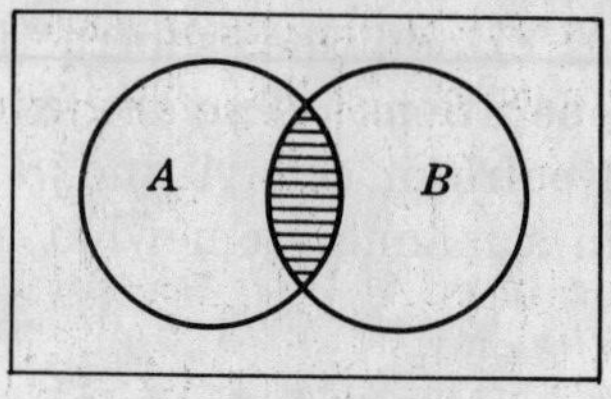

$A \cap B$ is shaded

Fig. 2-2

Example 2.2: Let $S = \{a, b, c, d\}$ and $T = \{f, b, d, g\}$. Then

$$S \cap T = \{b, d\}$$

Example 2.3: Let $V = \{2, 4, 6, \ldots\}$, i.e. the multiples of 2; and let $W = \{3, 6, 9, \ldots\}$, i.e. the multiples of 3. Then

$$V \cap W = \{6, 12, 18, \ldots\}$$

The intersection of A and B may also be defined concisely by

$$A \cap B = \{x \mid x \varepsilon A, \ x \varepsilon B\}$$

Here, the comma has the same meaning as "and".

Remark 2.3: It follows directly from the definition of the intersection of two sets that

$$A \cap B = B \cap A$$

Remark 2.4: Each of the sets A and B contains $A \cap B$ as a subset, i.e.,

$$(A \cap B) \subset A \quad \text{and} \quad (A \cap B) \subset B$$

Remark 2.5: If sets A and B have no elements in common, i.e. if A and B are disjoint, then the intersection of A and B is the null set, i.e. $A \cap B = \emptyset$.

In some books, especially on probability, the intersection of A and B is denoted by AB and is called the set-theoretic product of A and B or, simply, A times B.

DIFFERENCE

The *difference* of sets A and B is the set of elements which belong to A but which do not belong to B. We denote the difference of A and B by

$$A - B$$

which is read "A difference B" or, simply, "A minus B".

Example 3.1: In the Venn diagram in Fig. 2-3, we have shaded $A - B$, the area in A which is not part of B.

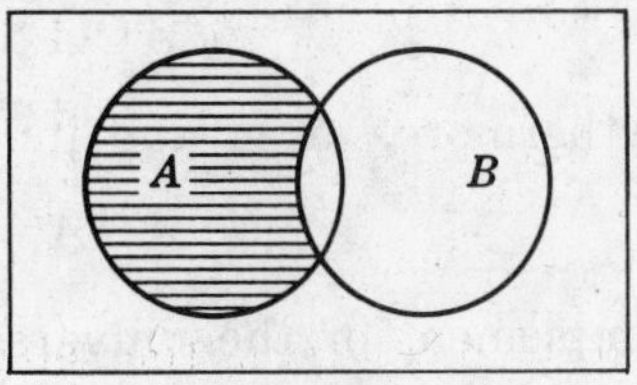

$A - B$ is shaded

Fig. 2-3

Example 3.2: Let $S = \{a, b, c, d\}$ and $T = \{f, b, d, g\}$. Then

$$S - T = \{a, c\}$$

Example 3.3: Let R be the set of real numbers and let Q be the set of rational numbers. Then $R - Q$ consists of the irrational numbers.

The difference of A and B may also be defined concisely by

$$A - B = \{x \mid x \in A, x \notin B\}$$

Remark 2.6: Set A contains $A - B$ as a subset, i.e.,

$$(A - B) \subset A$$

Remark 2.7: The sets $(A - B)$, $A \cap B$ and $(B - A)$ are mutually disjoint, that is, the intersection of any two is the null set.

The difference of A and B is sometimes denoted by A/B or $A \sim B$.

COMPLEMENT

The complement of a set A is the set of elements which do not belong to A, that is, the difference of the universal set U and A. We denote the complement of A by

$$A'$$

Example 4.1: In the Venn diagram in Fig. 2-4, we shaded the complement of A, i.e. the area outside of A. Here we assume that the universal set U consists of the area in the rectangle.

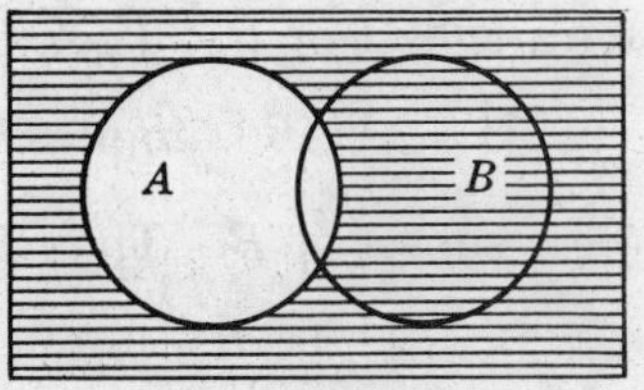

A' is shaded

Fig. 2-4

Example 4.2: Let the universal set U be the English alphabet and let $T = \{a, b, c\}$. Then

$$T' = \{d, e, f, \ldots, y, z\}$$

Example 4.3: Let $E = \{2, 4, 6, \ldots\}$, that is, the even numbers. Then $E' = \{1, 3, 5, \ldots\}$, the odd numbers. Here we assume that the universal set is the natural numbers, $1, 2, 3, \ldots$.

The complement of A may also be defined concisely by

$$A' = \{x \mid x \in U, x \notin A\}$$

or, simply,

$$A' = \{x \mid x \notin A\}$$

We state some facts about sets which follow directly from the definition of the complement of a set.

Remark 2.8: The union of any set A and its complement A' is the universal set, i.e.,

$$A \cup A' = U$$

Furthermore, set A and its complement A' are disjoint, i.e.

$$A \cap A' = \emptyset$$

Remark 2.9: The complement of the universal set U is the null set $\emptyset$, and vice versa, that is,

$$U' = \emptyset \quad \text{and} \quad \emptyset' = U$$

Remark 2.10: The complement of the complement of a set A is the set A itself. More briefly

$$(A')' = A$$

Our next remark shows how the difference of two sets can be defined in terms of the complement of a set and the intersection of two sets. More specifically, we have the following basic relationship:

Remark 2.11: The difference of A and B is equal to the intersection of A and the complement of B, that is,

$$A - B = A \cap B'$$

The proof of Remark 2.11 follows directly from definitions:

$$A - B = \{x \mid x \in A, x \notin B\} = \{x \mid x \in A, x \in B'\} = A \cap B'$$

OPERATIONS ON COMPARABLE SETS

The operations of union, intersection, difference and complement have simple properties when the sets under investigation are comparable. The following theorems can be proved.

Theorem 2.1: Let A be a subset of B. Then the intersection of A and B is precisely A, that is,

$$A \subset B \quad \text{implies} \quad A \cap B = A$$

Theorem 2.2: Let A be a subset of B. Then the union of A and B is precisely B, that is,

$$A \subset B \quad \text{implies} \quad A \cup B = B$$

Theorem 2.3: Let A be a subset of B. Then B' is a subset of A', that is,

$$A \subset B \quad \text{implies} \quad B' \subset A'$$

We illustrate Theorem 2.3 by the Venn diagrams in Fig. 2-5 and 2-6. Notice how the area of B' is included in the area of A'.

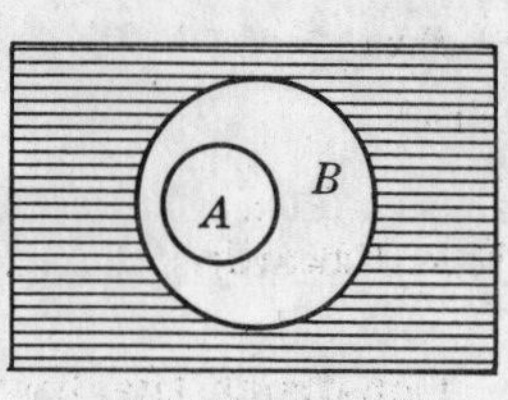

B' is shaded

Fig. 2-5

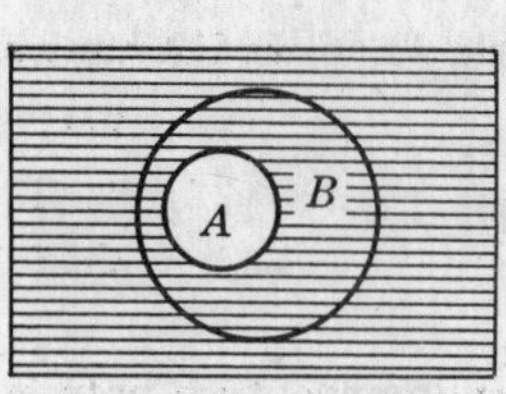

A' is shaded

Fig. 2-6

Theorem 2.4: Let A be a subset of B. Then the union of A and $(B - A)$ is precisely B, that is,

$$A \subset B \quad \text{implies} \quad A \cup (B - A) = B$$

Solved Problems

UNION

1. In the Venn diagrams below, shade A union B, that is, $A \cup B$:

(a) (b) (c) (d)

Solution:

The union of A and B is the set of all elements which belong to A or to B or to both. We therefore shade the area in A and B as follows:

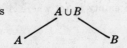

(a) (b) (c) (d)

$A \cup B$ is shaded

2. Let $A = \{1, 2, 3, 4\}$, $B = \{2, 4, 6, 8\}$ and $C = \{3, 4, 5, 6\}$. Find (a) $A \cup B$, (b) $A \cup C$, (c) $B \cup C$, (d) $B \cup B$.

Solution:

To form the union of A and B we put all the elements from A together with all the elements from B. Accordingly,

$$A \cup B = \{1, 2, 3, 4, 6, 8\}$$

Similarly,

$$A \cup C = \{1, 2, 3, 4, 5, 6\}$$
$$B \cup C = \{2, 4, 6, 8, 3, 5\}$$
$$B \cup B = \{2, 4, 6, 8\}$$

Notice that $B \cup B$ is precisely B.

3. Let A, B and C be the sets in Problem 2. Find (1) $(A \cup B) \cup C$, (2) $A \cup (B \cup C)$.

Solution:

(1) We first find $(A \cup B) = \{1, 2, 3, 4, 6, 8\}$. Then the union of $(A \cup B)$ and C is

$$(A \cup B) \cup C = \{1, 2, 3, 4, 6, 8, 5\}$$

(2) We first find $(B \cup C) = \{2, 4, 6, 8, 3, 5\}$. Then the union of A and $(B \cup C)$ is

$$A \cup (B \cup C) = \{1, 2, 3, 4, 6, 8, 5\}$$

Notice that $(A \cup B) \cup C = A \cup (B \cup C)$.

4. Let $X = \{\text{Tom, Dick, Harry}\}$, $Y = \{\text{Tom, Marc, Eric}\}$ and $Z = \{\text{Marc, Eric, Edward}\}$. Find (a) $X \cup Y$, (b) $Y \cup Z$, (c) $X \cup Z$.

Solution:

To find $X \cup Y$ we list the names of X with the names of Y; thus

$$X \cup Y = \{\text{Tom, Dick, Harry, Marc, Eric}\}$$

Similarly,

$$Y \cup Z = \{\text{Tom, Marc, Eric, Edward}\}$$
$$X \cup Z = \{\text{Tom, Dick, Harry, Marc, Eric, Edward}\}$$

5. Let A and B be two sets which are not comparable. Construct the line diagram for the sets A, B and $A \cup B$.

Solution:

Notice first, by Remark 2.2, that A and B are both subsets of $A \cup B$, that is,

$$A \subset (A \cup B) \quad \text{and} \quad B \subset (A \cup B)$$

Accordingly, the line diagram of A, B and $A \cup B$ is

6. Prove Remark 2.2: A and B are subsets of $A \cup B$.

Solution:

Since $A \cup B = B \cup A$ we need only show that A is a subset of $A \cup B$, that is, $x \varepsilon A$ implies $x \varepsilon A \cup B$.

Let x be a member of A. Then it follows that x is a member of A or B, i.e. $x \varepsilon A \cup B$. Thus $A \subset (A \cup B)$.

7. Prove: $A = A \cup A$.

Solution:

By Definition 1.1, we must show that $A \subset (A \cup A)$ and $(A \cup A) \subset A$. By Remark 2.2, $A \subset (A \cup A)$. Now let $x \varepsilon (A \cup A)$. Then, by the definition of union, $x \varepsilon A$ or $x \varepsilon A$; thus x belongs to A. Hence, $(A \cup A) \subset A$ and by Definition 1.1, $A = (A \cup A)$.

8. Prove: $U \cup A = U$, where U is the universal set.

Solution:

By Remark 2.2, $U \subset (U \cup A)$. Since every set is a subset of the universal set, $(U \cup A) \subset U$ and the conclusion follows from Definition 1.1.

9. Prove: $\emptyset \cup A = A$.

Solution:

By Remark 2.2, $A \subset (A \cup \emptyset)$. Now let $x \varepsilon (A \cup \emptyset)$; then $x \varepsilon A$ or $x \varepsilon \emptyset$. By definition of the null set, $x \not\varepsilon \emptyset$; hence $x \varepsilon A$. We have shown that $x \varepsilon (A \cup \emptyset)$ implies $x \varepsilon A$, i.e. $(A \cup \emptyset) \subset A$. By Definition 1.1, $A = \emptyset \cup A$.

10. Prove: $A \cup B = \emptyset$ implies $A = \emptyset$ and $B = \emptyset$.

Solution:

By Remark 2.2, $A \subset (A \cup B)$, that is, $A \subset \emptyset$. But $\emptyset$ is a subset of every set; in particular, $\emptyset \subset A$. Hence by Definition 1.1, $A = \emptyset$. In a similar manner we can show that $B = \emptyset$.

INTERSECTION

11. In the Venn diagrams in Problem 1, shade the intersection of A and B, that is, $A \cap B$.

Solution:

The intersection of A and B consists of the area that is common to both A and to B. To compute $A \cap B$, we first shade A with strokes slanting upward to the right (////) and we shade B with strokes slanting downward to the right (\\\\), as follows:

(a) (b) (c) (d)

Then $A \cap B$ consists of the cross-hatched area. We shade our final result, $A \cap B$, with horizontal lines, as follows:

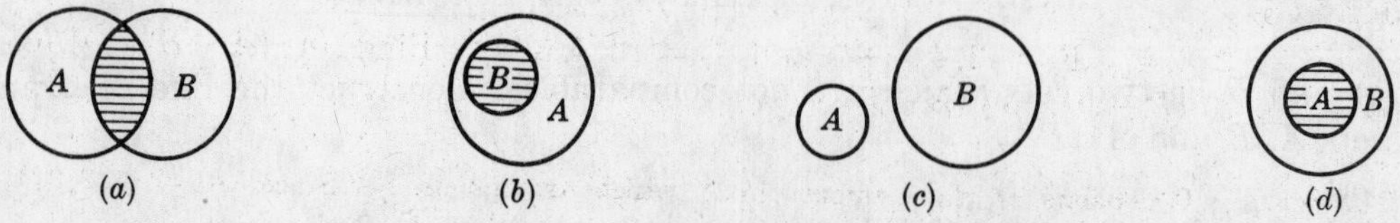

(a) (b) (c) (d)

$A \cap B$ is shaded

Notice that $A \cap B$ is empty in (c) where A and B are disjoint.

12. Let $A = \{1,2,3,4\}$, $B = \{2,4,6,8\}$ and $C = \{3,4,5,6\}$. Find (*a*) $A \cap B$, (*b*) $A \cap C$, (*c*) $B \cap C$, (*d*) $B \cap B$.

Solution:

To form the intersection of A and B, we list all the elements which are common to A and B; thus $A \cap B = \{2,4\}$. Similarly, $A \cap C = \{3,4\}$, $B \cap C = \{4,6\}$ and $B \cap B = \{2,4,6,8\}$. Notice that $B \cap B$ is, in fact, B.

13. Let A, B and C be the sets in Problem 12. Find (*a*) $(A \cap B) \cap C$, (*b*) $A \cap (B \cap C)$.

Solution:

(*a*) $A \cap B = \{2,4\}$. Then the intersection of $\{2,4\}$ with C is $(A \cap B) \cap C = \{4\}$.

(*b*) $B \cap C = \{4,6\}$. The intersection of this set with A is $\{4\}$, that is, $A \cap (B \cap C) = \{4\}$.

Notice that $(A \cap B) \cap C = A \cap (B \cap C)$.

14. Let A and B be two sets which are not comparable. Construct the line diagram of A, B and $A \cap B$.

Solution:

By Remark 2.4, $A \cap B$ is a subset of both A and B, that is, $(A \cap B) \subset A$ and $(A \cap B) \subset B$. Accordingly, we have the following line diagram

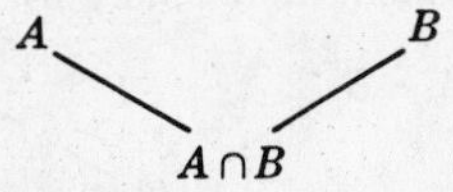

15. Prove Remark 2.4: $(A \cap B)$ is a subset of A and of B.

Solution:

Let x be any element in $A \cap B$. By definition of intersection, x belongs to both A and B; in particular, $x \in A$. We have shown that $x \in A \cap B$ implies $x \in A$, i.e. $(A \cap B) \subset A$. Similarly, $(A \cap B) \subset B$.

16. Prove: $A \cap A = A$.

Solution:

By Remark 2.4, $(A \cap A) \subset A$. Let x be any element in A; then, obviously, x belongs to both A and A, i.e. x belongs to $A \cap A$. We have proven that $x \in A$ implies $x \in (A \cap A)$, i.e. $A \subset (A \cap A)$. By Definition 1.1, $A \cap A = A$.

17. Prove: $U \cap A = A$, where U is the universal set.

Solution:

By Remark 2.4, $(U \cap A) \subset A$. Let x be any element in A. Since U is the universal set, x also belongs to U. Since $x \in A$ and $x \in U$, by definition of intersection, $x \in (U \cap A)$. We have shown that $x \in A$ implies $x \in (U \cap A)$, i.e. we have proven that $A \subset (U \cap A)$. By Definition 1.1, $(U \cap A) = A$.

18. Prove: $A \cap \emptyset = \emptyset$.

Solution:

By Remark 2.4, $(A \cap \emptyset) \subset \emptyset$. But the null set is a subset of every set; in particular, $\emptyset \subset (A \cap \emptyset)$. Therefore $A \cap \emptyset = \emptyset$.

DIFFERENCE

19. Let $A = \{1,2,3,4\}$, $B = \{2,4,6,8\}$ and $C = \{3,4,5,6\}$. Find (*a*) $(A - B)$, (*b*) $(C - A)$, (*c*) $(B - C)$, (*d*) $(B - A)$, (*e*) $(B - B)$.

Solution:

(*a*) The set $A - B$ consists of the elements in A which are not in B. Since $A = \{1,2,3,4\}$ and $2, 4 \in B$, then $A - B = \{1,3\}$.

(*b*) The only elements in C which are not in A are 5 and 6; hence $C - A = \{5,6\}$.

(*c*) $B - C = \{2,8\}$ (*d*) $B - A = \{6,8\}$ (*e*) $B - B = \emptyset$.

20. In the Venn diagrams exhibited in Problem 1, shade A minus B, that is, $A-B$.

Solution:

In each case the set $A-B$ consists of the elements in A which are not in B, that is, the area in A which does not lie in B.

(a) (b) (c) (d)

$A-B$ is shaded

Notice, as in (c), that $A-B=A$ if A and B are disjoint. Notice also, as in (d), that $A-B=\emptyset$ if A is a subset of B.

21. Suppose sets A and B are not comparable. Construct a line diagram for the sets A, B, $(A-B)$, $(B-A)$, $\emptyset$ and the universal set U.

Solution:

Notice first, by Remark 2.6, that $(A-B)\subset A$ and $(B-A)\subset B$.

Since $\emptyset$ is a subset of every element and since, by Remark 2.7, $(A-B)$ and $(B-A)$ are not comparable, we can first draw

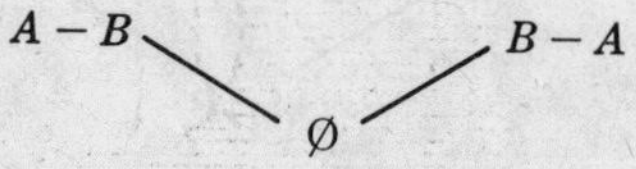

Since $A\supset(A-B)$ and $B\supset B-A$, we add A and B to the diagram as follows:

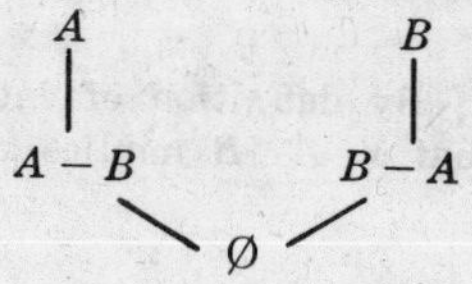

Since U contains every set, we complete the diagram:

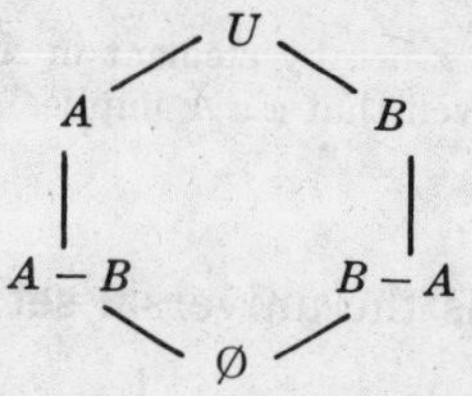

If we did not include U or $\emptyset$ to the diagram, then the line diagram would not be connected.

22. Prove Remark 2.6: $(A-B)\subset A$.

Solution:

Let x be any element of the set $A-B$. By definition of difference, $x \,\varepsilon\, A$ and $x \notin B$; in particular, x belongs to A. We have shown that $x \,\varepsilon\, (A-B)$ implies $x \,\varepsilon\, A$; in other words, $(A-B)\subset A$.

23. Prove: $(A-B)\cap B=\emptyset$.

Solution:

Let x belong to $(A-B)\cap B$. By definition of intersection, $x \,\varepsilon\, (A-B)$ and $x \,\varepsilon\, B$. But by definition of difference, $x \,\varepsilon\, A$ and $x \notin B$. Since there is no element satisfying both $x \,\varepsilon\, B$ and $x \notin B$, then $(A-B)\cap B=\emptyset$.

COMPLEMENT

24. Let $U=\{1,2,3,\ldots,8,9\}$, $A=\{1,2,3,4\}$, $B=\{2,4,6,8\}$ and $C=\{3,4,5,6\}$. Find (a) A', (b) B', (c) $(A\cap C)'$, (d) $(A\cup B)'$, (e) $(A')'$, (f) $(B-C)'$.

Solution:

(a) The set A' consists of the elements that are in U but not in A. Hence $A'=\{5,6,7,8,9\}$.

(b) The set which consists of the elements that are in U but not in B is $B' = \{1, 3, 5, 7, 9\}$.

(c) $(A \cap C) = \{3, 4\}$ and hence $(A \cap C)' = \{1, 2, 5, 6, 7, 8, 9\}$.

(d) $(A \cup B) = \{1, 2, 3, 4, 6, 8\}$ and hence $(A \cup B)' = \{5, 7, 9\}$.

(e) $A' = \{5, 6, 7, 8, 9\}$ and hence $(A')' = \{1, 2, 3, 4\}$, i.e. $(A')' = A$.

(f) $(B - C) = \{2, 8\}$ and hence $(B - C)' = \{1, 3, 4, 5, 6, 7, 9\}$.

25. In the Venn diagram below, shade (a) B', (b) $(A \cup B)'$, (c) $(B - A)'$, (d) $A' \cap B'$.

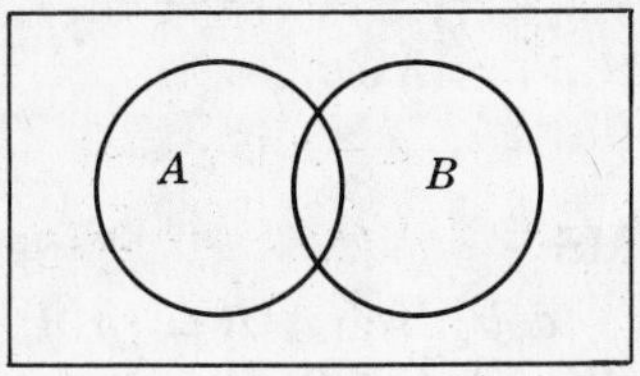

Solution:

(a) Since B', the complement of B, consists of the elements which do not belong to B, shade the area outside of B.

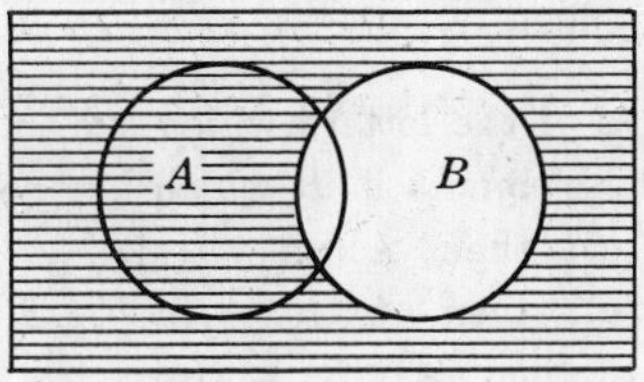

B' is shaded

(b) First shade the area $A \cup B$; then $(A \cup B)'$ is the area outside of $(A \cup B)$.

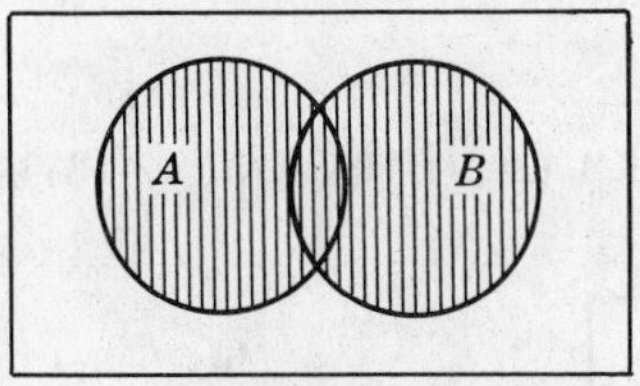

$A \cup B$ is shaded

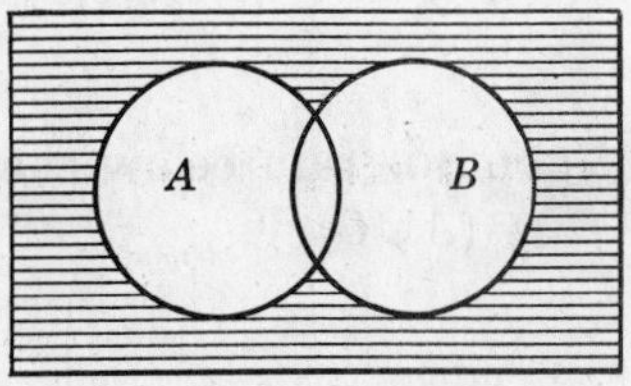

$(A \cup B)'$ is shaded

(c) First shade $B - A$; then $(B - A)'$ is the area outside of $B - A$.

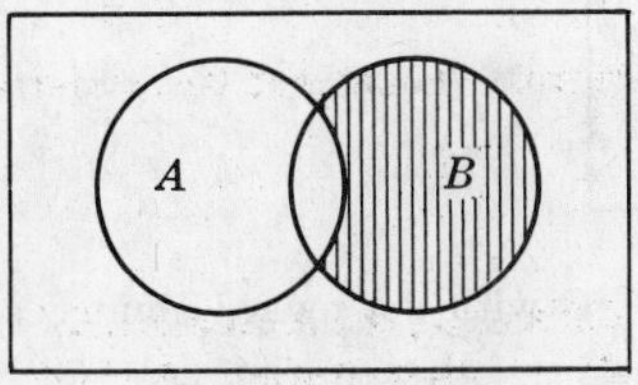

$B - A$ is shaded

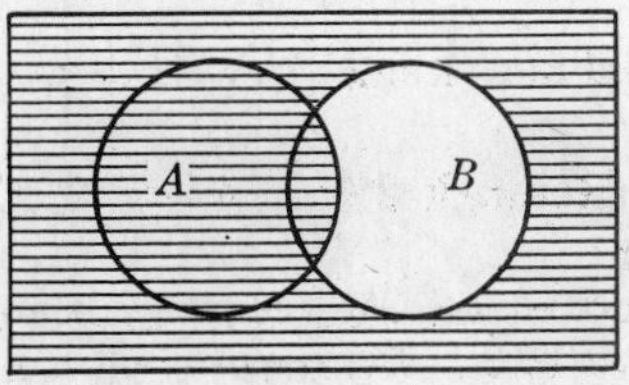

$(B - A)'$ is shaded

(d) First shade A', the area outside of A, with strokes that slant upward to the right (////) and shade B' with strokes that slant downward to the right (\\\\): then $A' \cap B'$ is the cross-hatched area.

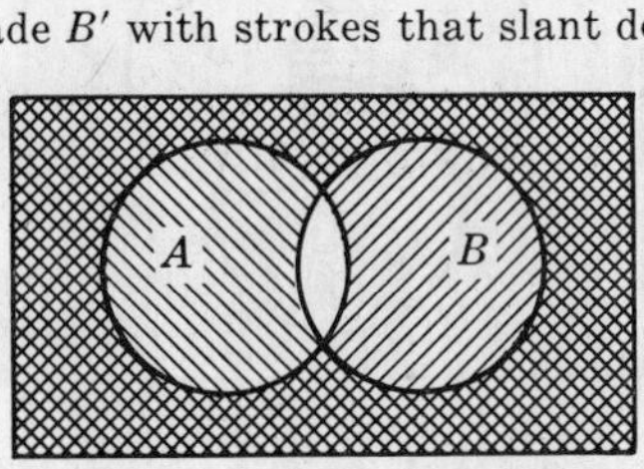

A' and B' are shaded

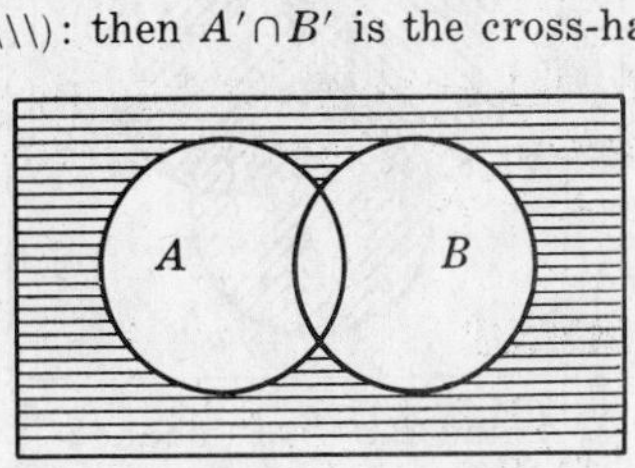

$A' \cap B'$ is shaded

Notice that the area of $(A \cup B)'$ is the same as the area of $A' \cap B'$.

26. Prove De Morgan's Theorem: $(A \cup B)' = A' \cap B'$.

Solution:

Let $x \in (A \cup B)'$; then x does not belong to $A \cup B$. Therefore $x \notin A$ and $x \notin B$, i.e. $x \in A'$ and $x \in B'$, and by the definition of intersection, x belongs to $A' \cap B'$. We have shown that $x \in (A \cup B)'$ implies $x \in (A' \cap B')$, i.e.,

$$(A \cup B)' \subset (A' \cap B')$$

Now let $y \in A' \cap B'$; then y belongs to A' and y belongs to B'. Thus $y \notin A$ and $y \notin B$ and hence $y \notin (A \cup B)$, i.e. $y \in (A \cup B)'$. We have shown that $y \in (A' \cap B')$ implies $y \in (A \cup B)'$, i.e.,

$$(A' \cap B') \subset (A \cup B)'$$

Hence by Definition 1.1, $(A' \cap B') = (A \cup B)'$.

MISCELLANEOUS PROBLEMS

27. Let $U = \{a, b, c, d, e\}$, $A = \{a, b, d\}$ and $B = \{b, d, e\}$. Find (*a*) $A \cup B$, (*b*) $B \cap A$, (*c*) B', (*d*) $B - A$, (*e*) $A' \cap B$, (*f*) $A \cup B'$, (*g*) $A' \cap B'$, (*h*) $B' - A'$, (*i*) $(A \cap B)'$, (*j*) $(A \cup B)'$.

Solution:

(*a*) The union of A and B consists of the elements in A and the elements in B, i.e. $A \cup B = \{a, b, d, e\}$.

(*b*) The intersection of A and B consists of the elements which are common to A and B, i.e. $A \cap B = \{b, d\}$.

(*c*) The complement of B consists of the letters which are in U but not in B; thus $B' = \{a, c\}$.

(*d*) The set $B - A$ consists of the elements in B which are not in A, i.e. $B - A = \{e\}$.

(*e*) $A' = \{c, e\}$ and $B = \{b, d, e\}$; then $A' \cap B = \{e\}$.

(*f*) $A = \{a, b, d\}$ and $B' = \{a, c\}$; then $A \cup B' = \{a, b, c, d\}$.

(*g*) $A' = \{c, e\}$ and $B' = \{a, c\}$; then $A' \cap B' = \{c\}$.

(*h*) $B' - A' = \{a\}$.

(*i*) From (*b*), $A \cap B = \{b, d\}$; hence $(A \cap B)' = \{a, c, e\}$.

(*j*) From (*a*), $A \cup B = \{a, b, d, e\}$; hence $(A \cup B)' = \{c\}$.

28. In the Venn diagram below shade (1) $A \cap (B \cup C)$, (2) $(A \cap B) \cup (A \cap C)$, (3) $A \cup (B \cap C)$, (4) $(A \cup B) \cap (A \cup C)$.

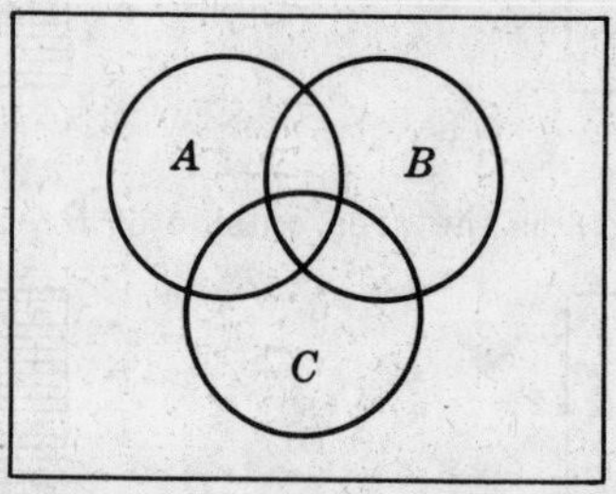

Solution:

(1) First shade A with upward slanted strokes and shade $B \cup C$ with downward slanted strokes; then $A \cap (B \cup C)$ is the cross-hatched area.

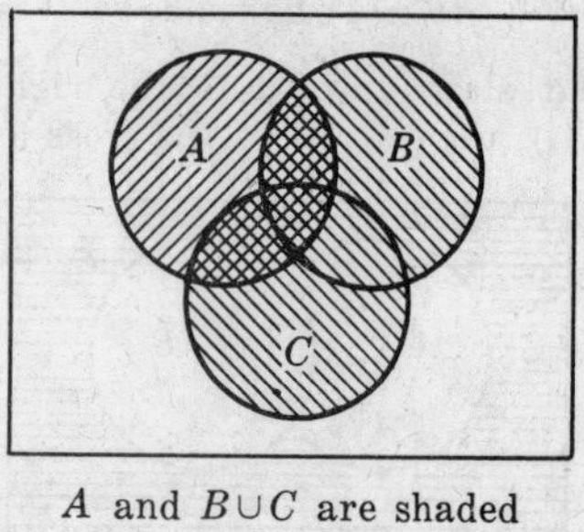

A and $B \cup C$ are shaded

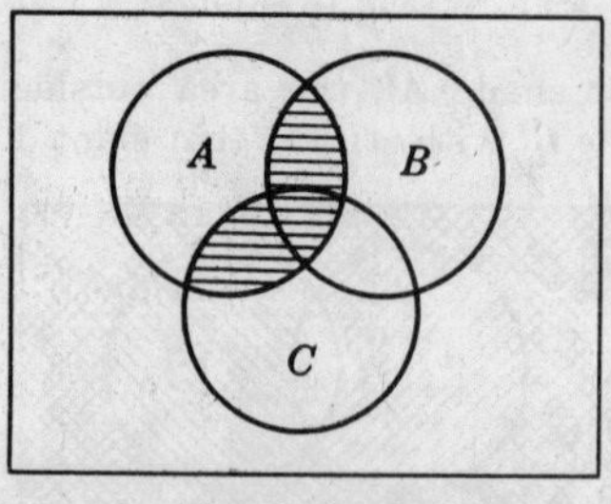

$A \cap (B \cup C)$ is shaded

(2) First shade $A \cap B$ with upward slanted strokes and then shade $A \cap C$ with downward slanted strokes; then $(A \cap B) \cup (A \cap C)$ is the total area shaded, as shown below.

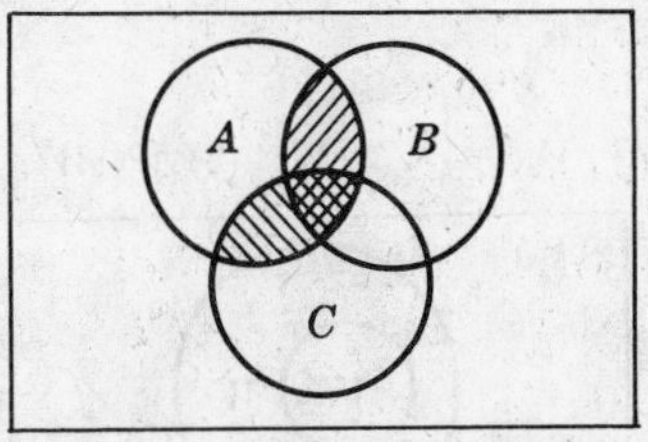

$A \cap B$ and $A \cap C$ are shaded

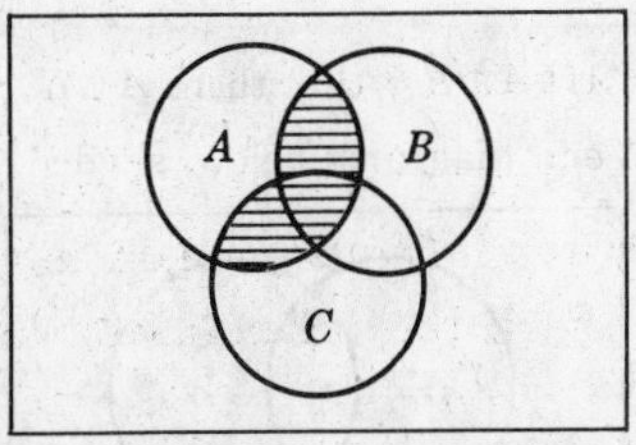

$(A \cap B) \cup (A \cap C)$ is shaded

Notice that $A \cap (B \cup C) = (A \cap B) \cup (A \cap C)$.

(3) First shade A with upward slanted strokes and shade $B \cap C$ with downward slanted strokes; then $A \cup (B \cap C)$ is the total area shaded.

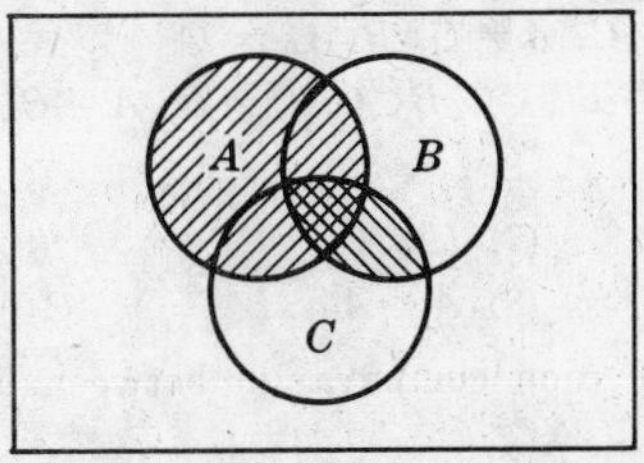

A and $B \cap C$ are shaded

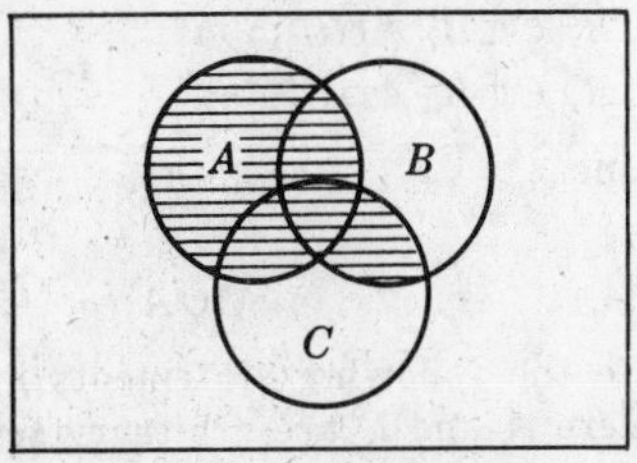

$A \cup (B \cap C)$ is shaded

(4) First shade $A \cup B$ with upward slanted strokes and shade $A \cup C$ with downward slanted strokes; then the cross-hatched area is $(A \cup B) \cap (A \cup C)$.

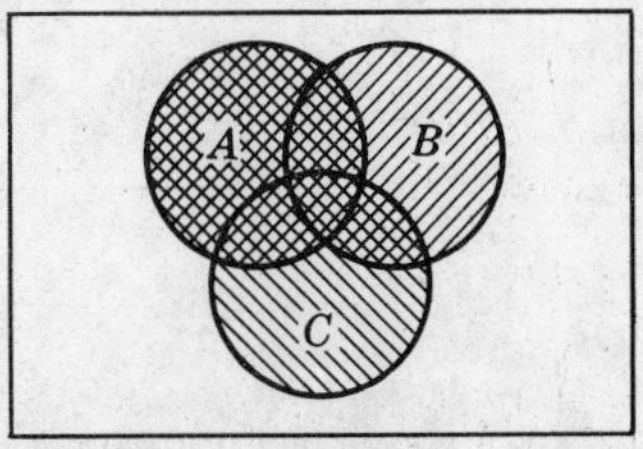

$A \cup B$ and $A \cup C$ are shaded

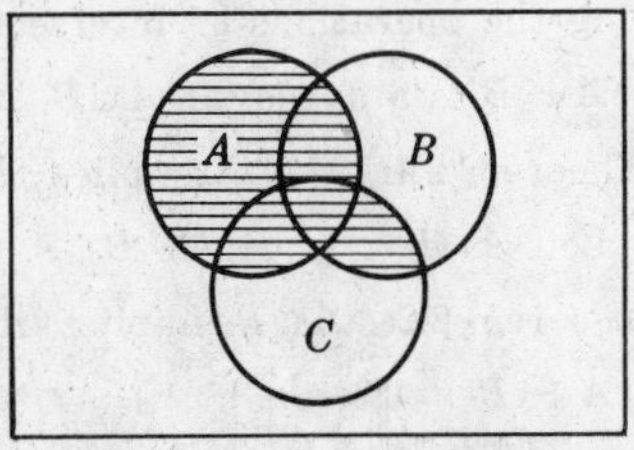

$(A \cup B) \cap (A \cup C)$ is shaded

Notice that $A \cup (B \cap C) = (A \cup B) \cap (A \cup C)$.

29. Prove: $B - A$ is a subset of A'.

Solution:

Let x belong to $B - A$. Then $x \in B$ and $x \notin A$; hence x is a member of A'.

Since $x \in B - A$ implies $x \in A'$, $B - A$ is a subset of A'.

30. Prove: $B - A' = B \cap A$.

Solution:

$B - A' = \{x \mid x \in B,\ x \notin A'\} = \{x \mid x \in B,\ x \in A\} = B \cap A$.

Supplementary Problems

31. Let the universal set $U = \{a, b, c, d, e, f, g\}$, and let $A = \{a, b, c, d, e\}$, $B = \{a, c, e, g\}$ and $C = \{b, e, f, g\}$. Find:

(1) $A \cup C$ (3) $C - B$ (5) $A' - B$ (7) $(A - C)'$ (9) $(A - B')'$

(2) $B \cap A$ (4) B' (6) $B' \cup C$ (8) $C' \cap A$ (10) $(A \cap A')'$

32. Prove: If $A \cap B = \emptyset$, then $A \subset B'$.

33. In the Venn diagrams below, shade (1) $V \cap W$, (2) W', (3) $W - V$, (4) $V' \cup W$, (5) $V \cap W'$, (6) $V' - W'$.

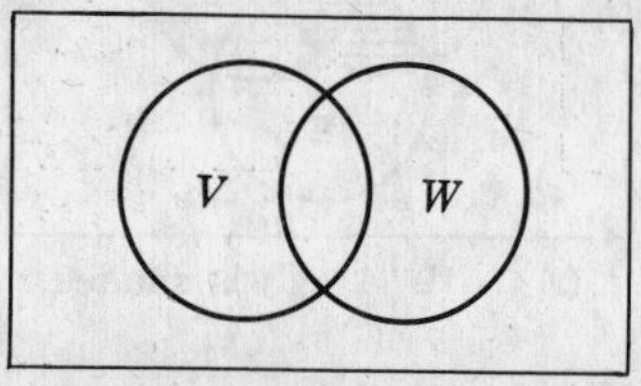

(a)

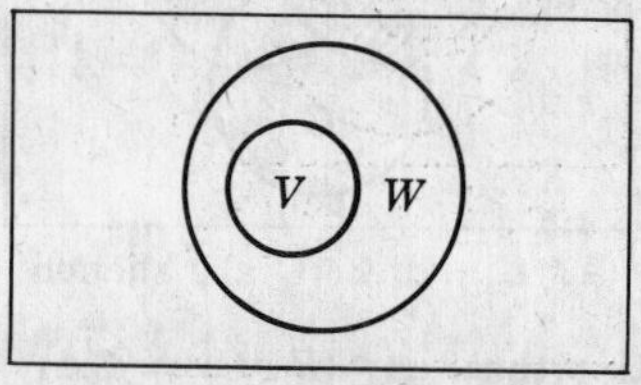

(b)

34. Draw a Venn diagram for three non-empty sets A, B and C so that A, B and C will have the following properties:

(1) $A \subset B$, $C \subset B$, $A \cap C = \emptyset$
(2) $A \subset B$, $C \not\subset B$, $A \cap C \neq \emptyset$
(3) $A \subset C$, $A \neq C$, $B \cap C = \emptyset$
(4) $A \subset (B \cap C)$, $B \subset C$, $C \neq B$, $A \neq C$

35. Determine:

(1) $U \cap A$
(2) $A \cup A$
(3) $\emptyset'$
(4) $\emptyset \cup A$
(5) $A' \cap A$
(6) U'
(7) $U \cup A$
(8) $A' \cup A$
(9) $A \cap A$
(10) $\emptyset \cap A$.

36. Complete the following statements by inserting $\subset$, $\supset$ or nc (non comparable) between each pair of sets. Here A and B are arbitrary sets.

(1) $A \ldots . A - B$
(2) $A \ldots . A \cap B$
(3) $A' \ldots . B - A$
(4) $A \ldots . A \cup B$
(5) $A' \ldots . A - B$
(6) $A \ldots . B - A$

37. The formula $A - B = A \cap B'$ can define the difference of two sets using only the operations of intersection and complement. Find a formula that can define the union of two sets, $A \cup B$, again only using the operations of intersection and complement.

38. Prove: $A - B$ is a subset of $A \cup B$.

39. Prove Theorem 2.1: $A \subset B$ implies $A \cap B = A$.

40. Prove: Let $A \cap B = \emptyset$; then $B \cap A' = B$.

41. Prove Theorem 2.2: $A \subset B$ implies $A \cup B = B$.

42. Prove: $A' - B' = B - A$.

43. Prove Theorem 2.3: $A \subset B$ implies $B' \subset A'$.

44. Prove: Let $A \cap B = \emptyset$; then $A \cup B' = B'$.

45. Prove: $(A \cap B)' = A' \cup B'$.

46. Prove Theorem 2.4: $A \subset B$ implies $A \cup (B - A) = B$.

Answers to Supplementary Problems

31. (1) U (2) $\{a, c, e\}$ (3) $\{b, f\}$ (4) $\{b, d, f\}$ (5) $\{f\}$ (6) $\{b, d, f, e, g\}$ (7) $C = \{b, e, f, g\}$ (8) $\{a, c, d\}$ (9) $\{b, d, f, g\}$ (10) U

32. Proof. Let $x \in A$. Since A and B are disjoint, $x \notin B$; hence x belongs to B'. We have shown that $x \in A$ implies $x \in B'$, i.e. $A \subset B'$.

33. (a) (1)

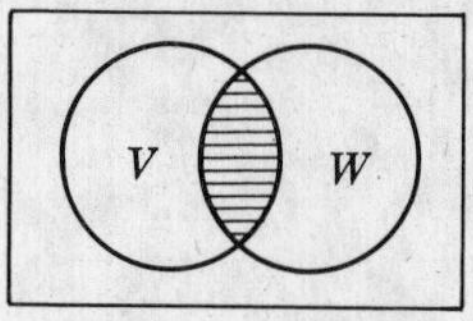

$V \cap W$ is shaded

(3)

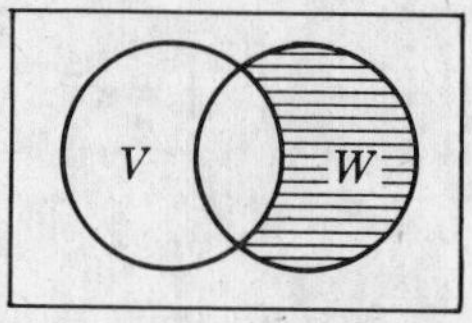

$W - V$ is shaded

(5)

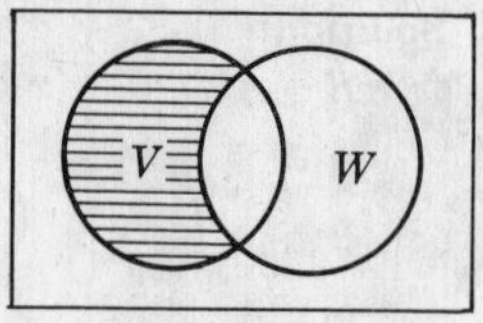

$V \cap W'$ is shaded

(2)

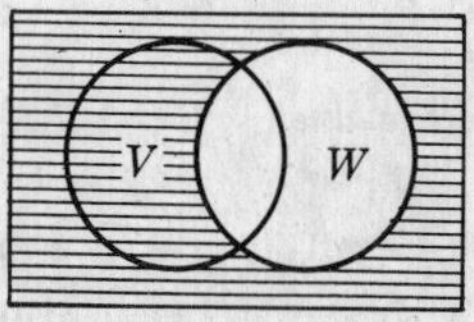

W' is shaded

(4)

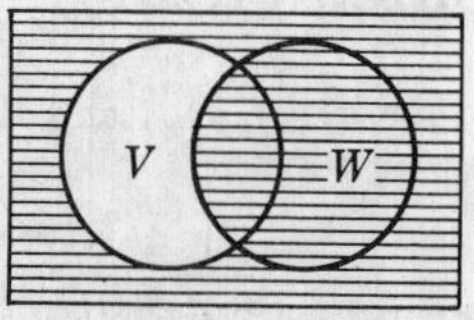

$V' \cup W$ is shaded

(6)

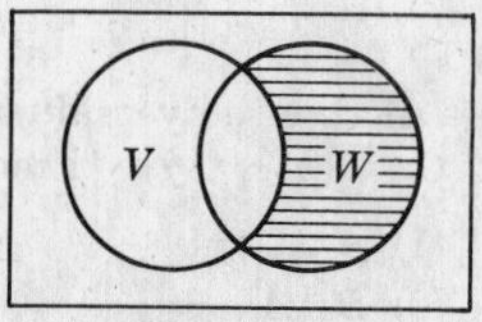

$V' - W'$ is shaded

(b) (1)

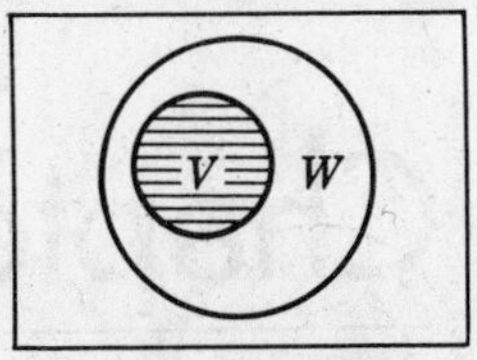

$V \cap W$ is shaded

(3)

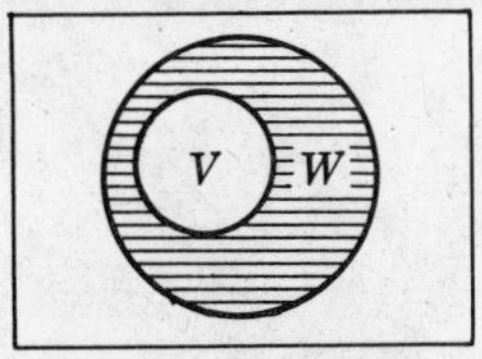

$W - V$ is shaded

(5)

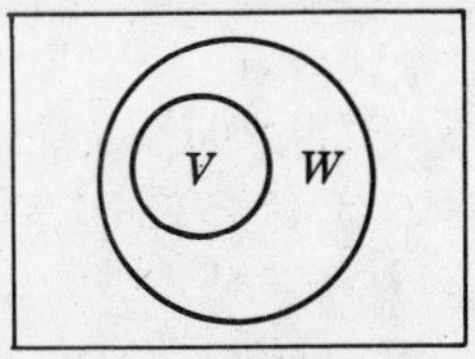

$V \cap W'$ is shaded

(2)

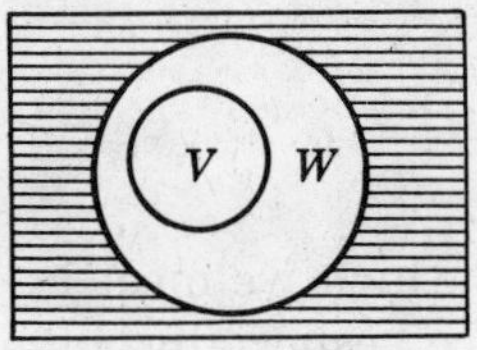

W' is shaded

(4)

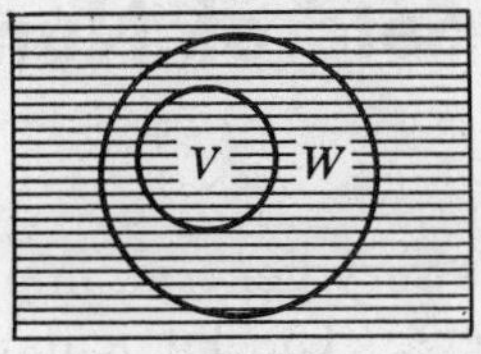

$V' \cup W$ is shaded

(6)

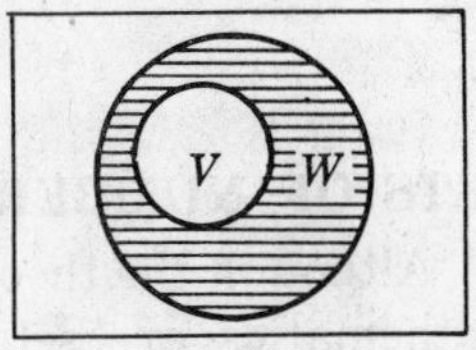

$V' - W'$ is shaded

34. (1)

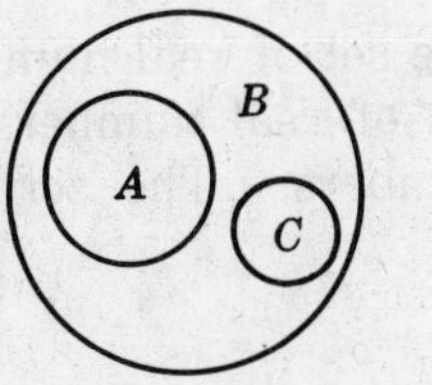

(3)

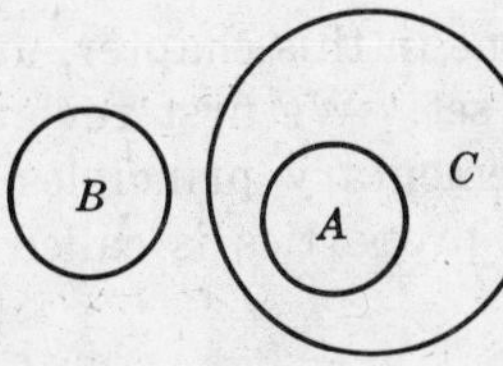

(2)

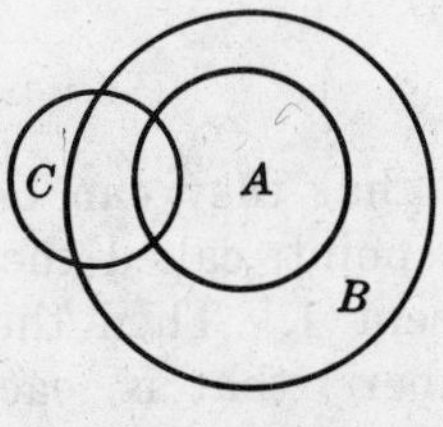

(4)

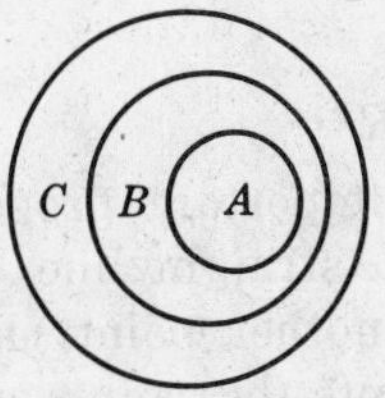

35. (1) A (2) A (3) U (4) A (5) $\emptyset$ (6) $\emptyset$ (7) U (8) U (9) A (10) $\emptyset$

36. (1) $\supset$ (2) $\supset$ (3) $\supset$ (4) $\subset$ (5) nc (6) nc

37. $A \cup B = (A' \cap B')'$.

Chapter 3

Sets of Numbers

SETS OF NUMBERS

Although the theory of sets is very general, important sets which we meet in elementary mathematics are sets of numbers. Of particular importance, especially in analysis, is the set of *real numbers* which we denote by

$$R^{\#}$$

In fact, we assume in this chapter, unless otherwise stated, that the set of real numbers $R^{\#}$ is our universal set. We first review some elementary properties of real numbers before applying our elementary principles of set theory to sets of numbers. The set of real numbers and its properties is called the *real number system.*

REAL NUMBERS, $R^{\#}$

One of the most important properties of the real numbers is that they can be represented by points on a straight line. As in Fig. 3-1, we choose a point, called the origin, to represent 0 and another point, usually to the right, to represent 1. Then there is a natural way to pair off the points on the line and the real numbers, that is, each point will represent a unique real number and each real number will be represented by a unique point. We refer to this line as the *real line.* Accordingly, we can use the words point and number interchangeably.

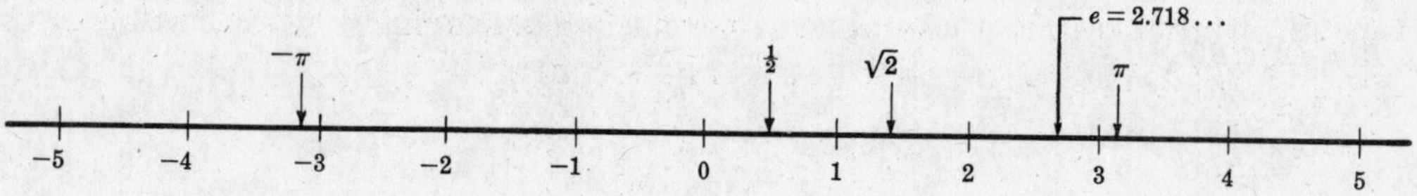

Fig. 3-1

Those numbers to the right of 0, i.e. on the same side as 1, are called the *positive numbers,* and those numbers to the left of 0 are called the *negative numbers.* The number 0 itself is neither positive nor negative.

INTEGERS, Z

The *integers* are those real numbers

$$\ldots, -3, -2, -1, 0, 1, 2, 3, \ldots$$

We denote the integers by Z; hence we can write

$$Z = \{\ldots, -2, -1, 0, 1, 2, \ldots\}$$

The integers are also referred to as the "whole" numbers.

One important property of the integers is that they are *"closed"* under the operations of addition, multiplication and subtraction; that is, the sum, product and difference of two integers is again an integer. Notice that the quotient of two integers, e.g. 3 and 7, need not be an integer; hence the integers are not closed under the operation of division.

RATIONAL NUMBERS, Q

The *rational numbers* are those real numbers which can be expressed as the ratio of two integers. We denote the set of rational numbers by Q. Accordingly,

$$Q = \{x \mid x = p/q \text{ where } p \in Z,\ q \in Z\}$$

Notice that each integer is also a rational number since, for example, $5 = 5/1$; hence Z is a subset of Q.

The rational numbers are closed not only under the operations of addition, multiplication and subtraction but also under the operation of division (except by 0). In other words, the sum, product, difference and quotient (except by 0) of two rational numbers is again a rational number.

NATURAL NUMBERS, N

The *natural numbers* are the positive integers. We denote the set of natural numbers by N; hence

$$N = \{1, 2, 3, \ldots\}$$

The natural numbers were the first number system developed and were used primarily, at one time, for counting. Notice the following relationships between the above number systems:

$$N \subset Z \subset Q \subset R^{\#}$$

The natural numbers are closed only under the operations of addition and multiplication. The difference and quotient of two natural numbers need not be a natural number.

The *prime numbers* are those natural numbers p, excluding 1, which are only divisible by 1 and p itself. We list the first few prime numbers:

$$2, 3, 5, 7, 11, 13, 17, 19, \ldots$$

IRRATIONAL NUMBERS, Q'

The irrational numbers are those real numbers which are not rational, that is, the set of irrational numbers is the complement of the set of rational numbers Q in the real numbers $R^{\#}$; hence Q' denotes the irrational numbers. Examples of irrational numbers are $\sqrt{3}$, π, $\sqrt{2}$, etc.

LINE DIAGRAM OF THE NUMBER SYSTEMS

Fig. 3-2 below is a line diagram of the various sets of numbers which we have investigated. (For completeness, the diagram includes the set of complex numbers, numbers of the form $a + bi$ where a and b are real. Notice that the set of complex numbers is a superset of the set of real numbers.)

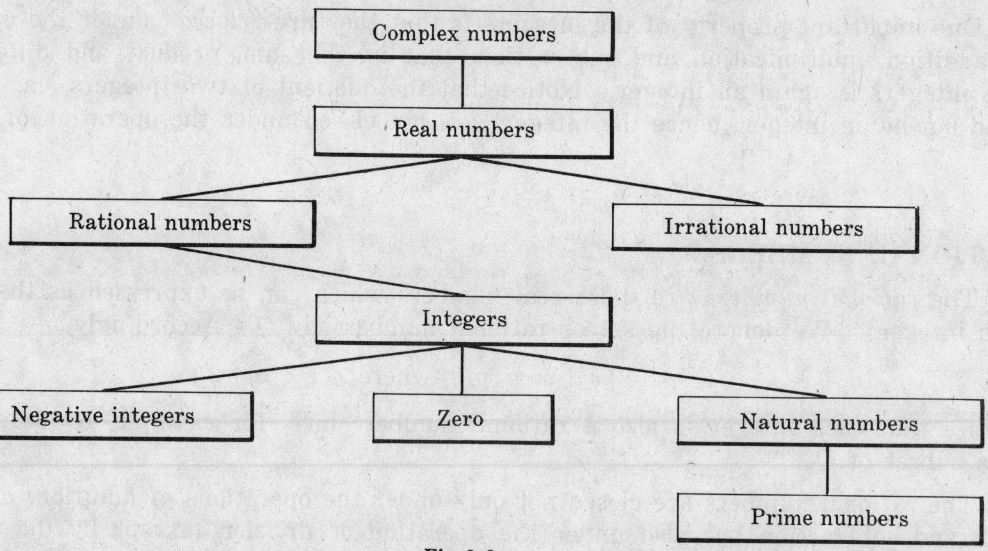

Fig. 3-2

DECIMALS AND REAL NUMBERS

Every real number can be represented by a "nonterminating decimal". The decimal representation of a rational number p/q can be found by "dividing the denominator q into the numerator p". If the indicated division terminates, as for

$$3/8 = .375$$

we write

$$3/8 = .375000\ldots$$

or

$$3/8 = .374999\ldots$$

If the indicated division of q into p does not terminate, then it is known that a block of digits will continually be repeated; for example,

$$2/11 = .181818\ldots$$

We now state the basic fact connecting decimals and real numbers. The rational numbers correspond precisely to those decimals in which a block of digits is continually repeated, and the irrational numbers correspond to the other nonterminating decimals.

INEQUALITIES

The concept of "order" is introduced in the real number system by the

Definition: The real number a is *less than* the real number b, written

$$a < b$$

if $b - a$ is a positive number.

The following properties of the relation $a < b$ can be proven. Let a, b and c be real numbers; then:

P_1: Either $a < b$, $a = b$ or $b < a$.

P_2: If $a < b$ and $b < c$, then $a < c$.

P_3: If $a < b$, then $a + c < b + c$.

P_4: If $a < b$ and c is positive, then $ac < bc$.

P_5: If $a < b$ and c is negative, then $bc < ac$.

Geometrically, if $a < b$ then the point a on the real line lies to the left of the point b.

We also denote $a < b$ by

$$b > a$$

which reads "b is *greater than* a". Furthermore, we write

$$a \leq b \quad \text{or} \quad b \geq a$$

if $a < b$ or $a = b$, that is, if a is not greater than b.

Example 1.1: $2 < 5$; $-6 \leq -3$ and $4 \leq 4$; $5 > -8$.

Example 1.2: The notation $x < 5$ means that x is a real number which is less than 5; hence x lies to the left of 5 on the real line.

The notation $2 < x < 7$ means $2 < x$ and also $x < 7$; hence x will lie between 2 and 7 on the real line.

Remark 3.1: Notice that the concept of order, i.e. the relation $a < b$, is defined in terms of the concept of positive numbers. The fundamental property of the positive numbers which is used to prove properties of the relation $a < b$ is that the positive numbers are closed under the operations of addition and multiplication. Moreover, this last fact is intimately connected with the fact that the natural numbers are also closed under the operations of addition and multiplication.

Remark 3.2: The following statements are true when a, b, c are any real numbers:

(1) $a \leq a$

(2) If $a \leq b$ and $b \leq a$ then $a = b$.

(3) If $a \leq b$ and $b \leq c$ then $a \leq c$.

ABSOLUTE VALUE

The absolute value of a real number x, denoted by

$$|x|$$

is defined by the formula

$$|x| = \begin{cases} x \text{ if } x \geq 0 \\ -x \text{ if } x < 0 \end{cases}$$

that is, if x is positive or zero then $|x|$ equals x, and if x is negative then $|x|$ equals $-x$. Consequently, the absolute value of any number is always non-negative, i.e. $|x| \geq 0$ for every $x \in R^\#$.

Geometrically speaking, the absolute value of x is the distance between the point x on the real line and the origin, i.e. the point 0. Moreover, the distance between any two points, i.e. real numbers, a and b is $|a - b| = |b - a|$.

Example 2.1: $|-2| = 2$, $|7| = 7$, $|-\pi| = \pi$

$|3 - 8| = |-5| = 5$, $|8 - 3| = |5| = 5$, $|-3 - 4| = |-7| = 7$.

Example 2.2: The statement

$$|x| < 5$$

can be interpreted to mean that the distance between x and the origin is less than 5, i.e. x must lie between -5 and 5 on the real line. In other words,

$$|x| < 5 \quad \text{and} \quad -5 < x < 5$$

have identical meaning. Similarly,

$$|x| \leq 5 \quad \text{and} \quad -5 \leq x \leq 5$$

have identical meaning.

INTERVALS

Consider the following sets of numbers:

$$A_1 = \{x \mid 2 < x < 5\}$$
$$A_2 = \{x \mid 2 \leqq x \leqq 5\}$$
$$A_3 = \{x \mid 2 < x \leqq 5\}$$
$$A_4 = \{x \mid 2 \leqq x < 5\}$$

Notice that the four sets contain only the points which lie between 2 and 5 with the possible exceptions of 2 and/or 5. We call these sets intervals, the numbers 2 and 5 being the endpoints of each interval. Moreover, A_1 is an *open interval* as it does not contain either endpoint; A_2 is a *closed interval* as it contains both endpoints; and A_3 and A_4 are *open-closed* and *closed-open* respectively.

We display, i.e. graph, these sets on the real line as follows:

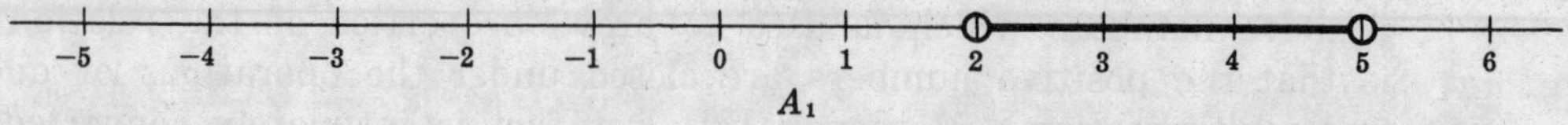

A_1

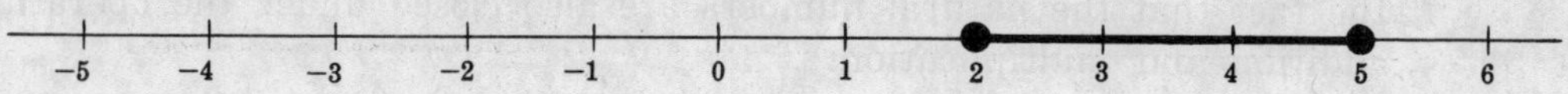

A_2

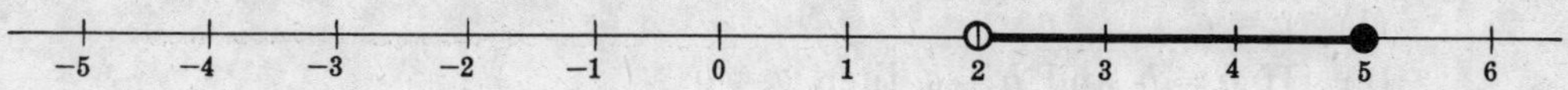

A_3

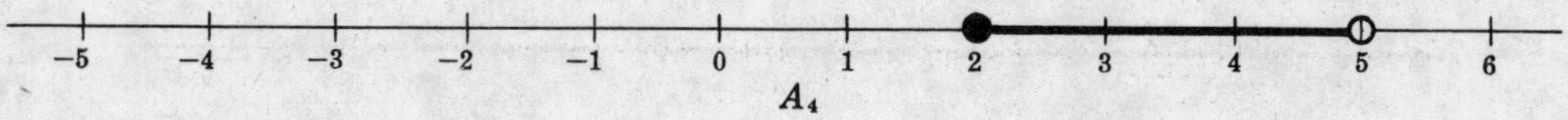

A_4

Notice that in each diagram we circle the endpoints 2 and 5 and thicken (or shade) the line segment between the points. If an interval includes an endpoint, then this is denoted by shading the circle about the endpoint.

Since intervals appear very often in mathematics, a shorter notation is frequently used to designate intervals. Specifically, the above intervals are sometimes denoted by

$$A_1 = (2, 5)$$
$$A_2 = [2, 5]$$
$$A_3 = (2, 5]$$
$$A_4 = [2, 5)$$

Notice that a parenthesis is used to designate an open endpoint, i.e. an endpoint that is not in the interval, and a bracket is used to designate a closed endpoint.

PROPERTIES OF INTERVALS

Let $\mathfrak{I}$ be the family of all intervals on the real line. We include in $\mathfrak{I}$ the null set $\emptyset$ and single points $a = [a, a]$. Then the intervals have the following properties:

(1) The intersection of two intervals is an interval, that is,

$$A \,\varepsilon\, \mathfrak{I},\; B \,\varepsilon\, \mathfrak{I} \quad \text{implies} \quad A \cap B \,\varepsilon\, \mathfrak{I}$$

(2) The union of two non-disjoint intervals is an interval, that is,

$$A \,\varepsilon\, \mathfrak{I},\; B \,\varepsilon\, \mathfrak{I},\; A \cap B \neq \emptyset \quad \text{implies} \quad A \cup B \,\varepsilon\, \mathfrak{I}$$

(3) The difference of two non-comparable intervals is an interval, that is,

$$A \,\varepsilon\, \mathfrak{I},\; B \,\varepsilon\, \mathfrak{I},\; A \not\subset B,\; B \not\subset A \quad \text{implies} \quad A - B \,\varepsilon\, \mathfrak{I}$$

Example 3.1: Let $A = [2, 4)$, $B = (3, 8)$. Then

$$A \cap B = (3, 4), \qquad A \cup B = [2, 8)$$
$$A - B = [2, 3], \qquad B - A = [4, 8)$$

INFINITE INTERVALS

Sets of the form

$$A = \{x \mid x > 1\},$$
$$B = \{x \mid x \geqq 2\},$$
$$C = \{x \mid x < 3\},$$
$$D = \{x \mid x \leqq 4\},$$
$$E = \{x \mid x \in R^{\#}\}$$

are called *infinite intervals* and are also denoted by

$$A = (1, \infty), \quad B = [2, \infty), \quad C = (-\infty, 3) \quad D = (-\infty, 4], \quad E = (-\infty, \infty)$$

We plot these infinite intervals on the real line as follows:

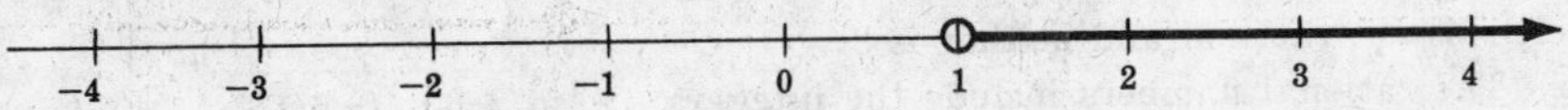

A is shaded

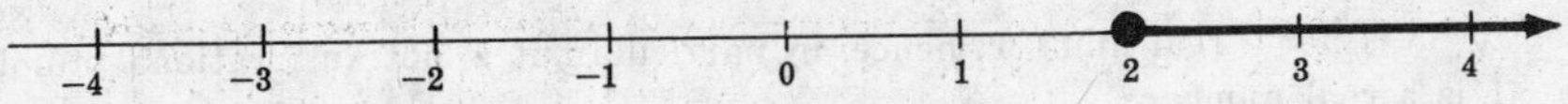

B is shaded

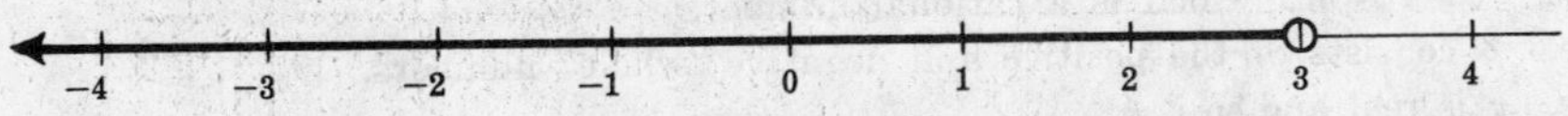

C is shaded

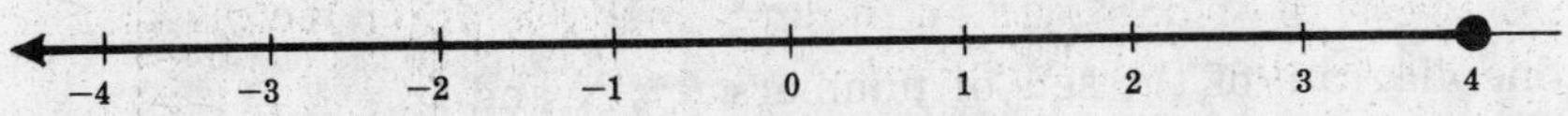

D is shaded

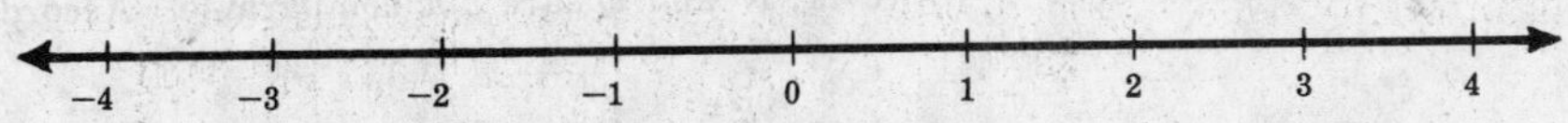

E is shaded

BOUNDED AND UNBOUNDED SETS

Let A be a set of numbers; then A is called a *bounded* set if A is the subset of a finite interval. An equivalent definition of boundedness is

Definition 3.1: Set A is *bounded* if there exists a positive number M such that

$$|x| \leqq M$$

for all $x \in A$. A set is called *unbounded* if it is not bounded.

Notice, then, that A is a subset of the finite interval $[-M, M]$.

Example 4.1: Let $A = \{1, 1/2, 1/3, \ldots\}$. Then A is bounded since A is certainly a subset of the closed interval $[0, 1]$.

Example 4.2: Let $A = \{2, 4, 6, \ldots\}$. Then A is an unbounded set.

Example 4.3: Let $A = \{7, 350, -473, 2322, 42\}$. Then A is bounded.

Remark 3.3: If a set A is finite then it is necessarily bounded. If a set is infinite then it can be either bounded as in Example 4.1 or unbounded as in Example 4.2.

Solved Problems

SETS OF NUMBERS

In the following problems, let $R^\#, Q, Q', Z, N$ and P denote respectively the real numbers, rational numbers, irrational numbers, integers, natural numbers and prime numbers.

1. State whether each of the following is true or false.

(1) $-7 \in N$
(2) $\sqrt{2} \in Q'$
(3) $4 \in Z$
(4) $9 \in P$
(5) $3\pi \in Q$
(6) $-6 \in Q$
(7) $11 \in P$
(8) $\frac{1}{2} \in Z$
(9) $\sqrt{-5} \in Q'$
(10) $1 \in R^\#$
(11) $\sqrt[3]{8} \in N$
(12) $\sqrt{9/4} \in Q'$
(13) $-2 \in Z$
(14) $\pi^2 \in R^\#$
(15) $\sqrt{-4} \in R^\#$

Solution:

(1) False. N contains only the positive integers; -7 is negative.
(2) True. $\sqrt{2}$ cannot be expressed as the ratio of two integers; hence $\sqrt{2}$ is not rational.
(3) True. Z, the set of integers, contains all the "whole" numbers, positive and negative.
(4) False. 3 divides 9; so 9 is not prime.
(5) False. π is not rational and neither is 3π.
(6) True. The rational numbers include the integers. Also, $-6 = (-6/1)$.
(7) True. 11 has no divisors except 11 and 1; hence 11 is prime.
(8) False. $\frac{1}{2}$ is not an integer.
(9) False. $\sqrt{-5}$ is not a real number; hence, in particular, it is not an irrational number.
(10) True. 1 is a real number.
(11) True. $\sqrt[3]{8} = 2$ which is a positive integer.
(12) False. $\sqrt{9/4} = 3/2$ which is a rational number.
(13) True. Z consists of the positive and negative "whole" numbers.
(14) True. π is real and so is π^2.
(15) False. $\sqrt{-4} = 2i$ is not real.

2. Draw a line diagram of the sets of numbers $R^\#, N$ and Q'.

Solution:

Both N and Q' are subsets of $R^\#$. However, N and Q' are not comparable. Accordingly, the line diagram is

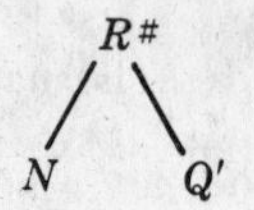

3. To which of the sets $R^\#, Q, Q', Z, N$ and P does each of the following numbers belong? (1) $-3/4$, (2) 13, (3) $\sqrt{-7}$.

Solution:

(1) $-3/4 \in Q$, the rational numbers, since it is the ratio of the two integers -3 and 4. Also, $-3/4 \in R^\#$ since $Q \subset R^\#$.
(2) $13 \in P$, since the only divisors of 13 are 13 and 1. 13 also belongs to N, Z, Q and $R^\#$, since P is a subset of each of them.
(3) $\sqrt{-7}$ is not a real number; hence it does not belong to any of the given sets.

4. Let $E = \{2, 4, 6, \ldots\}$ and $F = \{1, 3, 5, \ldots\}$. Are E and F closed under the operations of (1) addition, (2) multiplication?

Solution:

(1) The sum of two even numbers is even; hence E is closed under the operation of addition. The sum of two odd numbers is not odd; hence F is not closed under the operation of addition.
(2) The product of two even numbers is even, and the product of two odd numbers is odd; hence both E and F are closed under the operation of multiplication.

5. Consider the sets $R^{\#}, Q, Q', Z, N$ and P. Which of these are *not* closed under the operations of (1) addition, (2) subtraction?

Solution:

(1) Q' and P. For example, $-\sqrt{2} \in Q'$ and $\sqrt{2} \in Q'$, but $-\sqrt{2}+\sqrt{2} = 0 \notin Q'$; $3 \in P$ and $5 \in P$, but $3+5 = 8 \notin P$.

(2) Q', N and P. For example, $\sqrt{2} \in Q'$ but $\sqrt{2}-\sqrt{2} = 0 \notin Q'$; $3 \in N$ and $7 \in N$, but $3-7 = -4 \notin N$; $7 \in P$ and $3 \in P$, but $7-3 = 4 \notin P$.

INEQUALITIES AND ABSOLUTE VALUES

6. Write the following statements in notational form:

(1) a is less than b.
(2) a is not greater than or equal to b.
(3) a is less than or equal to b.
(4) a is not less than b.
(5) a is greater than or equal to b.
(6) a is not greater than b.

Solution:

Recall that a vertical or slant line through a symbol designates the opposite meaning of the symbol. We write:

(1) $a < b$, (2) $a \not\geqq b$, (3) $a \leqq b$, (4) $a \nless b$, (5) $a \geqq b$, (6) $a \ngtr b$.

7. Insert between the following pairs of numbers the correct symbol: $<, >$ or $=$.

(1) $3 \ldots\ldots -9$
(2) $-4 \ldots\ldots -8$
(3) $3^2 \ldots\ldots 7$
(4) $-5 \ldots\ldots 3$
(5) $3^2 \ldots\ldots 9$
(6) $-\pi \ldots\ldots \pi/2$

Solution:

We write $a < b$ if $b-a$ is positive, $a > b$ if $b-a$ is negative, and $a = b$ if $b-a = 0$. Then

(1) $3 > -9$, (2) $-4 > -8$, (3) $3^2 > 7$, (4) $-5 < 3$, (5) $3^2 = 9$, (6) $-\pi < \pi/2$.

8. Prove: If $a < b$ and $b < c$, then $a < c$.

Solution:

By definition, $a < b$ and $b < c$ means $b-a$ is positive and $c-b$ is positive. Since the sum of two positive numbers is positive,

$$(b-a) + (c-b) = c-a$$

is positive. Therefore, by definition, $a < c$.

9. Prove: If $a < b$, then $a+c < b+c$.

Solution:

Notice that

$$(b+c) - (a+c) = b-a$$

which, by hypothesis, is positive. Hence $a+c < b+c$.

10. Rewrite the following geometric relationships between real numbers using the inequality notation:

(1) y lies to the right of 8
(2) z lies to the left of 0.
(3) x lies between -3 and 7.
(4) w lies between 5 and 1.

Solution:

Recall that $a < b$ means a lies to the left of b on the real line. Accordingly,

(1) $y > 8$ or $8 < y$.

(2) $z < 0$.

(3) $-3 < x$ and $x < 7$ or, more briefly, $-3 < x < 7$.

(4) $5 > w$ and $w > 1$, or $w < 5$ and $1 < w$. Also, $1 < w < 5$. It is not customary to write $5 > w > 1$.

11. Give the precise definition of the absolute value function, that is, define $|x|$.

Solution:

By definition, $|x| = \begin{cases} x & \text{if } x \geqq 0 \\ -x & \text{if } x < 0. \end{cases}$

12. Suppose $|x| \leqq 0$. Find x.

Solution:

Note that the absolute value of a number is always non-negative, that is, for every number x, $|x| \geqq 0$. By hypothesis, $|x| \leqq 0$; hence $|x| = 0$. Accordingly, $x = 0$.

13. Evaluate:

(1) $|3-5|$ (6) $|-8| + |3-1|$

(2) $|-3+5|$ (7) $|2-5| - |4-7|$

(3) $|-3-5|$ (8) $13 + |-1-4| - 3 - |-8|$

(4) $|-2| - |-6|$ (9) $||-2| - |-6||$

(5) $|3-7| - |-5|$ (10) $|-|-5||$

Solution:

(1) $|3-5| = |-2| = 2$

(2) $|-3+5| = |2| = 2$

(3) $|-3-5| = |-8| = 8$

(4) $|-2| - |-6| = 2 - 6 = -4$

(5) $|3-7| - |-5| = |-4| - |-5| = 4 - 5 = -1$

(6) $|-8| + |3-1| = |-8| + |2| = 8 + 2 = 10$

(7) $|2-5| - |4-7| = |-3| - |-3| = 3 - 3 = 0$

(8) $13 + |-1-4| - 3 - |-8| = 13 + |-5| - 3 - |-8| = 13 + 5 - 3 - 8 = 7$

(9) $||-2| - |-6|| = |2-6| = |-4| = 4$

(10) $|-|-5|| = |-5| = 5$

14. Rewrite so that x is alone between the inequality signs:

(1) $3 < x-4 < 8$ (3) $-9 < 3x < 12$ (5) $3 < 2x-5 < 7$

(2) $-1 < x+3 < 2$ (4) $-6 < -2x < 4$ (6) $-7 < -2x+3 < 5$

Solution:

(1) By P_3, add 4 to each side of $3 < x-4 < 8$ to get $7 < x < 12$.

(2) By P_3, add -3 to each side of $-1 < x+3 < 2$ to get $-4 < x < -1$.

(3) By P_4, multiply each side of $-9 < 3x < 12$ by $\frac{1}{3}$ to get $-3 < x < 4$.

(4) By P_5, multiply each side of $-6 < -2x < 4$ by $-\frac{1}{2}$ and then reverse the inequalities to get $-2 < x < 3$.

(5) Add 5 to each side of $3 < 2x-5 < 7$ to get $8 < 2x < 12$. Then multiply by $\frac{1}{2}$ to obtain $4 < x < 6$.

(6) Add -3 to each side of $-7 < -2x+3 < 5$ to get $-10 < -2x < 2$. Then multiply by $-\frac{1}{2}$ and reverse the inequalities to obtain $-1 < x < 5$.

15. Rewrite without the absolute value sign:

(1) $|x| < 3$, (2) $|x-2| < 5$, (3) $|2x+3| < 7$.

Solution:

(1) $-3 < x < 3$

(2) $-5 < x-2 < 5$ or $-3 < x < 7$

(3) $-7 < 2x+3 < 7$ or $-10 < 2x < 4$ or $-5 < x < 2$

16. Rewrite using an absolute value sign: (1) $-2 < x < 6$, (2) $4 < x < 10$.

Solution:

Note first that we rewrite the inequality so that a number and its negative appear at the ends of the inequality.

(1) Add -2 to each side of $-2 < x < 6$ to get

$$-4 < x - 2 < 4$$

which is equivalent to

$$|x-2| < 4$$

(2) Add -7 to each side of $4 < x < 10$ to get

$$-3 < x - 7 < 3$$

which is equivalent to

$$|x-7| < 3$$

17. Insert between the following pairs of numbers the correct symbol: $\leq$ or $\geq$.
(1) $1 \ldots . -7$, (2) $-2 \ldots . -9$, (3) $2^3 \ldots . 8$, (4) $3 \ldots . 7$, (5) $3^2 \ldots . 9$, (6) $3^2 \ldots . -11$.

Solution:

Note that $a \leq b$ is true if either $a < b$ or $a = b$, and $a \geq b$ is true if either $a > b$ or $a = b$.

(1) $1 \geq -7$, since $1 > -7$.
(2) $-2 \geq -9$, since $-2 > -9$.
(3) Both $2^3 \leq 8$ and $2^3 \geq 8$ are correct, since $2^3 = 8$.
(4) $3 \leq 7$, since $3 < 7$.
(5) Both $3^2 \leq 9$ and $3^2 \geq 9$ are correct, since $3^2 = 9$.
(6) $3^2 \geq -11$, since $3^2 > -11$.

INTERVALS

18. Rewrite the following intervals in set-builder form:
(1) $M = [-3, 5)$, (2) $S = (3, 8)$, (3) $T = [0, 4]$, (4) $W = (-7, -2]$.

Solution:

Recall that the parenthesis means that the endpoint does not belong to the interval; and the bracket means that the endpoint belongs to the interval. Accordingly,

$$M = \{x \mid -3 \leq x < 5\}$$
$$S = \{x \mid 3 < x < 8\}$$
$$T = \{x \mid 0 \leq x \leq 4\}$$
$$W = \{x \mid -7 < x \leq -2\}$$

19. Plot the intervals $R = (-1, 2]$, $S = [-2, 2)$, $T = (0, 1)$ and $W = [1, 3]$ on the real line.

Solution:

To plot R, first circle its endpoints -1 and 2:

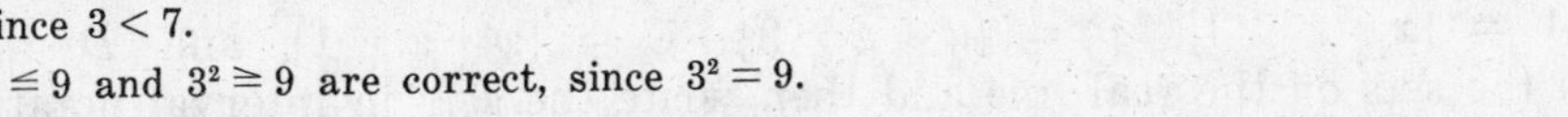

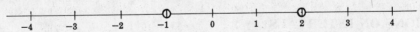

Since the endpoint 2 belongs to R, shade the circle around 2:

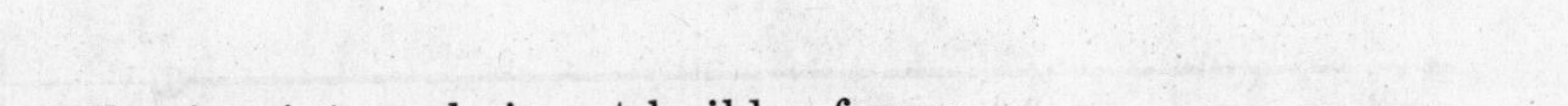

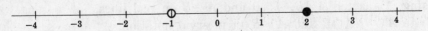

Finally, thicken the line between the endpoints:

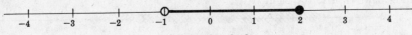

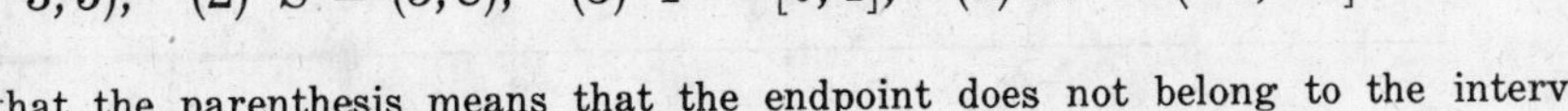

R is sketched

Similarly,

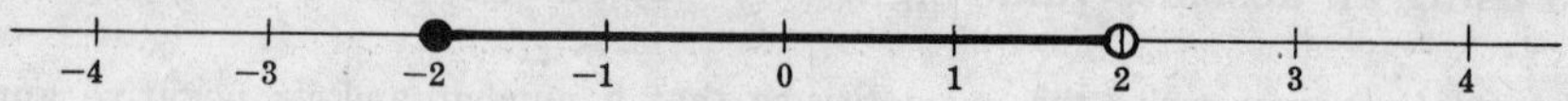

S is plotted

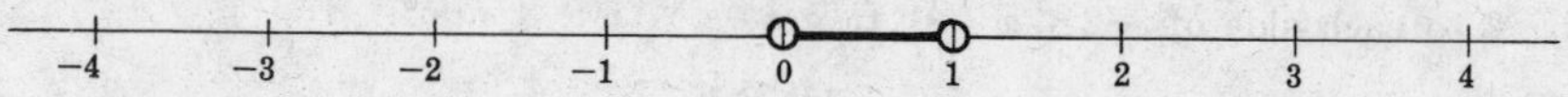

T is plotted

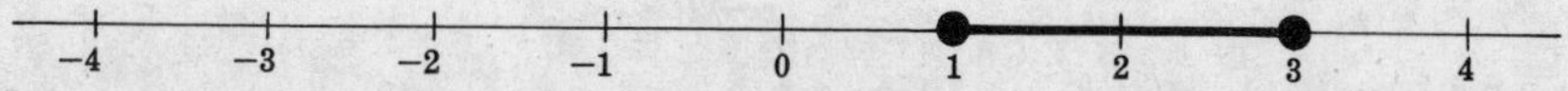

W is plotted

20. Let the real number a be less than b, i.e. $a < b$. Construct the line diagram of the four intervals (a, b), $[a, b]$, $(a, b]$ and $[a, b)$.

Solution:

The open interval (a, b) is a subset of both half-open intervals $[a, b)$ and $(a, b]$, which are subsets of the closed interval $[a, b]$. Thus

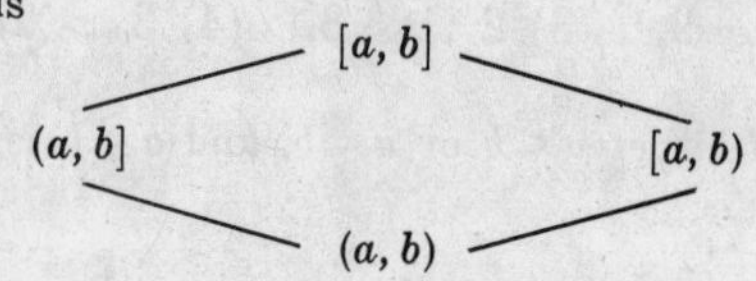

21. Let $A = \{x \mid x < 3\}$, $B = \{x \mid x \geqq 2\}$, $C = \{x \mid x \leqq 1\}$ and $D = \{x \mid x > -1\}$. Sketch the sets on the real line and then write the sets in interval notation.

Solution:

The sets are all infinite intervals. Circle the end point and draw a ray on the side of the end point where the set lies, as follows:

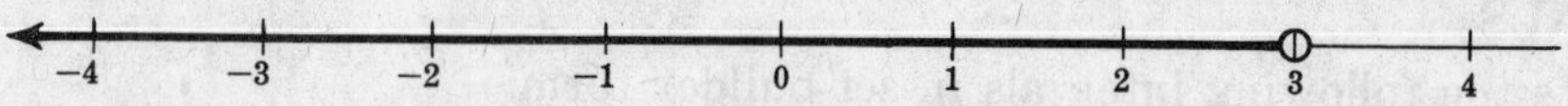

A

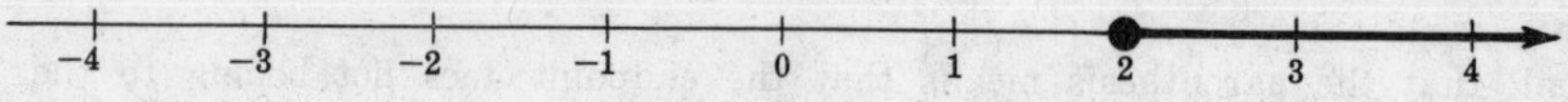

B

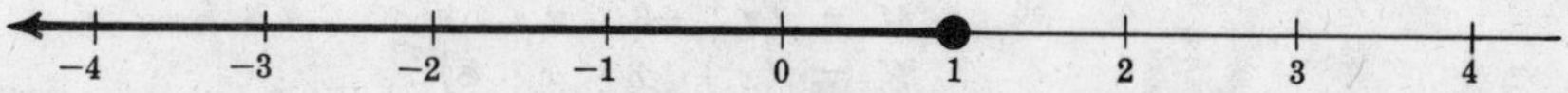

C

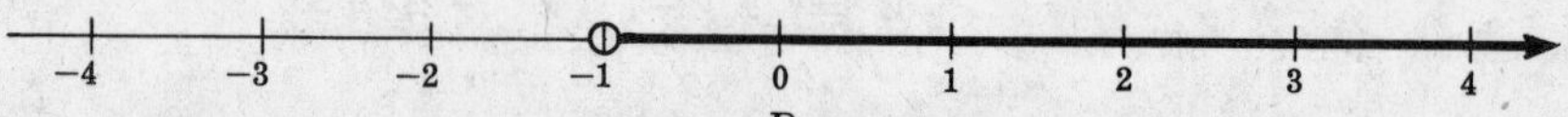

D

In interval notation, the sets are defined by $A = (-\infty, 3)$, $B = [2, \infty)$, $C = (-\infty, 1]$ and $D = (-1, \infty)$. Note that a parenthesis is used on the side with the infinity symbol.

OPERATIONS ON INTERVALS

22. Let $A = [-3, 1)$ and $B = [-1, 2]$.

(1) Sketch A and B on the same real line.

(2) Using (1), sketch $A \cup B$, $A \cap B$ and $A - B$ on real lines.

(3) Write $A \cup B$, $A \cap B$ and $A - B$ in interval notation.

Solution:

(1) On a real line, shade A with strokes slanting upward to the right (////) and shade B with strokes slanting downward to the right (\\\\):

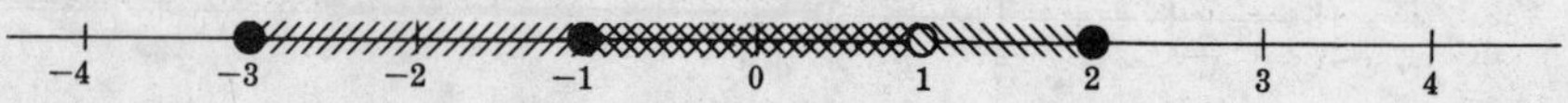

A and B are shaded

(2) $A \cup B$ consists of points in either interval, that is, points which are shaded:

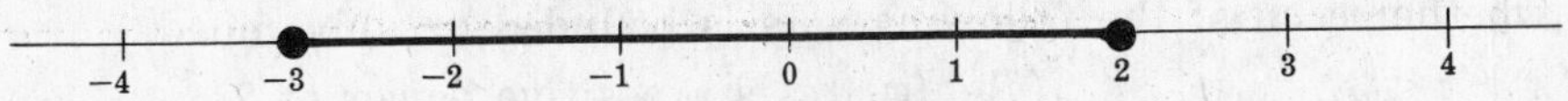

$A \cup B$ is shaded

$A \cap B$ consists only of those points which are in both A and B, i.e. only the cross-hatched points:

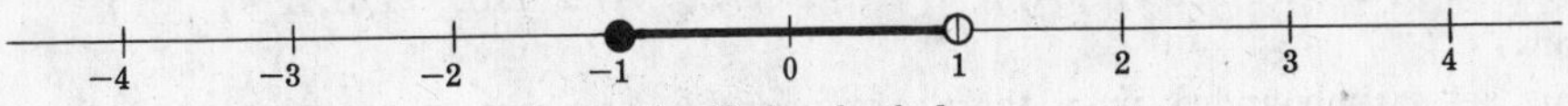

$A \cap B$ is shaded

$A - B$ consists of the points in A which are not in B, i.e. of the points which are shaded //// but not cross-hatched:

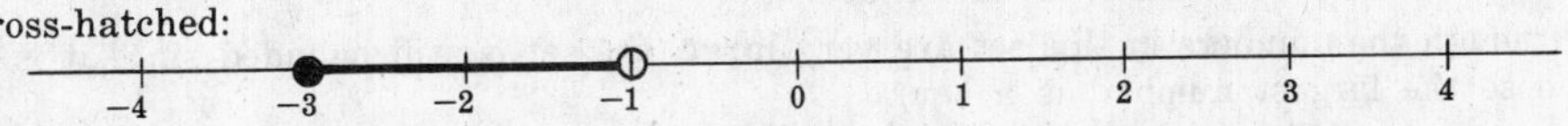

(3) The above sketches indicate that $A \cup B = [-3, 2]$, $A \cap B = [-1, 1)$, and $A - B = [-3, -1)$.

23. Under what conditions will the union of two disjoint intervals be an interval?

Solution:

First, the right endpoint of one of the intervals must be the left endpoint of the other. Secondly, at the common endpoint one interval must be closed and the other interval open. For example, let $M = [-3, 4)$ and $N = [4, 7)$. The right endpoint of M is the left endpoint of N; and M is open at 4 and N is closed at 4. $M \cup N = [-3, 7)$ and $M \cap N = \emptyset$.

24. In each case, sketch on a real line and write the resultant set in interval notation:

(*a*) $\{x \mid x \geqq -1\} \cap \{x \mid -3 < x < 2\}$

(*b*) $\{x \mid x < 2\} \cup \{x \mid x \geqq 0\}$

(*c*) $\{x \mid -3 < x \leqq 1\} \cap \{x \mid x > 2\}$

(*d*) $\{x \mid -2 < x \leqq 3\} \cup \{x \mid x < 1\}$

(*e*) $\{x \mid -3 \leqq x \leqq 0\} \cap \{x \mid -2 < x < 3\}$

Solution:

In each case, sketch the left set with strokes //// and the set on the right with strokes \\\\ .

(*a*)

The intersection consists of the cross-hatched points, i.e. is the set $[-1, 2)$.

(*b*)

The union consists of all points slashed; these constitute the set $(-\infty, \infty)$, the entire real line.

(*c*)

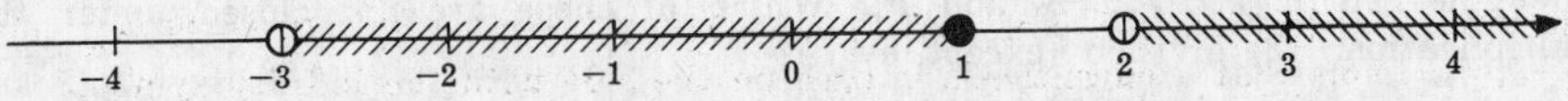

The intersection is the null set since there are no points which are cross-hatched, i.e. no point lies in both intervals.

(*d*)

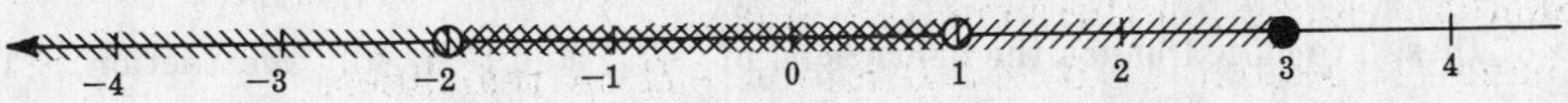

The union is the infinite interval $(-\infty, 3]$.

(*e*)

The intersection is the set of cross-hatched points, i.e. the set $(-2, 0]$.

BOUNDED AND UNBOUNDED SETS

25. State whether each of the following sets is bounded or unbounded.

(*a*) $\{x \mid x < 3\}$ (*d*) $\{x \mid x \text{ is a positive power of } 2\}$
(*b*) $\{1, 3, 5, 7, \ldots\}$ (*e*) $\{2, 5, 2, 5, \ldots\}$
(*c*) $\{2^{18}, 3^{-4}, 5, 0, 8^{35}\}$ (*f*) $\{1, -1, 1/2, -1/2, 1/3, -1/3, 1/4, \ldots\}$

Solution:

(*a*) This set is unbounded since there are negative numbers whose absolute values are arbitrarily large. In fact, this set is an infinite interval $(-\infty, 3)$, which cannot be contained in a finite interval.

(*b*) This set, the odd numbers, is unbounded.

(*c*) Although the numbers in this set are very large, the set is still bounded since it is finite. We can choose the largest number as a bound.

(*d*) The set of positive powers of 2, which are $2, 4, 8, 16, \ldots$, is an unbounded set.

(*e*) Since this set consists only of the two numbers 2 and 5, it is bounded.

(*f*) Although there are an infinite number of numbers in this set, the set is still bounded. It certainly is contained in the interval $[-1, 1]$.

26. If two sets W and V are bounded, what can be said about the union and intersection of these sets?

Solution:

Both the union and the intersection of bounded sets are bounded.

27. If two sets R and S are unbounded, what can be said about the union and intersection of these sets?

Solution:

The union of R and S must be unbounded, but the intersection of R and S could be either bounded or unbounded. For example if $R = (-\infty, 3)$ and $S = [-2, \infty)$, then the intersection of these infinite intervals is the finite, and therefore bounded, interval $[-2, 3)$. But if $R = (3, \infty)$ and $S = [-2, \infty)$, then the intersection is the infinite, and therefore unbounded, interval $(3, \infty)$.

Supplementary Problems

SETS OF NUMBERS

28. State whether each of the following is true or false:

(1) $\pi \in Q$ (3) $-3 \in N$ (5) $7 \in P$ (7) $-5 \in Z$ (9) $15 \in P$ (11) $2/3 \in Z$
(2) $3 \in Z$ (4) $\sqrt{-1} \in Q'$ (6) $\sqrt{9} \in N$ (8) $\sqrt{-3} \in R\#$ (10) $\sqrt[3]{2} \in Q'$ (12) $2 \in Q$

29. Draw a line diagram of the sets $R\#$, Z, Q' and P.

30. Consider the sets $R\#, Q, Q', Z, N$ and P. Which of these are not closed under the operations of (1) multiplication, (2) division (except by 0)?

31. Consider the sets

$$\begin{aligned} A &= \{x \mid x = 2^n,\ n \in N\} = \{2, 4, 8, 16, \ldots\} \\ B &= \{x \mid x = 3n,\ n \in N\} = \{3, 6, 9, 12, \ldots\} \\ C &= \{x \mid x = 3n,\ n \in Z\} = \{\ldots, -6, -3, 0, 3, 6, \ldots\} \end{aligned}$$

Which of these are closed under the operations of (1) addition, (2) subtraction, (3) multiplication?

32. State whether each of the following is (*a*) always true, (*b*) sometimes true, (*c*) never true. Here, $a \neq 0$ and $b \neq 0$.

(1) $a \in Z$, $b \in Q$ and $a - b \in N$. (5) $a \in P$, $b \in P$ and $a + b \in P$.
(2) $a \in P$, $b \in Q'$ and $ab \in Q$. (6) $a \in N$, $b \in Q'$ and $a + b \in Q'$.
(3) $a \in N$, $b \in Z$ and $ab \in Z$. (7) $a \in Z$, $b \in Q$ and $a/b \in N$.
(4) $a \in N$, $b \in Q'$ and $a/b \in Q$. (8) $a \in P$, $b \in Z$ and $b/a \in Q$.

INEQUALITIES AND ABSOLUTE VALUES

33. Write the following statements in notational form:

(1) x is not greater than y.
(2) The absolute value of x is less than 4.
(3) r is not less than y.
(4) r is greater than or equal to t.

34. Insert between the following pairs of numbers the correct symbol: <, > or =. Here, x denotes any real number.

(1) $5 \ldots . -8$
(2) $|x| \ldots . -3$
(3) $2^3 \ldots . 8$
(4) $-\pi \ldots . \pi/3$
(5) $2^3 \ldots . 19$
(6) $-|x| \ldots . 1$
(7) $-7 \ldots . 4$
(8) $-2 \ldots . -5$

35. Rewrite the geometric relationships between the numbers using the inequality notation.

(1) a lies to the right of b. (2) x lies to the left of y. (3) r lies between -5 and -8.

36. Evaluate:

(1) $|4-7|$
(2) $|-4-7|$
(3) $|-4+7|$
(4) $|3|-|-5|$
(5) $|2-3|+|-6|$
(6) $|-2|+|1-5|$
(7) $|3-8|-|2-1|$
(8) $||-3|-|-9||$
(9) $||2-6|-|1-9||$

37. Rewrite so that x is alone between the inequality signs:

(1) $-2 < x-3 < 4$
(2) $-5 < x+2 < 1$
(3) $-12 < 4x < -8$
(4) $4 < -2x < 10$
(5) $-1 < 2x-3 < 5$
(6) $-3 < 5-2x < 7$

38. Rewrite without the absolute value sign:

(1) $|x| \leq 8$, (2) $|x-3| < 8$, (3) $|2x+4| < 8$.

39. Rewrite using an absolute value sign:

(1) $-3 < x < 9$, (2) $2 \leq x \leq 8$, (3) $-7 < x < -1$.

40. Prove P_5: If $a < b$ and c is negative, then $bc < ac$. (*Note*: Assume that the product of a negative number and a positive number is negative.)

INTERVALS

41. Rewrite each of the following intervals in set-builder form:

$A = [-3,1)$, $B = [1,2]$, $C = (-1,3]$, $D = (-4,2)$.

42. Which of the sets in Problem 41 is (1) an open interval, (2) a closed interval?

43. Sketch each of the sets in Problem 41 on a real line.

44. Rewrite each of the following infinite intervals in interval notation:

$$R = \{x \mid x \leq 2\}, \quad S = \{x \mid x > -1\}, \quad T = \{x \mid x < -3\}.$$

45. Sketch each of the sets in Problem 44 on a real line.

46. Let $A = [-4,2)$, $B = (-1,6)$, $C = (-\infty,1]$. Find, and write in interval notation,

(1) $A \cup B$
(2) $A \cap B$
(3) $A - B$
(4) $B - A$
(5) $A \cup C$
(6) $A \cap C$
(7) $A - C$
(8) $C - A$
(9) $B \cup C$
(10) $B \cap C$
(11) $B - C$
(12) $C - B$

BOUNDED AND UNBOUNDED SETS

47. Write each of the following sets in tabular form and state whether it is bounded or unbounded.

$$E = \{x \mid x = (1/n),\ n \in N\} \qquad G = \{x \mid x = (\tfrac{1}{2})^n,\ n \in N\}$$
$$F = \{x \mid x = 3^n,\ n \in N\} \qquad H = \{x \mid x \in N,\ x < 2576\}$$

48. Are the following statements (*a*) always true, (*b*) sometimes true, (*c*) never true.

(1) If A is finite, A is bounded.
(2) If A is infinite, A is bounded.
(3) If A is a subset of $[-23,79]$, A is finite.
(4) If A is a subset of $[-23,79]$, A is unbounded.

Answers to Supplementary Problems

28. (1) F, (2) T, (3) F, (4) F, (5) T, (6) T, (7) T, (8) F, (9) F, (10) T, (11) F, (12) T.

29.

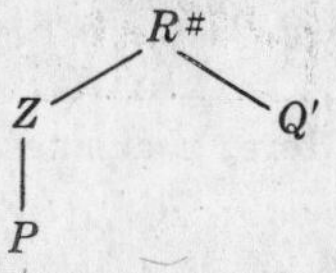

30. (1) Q' and P (2) Q', Z, N and P

31. (1) B and C (2) C (3) A, B and C

32. (1) Sometimes true. (2) Never true. (3) Always true. (4) Never true. (5) Sometimes true. (6) Always true. (7) Sometimes true. (8) Always true.

33. (1) $x \ngtr y$ (2) $|x| < 4$ (3) $r \nless y$ (4) $r \geqq t$

34. (1) $>$ (2) $>$ (3) $=$ (4) $<$ (5) $<$ (6) $<$ (7) $<$ (8) $>$

35. (1) $a > b$ or $b < a$ (2) $x < y$ (3) $-8 < r < -5$

36. (1) 3 (2) 11 (3) 3 (4) −2 (5) 7 (6) 6 (7) 4 (8) 6 (9) 4

37. (1) $1 < x < 7$ (2) $-7 < x < -1$ (3) $-3 < x < -2$ (4) $-5 < x < -2$ (5) $1 < x < 4$ (6) $-1 < x < 4$

38. (1) $-8 \leqq x \leqq 8$ (2) $-5 < x < 11$ (3) $-6 < x < 2$

39. (1) $|x-3| < 6$ (2) $|x-5| \leqq 3$ (3) $|x+4| < 3$

40. Since $a < b$, $b - a$ is positive. Since c is negative, the product $(b-a)c = bc - ac$ is also negative; hence $ac - bc$ is positive, i.e. $bc < ac$.

41. $A = \{x \mid -3 \leqq x < 1\}$ $\quad B = \{x \mid 1 \leqq x \leqq 2\}$ $\quad C = \{x \mid -1 < x \leqq 3\}$ $\quad D = \{x \mid -4 < x < 2\}$

42. D is an open interval and B is a closed interval.

43.

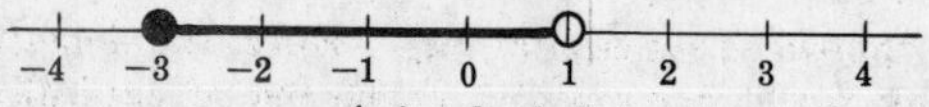

A is shaded

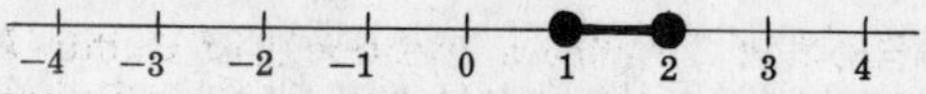

B is shaded

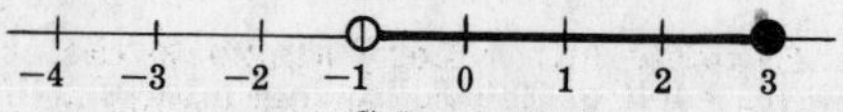

C is shaded

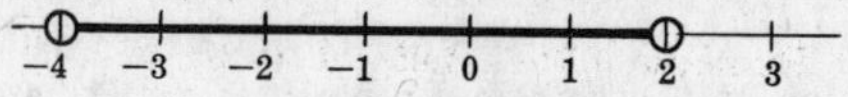

D is shaded

44. $R = (-\infty, 2]$, $S = (-1, \infty)$, $T = (-\infty, -3)$

45.

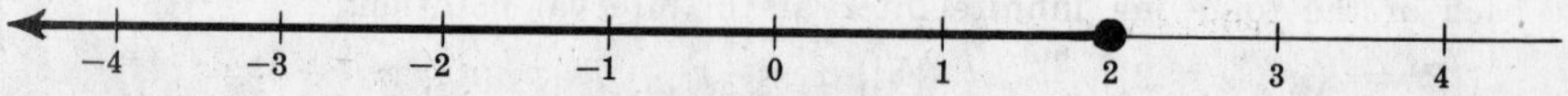

R is shaded

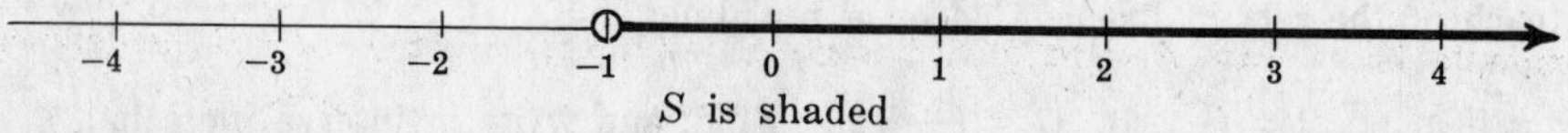

S is shaded

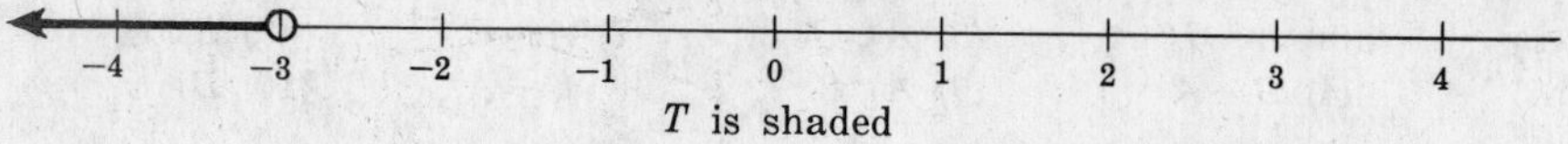

T is shaded

46.

$A \cup B = [-4, 6)$ $\quad A \cap B = (-1, 2)$ $\quad A - B = [-4, -1]$ $\quad B - A = [2, 6)$

$A \cup C = (-\infty, 2)$ $\quad A \cap C = [-4, 1]$ $\quad A - C = (1, 2)$ $\quad C - A = (-\infty, -4)$

$B \cup C = (-\infty, 6)$ $\quad B \cap C = (-1, 1]$ $\quad B - C = (1, 6)$ $\quad C - B = (-\infty, -1]$

47. $E = \{1, \frac{1}{2}, \frac{1}{3}, \frac{1}{4}, \ldots\}$. Bounded. $\quad F = \{3, 9, 27, 81, \ldots\}$. Unbounded.
$G = \{\frac{1}{2}, \frac{1}{4}, \frac{1}{8}, \ldots\}$. Bounded. $\quad H = \{1, 2, 3, \ldots, 2574, 2575\}$. Bounded.

48. (1) Always true. (2) Sometimes true. (3) Sometimes true. (4) Never true.

Chapter 4

Functions

DEFINITION OF A FUNCTION

Suppose that to each element in a set A there is assigned, by some manner or other, a unique element of a set B. We call such assignments a *function.* If we let f denote these assignments, then we write

$$f : A \rightarrow B$$

which reads "f is a function of A into B". The set A is called the *domain* of the function f, and B is called the co-domain of f. Further, if $a \in A$ then the element in B which is assigned to a is called the *image* of a and is denoted by

$$f(a)$$

which reads "f of a".

We list a number of instructive examples of functions.

Example 1.1: Let f assign to each real number its square, that is, for every real number x let $f(x) = x^2$. The domain and co-domain of f are both the real numbers, so we can write

$$f : R^{\#} \rightarrow R^{\#}$$

The image of -3 is 9; hence we can also write $f(-3) = 9$, or $f : -3 \rightarrow 9$.

Example 1.2: Let f assign to each country in the world its capital city. Here, the domain of f is the set of countries in the world; the co-domain of f is the list of capital cities in the world. The image of France is Paris, that is, $f(\text{France}) = \text{Paris}$.

Example 1.3: Let $A = \{a, b, c, d\}$ and $B = \{a, b, c\}$. Define a function f of A into B by the correspondence $f(a) = b$, $f(b) = c$, $f(c) = c$ and $f(d) = b$. By this definition, the image, for example, of b is c.

Example 1.4: Let $A = \{-1, 1\}$. Let f assign to each rational number in $R^{\#}$ the number 1, and to each irrational number in $R^{\#}$ the number -1. Then $f : R^{\#} \rightarrow A$, and f can be defined concisely by

$$f(x) = \begin{cases} 1 & \text{if } x \text{ is rational} \\ -1 & \text{if } x \text{ is irrational} \end{cases}$$

Example 1.5: Let $A = \{a, b, c, d\}$ and $B = \{x, y, z\}$. Let $f : A \rightarrow B$ be defined by the diagram:

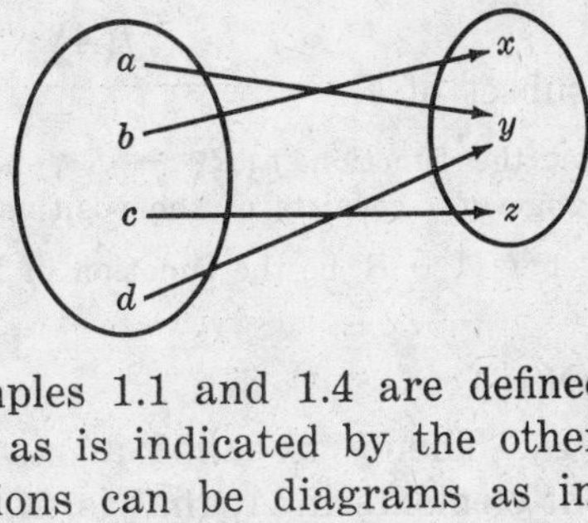

Notice that the functions in Examples 1.1 and 1.4 are defined by specific formulas. But this need not always be the case, as is indicated by the other examples. The rules of correspondence which define functions can be diagrams as in Example 1.5, can be geographical as in Example 1.2, or, when the domain is finite, the correspondence can be listed for each element in the domain as in Example 1.3.

MAPPINGS, OPERATORS, TRANSFORMATIONS

If A and B are sets in general, not necessarily sets of numbers, then a function f of A into B is frequently called a mapping of A into B; and the notation

$$f: A \rightarrow B$$

is then read "f maps A into B". We can also denote a mapping, or function, f of A into B by

$$A \xrightarrow{f} B$$

or by the diagram

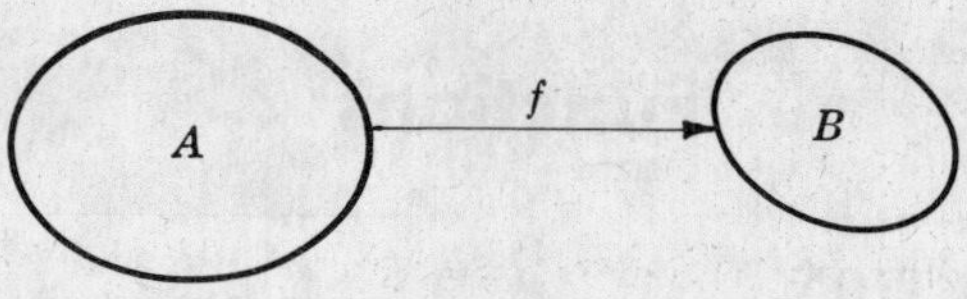

If the domain and co-domain of a function f are both the same set, say

$$f: A \rightarrow A$$

then f is frequently called an *operator* or *transformation* on A. As we will see later, operators are important special cases of functions.

EQUAL FUNCTIONS

If f and g are functions defined on the same domain D and if $f(a) = g(a)$ for every $a \in D$, then the functions f and g are equal and we write

$$f = g$$

Example 2.1: Let $f(x) = x^2$ where x is a real number. Let $g(x) = x^2$ where x is a complex number. Then the function f is not equal to g since they have different domains.

Example 2.2: Let the function f be defined by the diagram

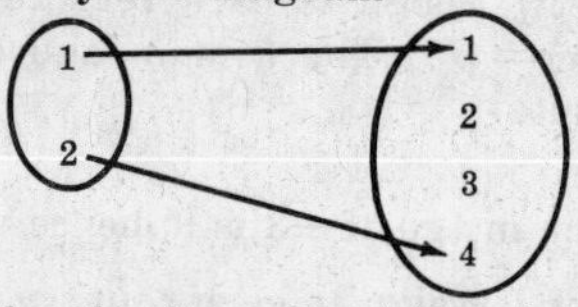

Let a function g be defined by the formula $g(x) = x^2$ where the domain of g is the set $\{1, 2\}$. Then $f = g$ since they both have the same domain and since f and g assign the same image to each element in the domain.

Example 2.3: Let $f: R^\# \rightarrow R^\#$ and $g: R^\# \rightarrow R^\#$. Suppose f is defined by $f(x) = x^2$ and g by $g(y) = y^2$. Then f and g are equal functions, that is, $f = g$. Notice that x and y are merely dummy variables in the formulas defining the functions.

RANGE OF A FUNCTION

Let f be a mapping of A into B, that is, let $f: A \rightarrow B$. Each element in B need not appear as the image of an element in A. We define the *range* of f to consist precisely of those elements in B which appear as the image of at least one element in A. We denote the range of $f: A \rightarrow B$ by

$$f(A)$$

Notice that $f(A)$ is a subset of B.

Example 3.1: Let the function $f: R^\# \rightarrow R^\#$ be defined by the formula $f(x) = x^2$. Then the range of f consists of the positive real numbers and zero.

Example 3.2: Let $f: A \rightarrow B$ be the function in Example 1.3. Then $f(A) = \{b, c\}$.

ONE-ONE FUNCTIONS

Let f map A into B. Then f is called a *one-one function* if different elements in B are assigned to different elements in A, that is, if no two different elements in A have the same image. More briefly, $f: A \rightarrow B$ is one-one if $f(a) = f(a')$ implies $a = a'$ or, equivalently, $a \neq a'$ implies $f(a) \neq f(a')$.

Example 4.1: Let the function $f: R\# \to R\#$ be defined by the formula $f(x) = x^2$. Then f is not a one-one function since $f(2) = f(-2) = 4$, that is, since the image of two different real numbers, 2 and -2, is the same number, 4.

Example 4.2: Let the function $f: R\# \to R\#$ be defined by the formula $f(x) = x^3$. Then f is a one-one mapping since the cubes of two different real numbers are themselves different.

Example 4.3: The function f which assigns to each country in the world its capital city is one-one since different countries have different capitals, that is, no city is the capital of two different countries.

ONTO FUNCTIONS

Let f be a function of A into B. Then the range $f(A)$ of the function f is a subset of B, that is, $f(A) \subset B$. If $f(A) = B$, that is, if every member of B appears as the image of at least one element of A, then we say "f is a function of A *onto* B", or "f maps A *onto* B", or "f is an *onto function*".

Example 5.1: Let the function $f: R\# \to R\#$ be defined by the formula $f(x) = x^2$. Then f is not an onto function since the negative numbers do not appear in the range of f, that is, no negative number is the square of a real number.

Example 5.2: Let $f: A \to B$ be the function in Example 1.3. Notice that $f(A) = \{b, c\}$. Since $B = \{a, b, c\}$, the range of f does not equal the co-domain, i.e. f is not onto.

Example 5.3: Let $f: A \to B$ be the function in Example 1.5. Notice that

$$f(A) = \{x, y, z\} = B$$

that is, the range of f is equal to the co-domain B. Thus f maps A onto B, i.e. f is an onto mapping.

IDENTITY FUNCTION

Let A be any set. Let the function $f: A \to A$ be defined by the formula $f(x) = x$, that is, let f assign to each element in A the element itself. Then f is called the identity function or the identity transformation on A. We denote this function by 1 or by 1_A.

CONSTANT FUNCTIONS

A function f of A into B is called a *constant* function if the same element $b \in B$ is assigned to every element in A. In other words, $f: A \to B$ is a constant function if the range of f consists of only one element.

Example 6.1: Let the function f be defined by the diagram:

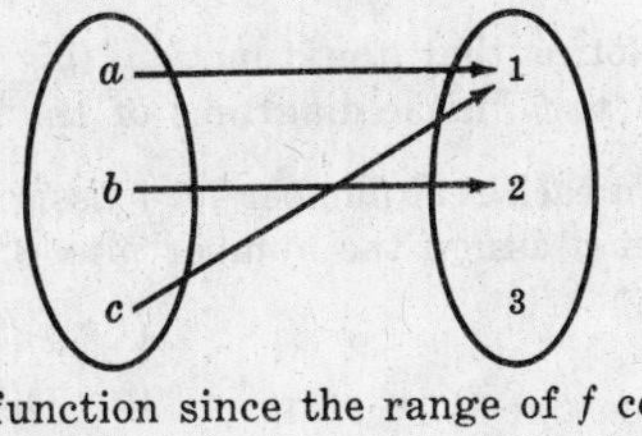

Then f is not a constant function since the range of f consists of both 1 and 2.

Example 6.2: Let the function f be defined by the diagram:

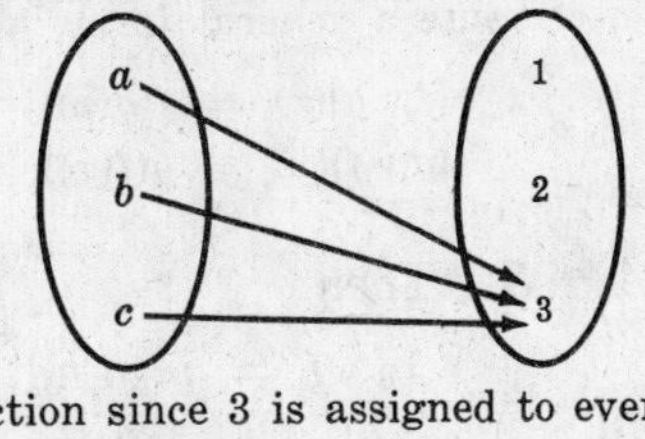

Then f is a constant function since 3 is assigned to every element in A.

Example 6.3: Let $f: R\# \to R\#$ be defined by the formula $f(x) = 5$. Then f is a constant function since 5 is assigned to every element.

PRODUCT FUNCTION

Let f be a function of A into B and let g be a function of B, the co-domain of f, into C. We illustrate the functions below.

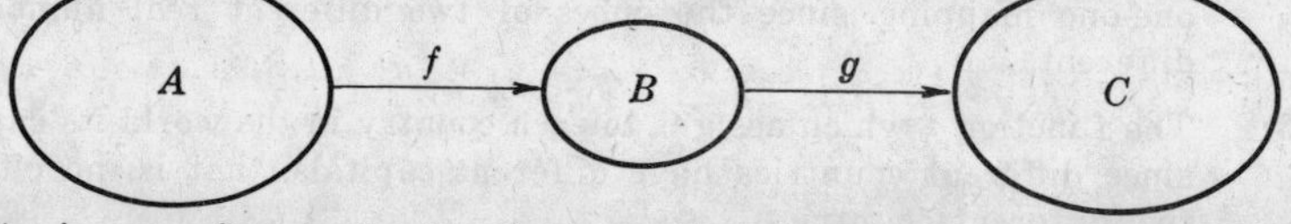

Let $a \in A$; then its image $f(a)$ is in B which is the domain of g. Accordingly, we can find the image of $f(a)$ under the mapping g, that is, we can find $g(f(a))$. Thus we have a rule which assigns to each element $a \in A$ a corresponding element $g(f(a)) \in C$. In other words, we have a function of A into C. This new function is called the *product function* or *composition function* of f and g and it is denoted by

$$(g \circ f) \quad \text{or} \quad (gf)$$

More briefly, if $f: A \to B$ and $g: B \to C$ then we define a function $(g \circ f): A \to C$ by

$$(g \circ f)(a) \equiv g(f(a))$$

Here $\equiv$ is used to mean equal by definition. We can now complete our diagram:

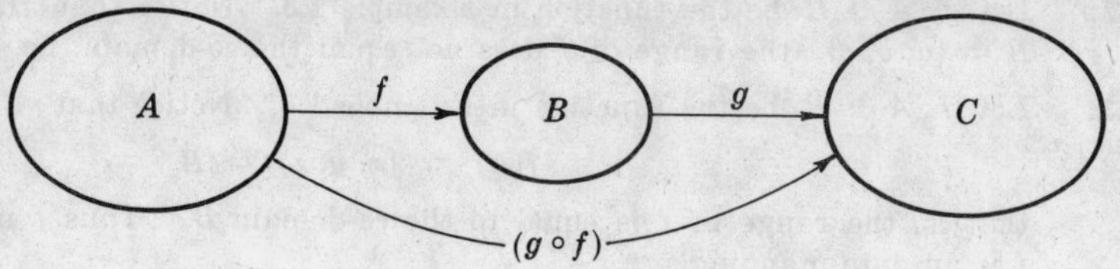

Example 7.1: Let $f: A \to B$ and $g: B \to C$ be defined by the diagrams

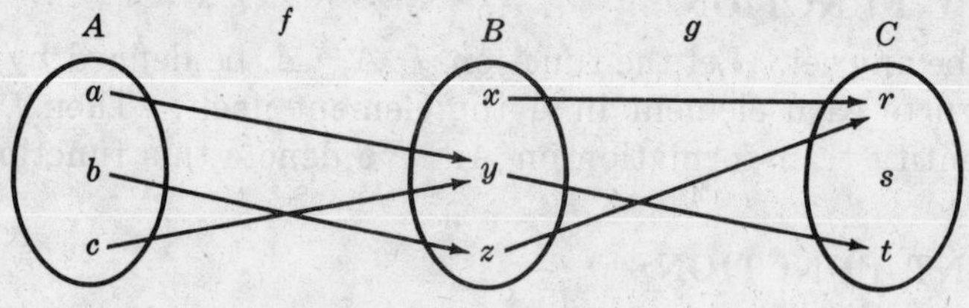

We compute $(g \circ f): A \to C$ by its definition:

$$(g \circ f)(a) \equiv g(f(a)) = g(y) = t$$
$$(g \circ f)(b) \equiv g(f(b)) = g(z) = r$$
$$(g \circ f)(c) \equiv g(f(c)) = g(y) = t$$

Notice that the function $(g \circ f)$ is equivalent to "following the arrows" from A to C in the diagrams of the functions f and g.

Example 7.2: To each real number let f assign its square, i.e. let $f(x) = x^2$. To each real number let g assign the number plus 3, i.e. let $g(x) = x + 3$. Then

$$(f \circ g)(2) \equiv f(g(2)) = f(5) = 25,$$
$$(g \circ f)(2) \equiv g(f(2)) = g(4) = 7$$

Notice that the product functions $(f \circ g)$ and $(g \circ f)$ are not the same function. We compute a general formula for these product functions:

$$(f \circ g)(x) \equiv f(g(x)) = f(x+3) = (x+3)^2 = x^2 + 6x + 9$$
$$(g \circ f)(x) \equiv g(f(x)) = g(x^2) = x^2 + 3$$

Remark 4.1: Let $f: A \to B$. Then

$$1_B \circ f = f \quad \text{and} \quad f \circ 1_A = f$$

that is, the product of any function and the identity function is the function itself.

ASSOCIATIVITY OF PRODUCTS OF FUNCTIONS

Let $f: A \to B$, $g: B \to C$ and $h: C \to D$. Then, as illustrated in Figure 4-1, we can form the product function $g \circ f: A \to C$, and then the function $h \circ (g \circ f): A \to D$.

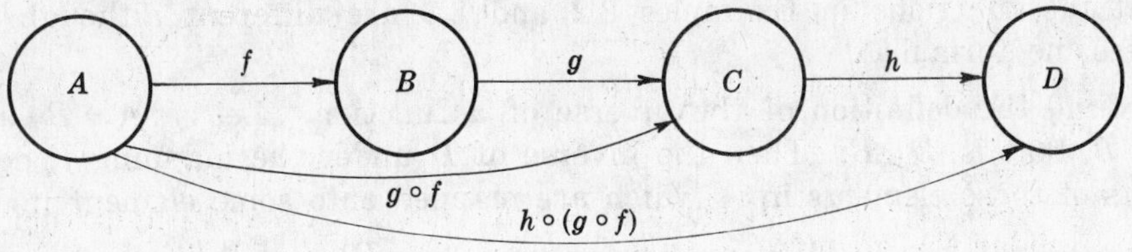

Fig. 4-1

Similarly, as illustrated in Figure 4-2, we can form the product function $h \circ g: B \to D$ and then the function $(h \circ g) \circ f: A \to D$.

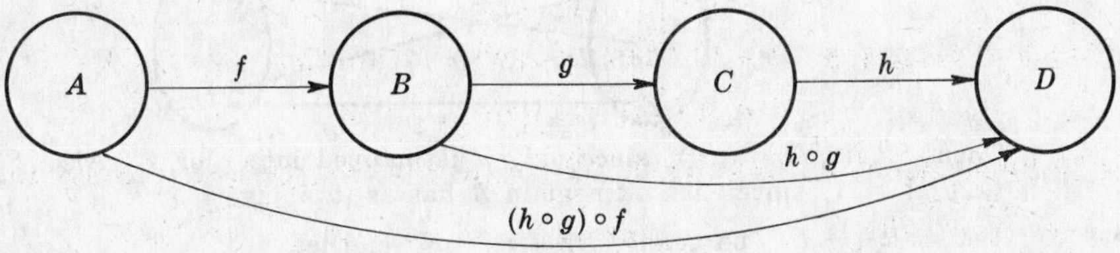

Fig. 4-2

Both $h \circ (g \circ f)$ and $(h \circ g) \circ f$ are functions of A into D. A basic theorem on functions states that these functions are equal. Specifically,

Theorem 4.1: Let $f: A \to B$, $g: B \to C$ and $h: C \to D$. Then

$$(h \circ g) \circ f = h \circ (g \circ f)$$

In view of Theorem 4.1, we can write

$$h \circ g \circ f: A \to D$$

without any parenthesis.

INVERSE OF A FUNCTION

Let f be a function of A into B, and let $b \in B$. Then the *inverse* of b, denoted by

$$f^{-1}(b)$$

consists of those elements in A which are mapped onto b, that is, those elements in A which have b as their image. More briefly, if $f: A \to B$ then

$$f^{-1}(b) = \{x \mid x \in A,\ f(x) = b\}$$

Notice that $f^{-1}(b)$ is always a subset of A. We read f^{-1} as "f inverse".

Example 8.1: Let the function $f: A \to B$ be defined by the diagram

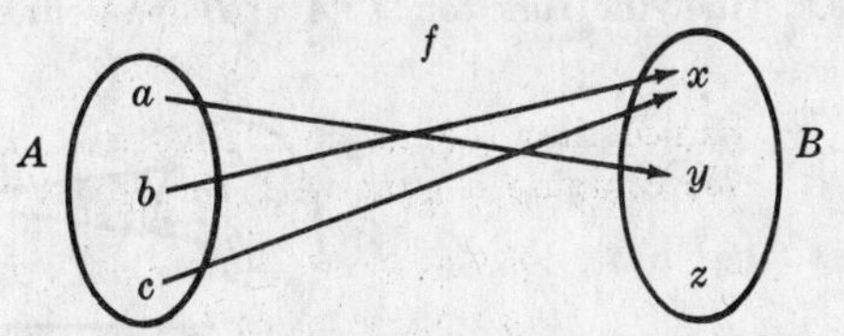

Then $f^{-1}(x) = \{b, c\}$, since both b and c have x as their image point. Also, $f^{-1}(y) = \{a\}$, as only a is mapped into y. The inverse of z, $f^{-1}(z)$, is the null set $\emptyset$, since no element of A is mapped into z.

Example 8.2: Let $f: R^\# \to R^\#$, the real numbers, be defined by the formula $f(x) = x^2$. Then $f^{-1}(4) = \{2, -2\}$, since 4 is the image of both 2 and -2 and there is no other real number whose square is four. Notice that $f^{-1}(-3) = \emptyset$, since there is no element in $R^\#$ whose square is -3.

Example 8.3: Let f be a function of the complex numbers into the complex numbers, where f is defined by the formula $f(x) = x^2$. Then $f^{-1}(-3) = \{\sqrt{3}\,i, -\sqrt{3}\,i\}$, as the square of each of these numbers is -3.

Notice that the functions in Examples 8.2 and 8.3 are different although they are defined by the same formula.

We now extend the definition of the inverse of a function. Let $f: A \to B$ and let D be a subset of B, that is, $D \subset B$. Then the inverse of D under the mapping f, denoted by $f^{-1}(D)$, consists of those elements in A which are mapped onto some element in D. More briefly,

$$f^{-1}(D) = \{x \mid x \,\varepsilon\, A,\ f(x) \,\varepsilon\, D\}$$

Example 9.1: Let the function $f: A \to B$ be defined by the diagram

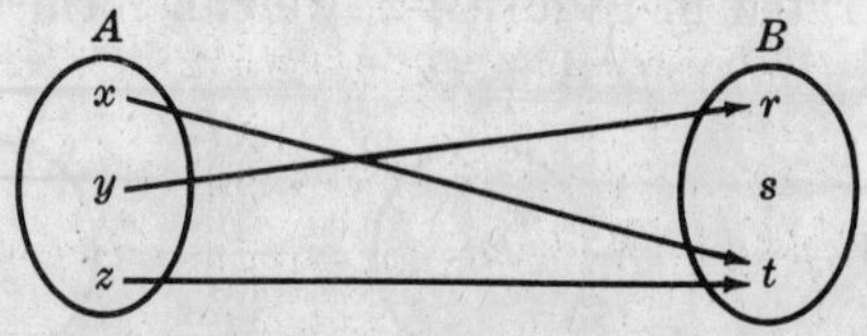

Then $f^{-1}(\{r, s\}) = \{y\}$, since only y is mapped into r or s Also $f^{-1}(\{r, t\}) = \{x, y, z\} = A$, since each element in A has as its image r or t.

Example 9.2: Let $f: R^\# \to R^\#$ be defined by $f(x) = x^2$, and let

$$D = [4, 9] = \{x \mid 4 \leq x \leq 9\}$$

Then

$$f^{-1}(D) = \{x \mid -3 \leq x \leq -2 \text{ or } 2 \leq x \leq 3\}$$

Example 9.3: Let $f: A \to B$ be any function. Then $f^{-1}(B) = A$, since every element in A has its image in B. If $f(A)$ denotes the range of the function f, then

$$f^{-1}(f(A)) = A$$

Further, if $b \,\varepsilon\, B$, then

$$f^{-1}(b) = f^{-1}(\{b\})$$

Here f^{-1} has two meanings, as the inverse of an element of B and as the inverse of a subset of B.

INVERSE FUNCTION

Let f be a function of A into B. In general, $f^{-1}(b)$ could consist of more than one element or might even be the empty set $\emptyset$. Now if $f: A \to B$ is a one-one function and an onto function, then for each $b \,\varepsilon\, B$ the inverse $f^{-1}(b)$ will consist of a single element in A. We therefore have a rule that assigns to each $b \,\varepsilon\, B$ a unique element $f^{-1}(b)$ in A. Accordingly, f^{-1} is a function of B into A and we can write

$$f^{-1}: B \to A$$

In this situation, when $f: A \to B$ is one-one and onto, we call f^{-1} the inverse function of f.

Example 10.1: Let the function $f: A \to B$ be defined by the diagram

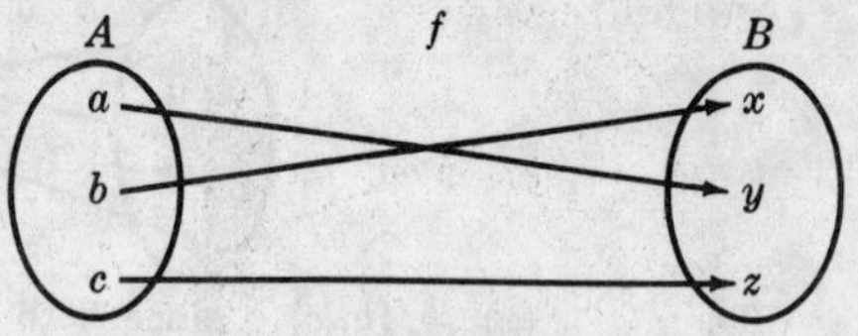

Notice that f is one-one and onto. Therefore f^{-1}, the inverse function, exists. We describe $f^{-1}: B \to A$ by the diagram

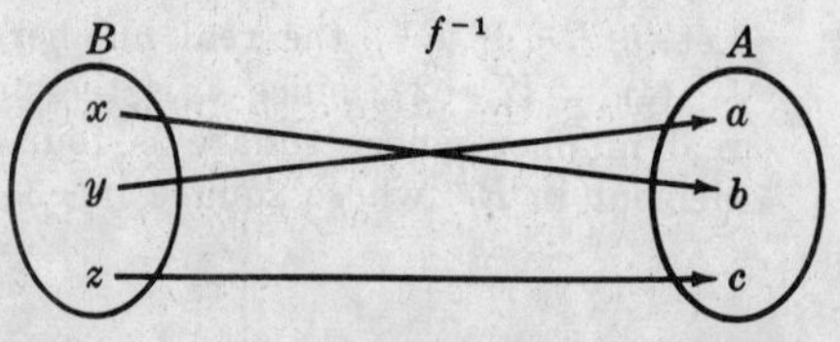

Notice, further, that if we send the arrows in the opposite direction in the first diagram of f we essentially have the diagram of f^{-1}.

Example 10.2: Let the function $f:A \to B$ be defined by the diagram

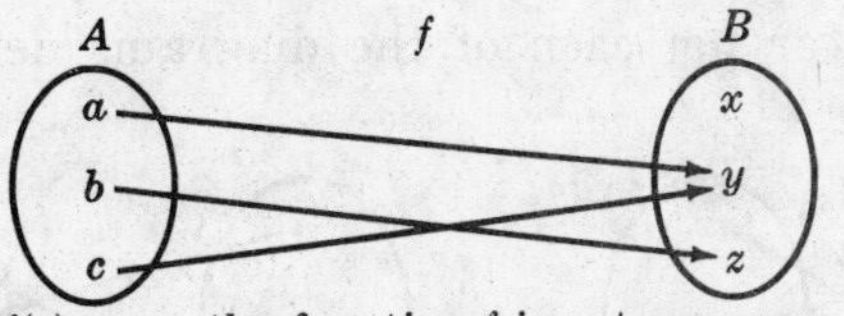

Since $f(a) = y$ and $f(c) = y$, the function f is not one-one. Therefore the inverse function f^{-1} does not exist. As $f^{-1}(y) = \{a, c\}$, we cannot assign both a and c to the element $y \in B$.

Example 10.3: Let $f: R\# \to R\#$, the real numbers, be defined by $f(x) = x^3$. Notice that f is one-one and onto. Hence $f^{-1}: R\# \to R\#$ exists. In fact we have a formula which defines the inverse function, $f^{-1}(x) = \sqrt[3]{x}$.

THEOREMS ON THE INVERSE FUNCTION

Let a function $f:A \to B$ have an inverse function $f^{-1}:B \to A$. Then we see by the diagram

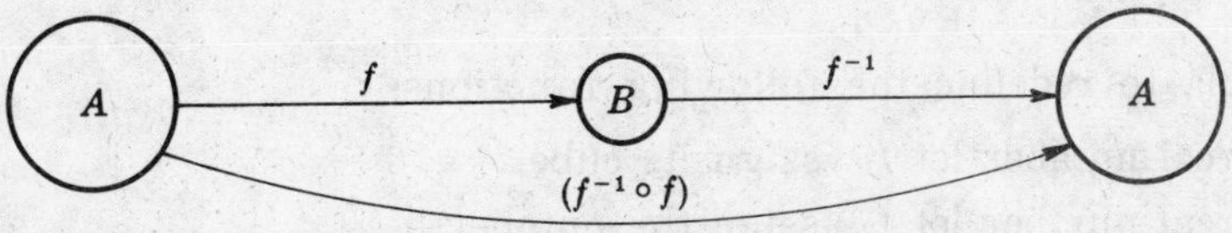

that we can form the product function $(f^{-1} \circ f)$ which maps A into A, and we see by the diagram

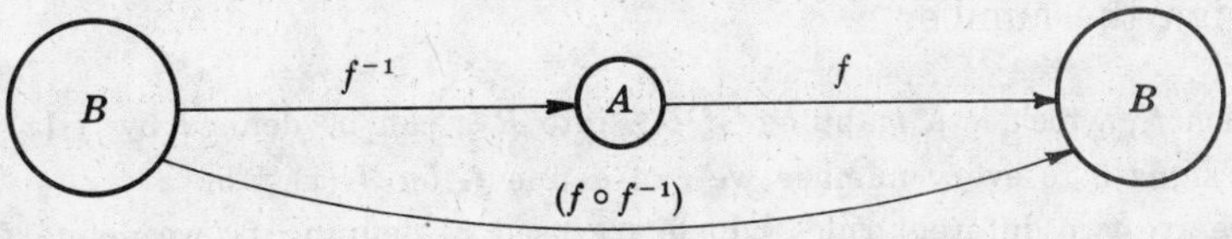

that we can form the product function $(f \circ f^{-1})$ which maps B into B. We now state the basic theorems on the inverse function:

Theorem 4.2: Let the function $f:A \to B$ be one-one and onto; i.e. the inverse function $f^{-1}:B \to A$ exists. Then the product function

$$(f^{-1} \circ f): A \to A$$

is the identity function on A, and the product function

$$(f \circ f^{-1}): B \to B$$

is the identity function on B.

Theorem 4.3: Let $f:A \to B$ and $g:B \to A$. Then g is the inverse function of f, i.e. $g = f^{-1}$, if the product function $(g \circ f): A \to A$ is the identity function on A and $(f \circ g): B \to B$ is the identity function on B.

Both conditions are necessary in Theorem 4.3 as we shall see from

Example 11.1: Let $A = \{x, y\}$ and let $B = \{a, b, c\}$. Define a function $f: A \to B$ by the diagram (a) below

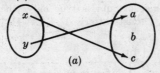

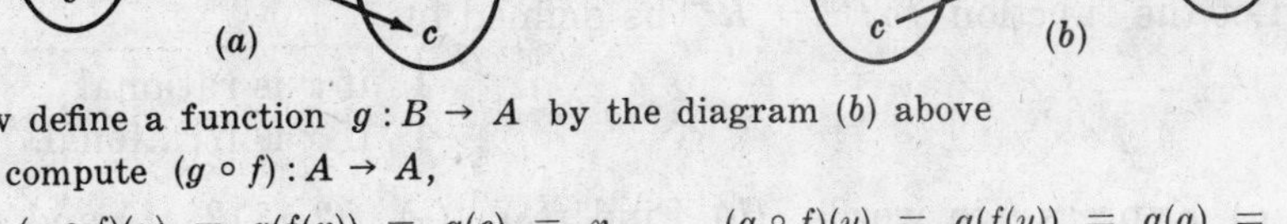

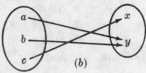

Now define a function $g: B \to A$ by the diagram (b) above

We compute $(g \circ f): A \to A$,

$$(g \circ f)(x) = g(f(x)) = g(c) = x \qquad (g \circ f)(y) = g(f(y)) = g(a) = y$$

Therefore the product function $(g \circ f)$ is the identity function on A. But g is not the inverse function of f because the product function $(f \circ g)$ is not the identity function on B, f not being an onto function.

Solved Problems

DEFINITION OF FUNCTION

1. State whether or not each of the diagrams defines a function of $A = \{a, b, c\}$ into $B = \{x, y, z\}$.

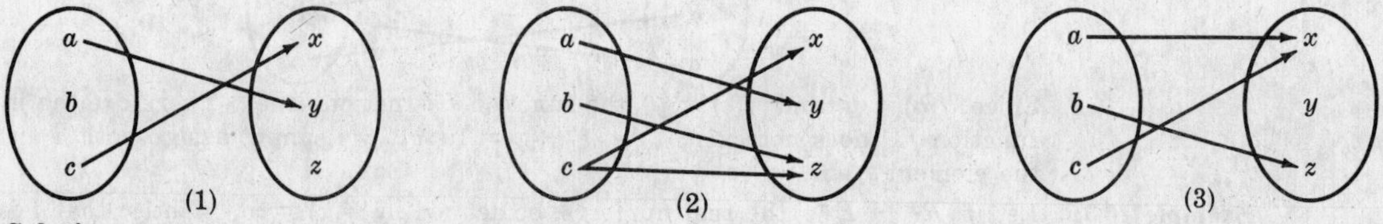

Solution:

(1) No. There is nothing assigned to the element $b \in A$.

(2) No. Two elements, x and z, are assigned to the element $c \in A$. In a function, only one element can be assigned to an element in the domain.

(3) Yes. It is possible in a function for the same element in the co-domain to be assigned to more than one element in the domain.

2. Use a formula to redefine the following functions:

(1) To each real number let f_1 assign its cube.

(2) To each real number let f_2 assign the number 5.

(3) To each positive number let f_3 assign its square, and to the other real numbers let f_3 assign the number 4.

Solution:

(1) The function f_1, which is a mapping of $R^\#$ into $R^\#$, can be defined by $f_1(x) = x^3$.

(2) Since f_2 assigns 5 to every number we can define f_2 *by* $f_2(x) = 5$.

(3) Since there are two different rules which are used in defining f_3, we define f_3 as follows:

$$f_3(x) = \begin{cases} x^2 & \text{if } x > 0 \\ 4 & \text{if } x \leqq 0 \end{cases}$$

3. Which of these statements is different from the others, and why?

(1) f is a function of A into B. (3) $f: x \to f(x)$ (5) f is a mapping of A into B.
(2) $f: A \to B$ (4) $A \xrightarrow{f} B$

Solution:

(3) is different from the others. We are not told what is the domain and the co-domain in (3), whereas in all the others we are told that A is the domain and B is the co-domain.

4. Let $f(x) = x^2$ define a function on the closed interval $-2 \leqq x \leqq 8$. Find (1) $f(4)$, (2) $f(-3)$, (3) $f(t-3)$.

Solution:

(1) $f(4) = 4^2 = 16$.

(2) $f(-3)$ has no meaning, i.e. is undefined, since -3 is not in the domain of the function.

(3) $f(t-3) = (t-3)^2 = t^2 - 6t + 9$. But this formula is true only when $t-3$ is in the domain, i.e. when $-2 \leqq t-3 \leqq 8$. In other words, t must satisfy $1 \leqq t \leqq 11$.

5. Let the function $f: R^\# \to R^\#$ be defined by

$$f(x) = \begin{cases} 1 & \text{if } x \text{ is rational} \\ -1 & \text{if } x \text{ is irrational.} \end{cases}$$

(*a*) Express f in words. (*b*) Find $f(\frac{1}{2})$, $f(\pi)$, $f(2.1313\ldots)$ and $f(\sqrt{2})$.

Solution:

(*a*) The function f assigns the number 1 to each rational number and the number -1 to each irrational number.

(*b*) Since $\frac{1}{2}$ is a rational number, $f(\frac{1}{2}) = 1$. Since π is an irrational number, $f(\pi) = -1$. As $2.1313\ldots$ is a repeating decimal, which represents a rational number, $f(2.1313\ldots) = 1$. Since $\sqrt{2}$ is irrational, $f(\sqrt{2}) = -1$.

6. Let the function $f: R^\# \to R^\#$ be defined by

$$f(x) = \begin{cases} 3x - 1 & \text{if } x > 3 \\ x^2 - 2 & \text{if } -2 \leqq x \leqq 3 \\ 2x + 3 & \text{if } x < -2 \end{cases}$$

Find (*a*) $f(2)$, (*b*) $f(4)$, (*c*) $f(-1)$, (*d*) $f(-3)$.

Solution:

(*a*) Since 2 belongs to the closed interval $[-2, 3]$, we use the formula $f(x) = x^2 - 2$. Hence $f(2) = 2^2 - 2 = 4 - 2 = 2$.

(*b*) Since 4 belongs to $(3, \infty)$, we use the formula $f(x) = 3x - 1$. Thus $f(4) = 3(4) - 1 = 12 - 1 = 11$.

(*c*) Since -1 is in the interval $[-2, 3]$, we use the formula $f(x) = x^2 - 2$. Computing, $f(-1) = (-1)^2 - 2 = 1 - 2 = -1$.

(*d*) Since -3 is less than -2, i.e. -3 belongs to $(-\infty, -2)$, we use the formula $f(x) = 2x + 3$. Thus $f(-3) = 2(-3) + 3 = -6 + 3 = -3$.

Notice that there is only one function f defined even though there are three formulas which are used to define f. The reader should not confuse formulas and functions.

7. Let $A = \{a, b, c\}$ and $B = \{1, 0\}$. How many different functions are there from A into B, and what are they?

Solution:

We list all the functions of A into B by diagrams. In each function we assign either 1 or 0, but not both, to each element in A.

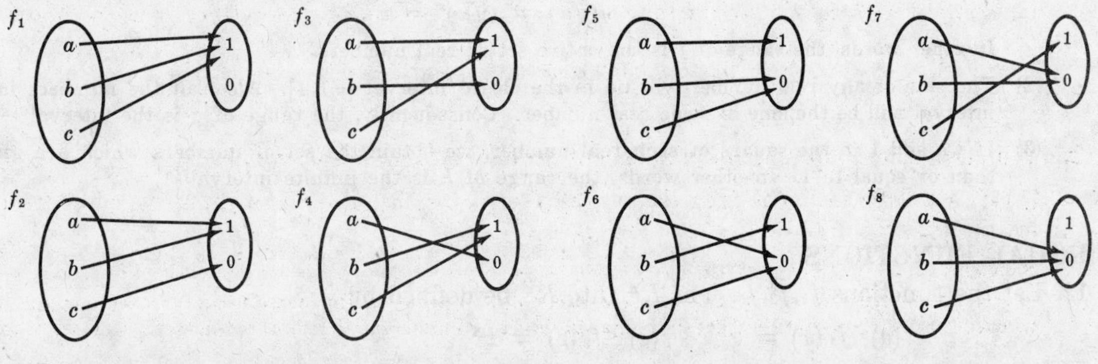

Notice that there are eight functions.

RANGE OF A FUNCTION

8. Let $A = \{1, 2, 3, 4, 5\}$. Define a function $f: A \to A$ by the diagram

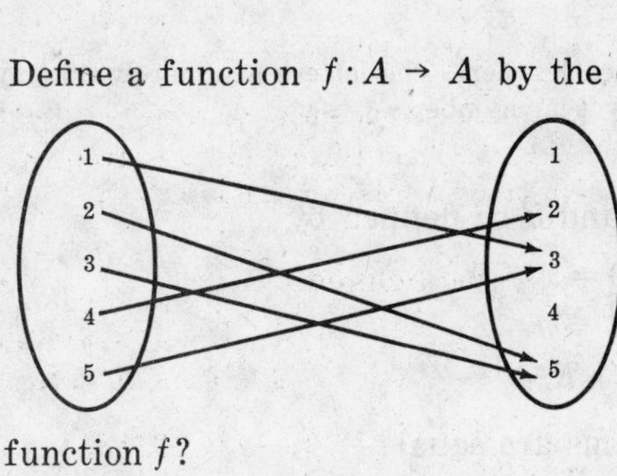

What is the range of the function f?

Solution:

The range consists of all the image points. Since only the numbers 2, 3 and 5 appear as image points, the range of f is the set $\{2, 3, 5\}$.

9. Let $W = \{a, b, c, d\}$. Let a function f of W into W be defined by $f(a) = a$, $f(b) = c$, $f(c) = a$, $f(d) = a$. Find the range of the function $f: W \to W$.

Solution:

The range of f consists of those elements which appear as image points. Only a and c appear as the image points of elements in W. Accordingly, the range of f is $\{a, c\}$.

10. Let $V = \{-2, -1, 0, 1, 2\}$. Let the function $g: V \to R^\#$ be defined by the formula

$$g(x) = x^2 + 1$$

Find the range of g.

Solution:

We compute the image of each element of V.

$$\begin{aligned} g(-2) &= (-2)^2 + 1 = 4 + 1 = 5 \\ g(-1) &= (-1)^2 + 1 = 1 + 1 = 2 \\ g(0) &= (0)^2 + 1 = 0 + 1 = 1 \\ g(1) &= (1)^2 + 1 = 1 + 1 = 2 \\ g(2) &= (2)^2 + 1 = 4 + 1 = 5 \end{aligned}$$

Thus the range of g is the set of image points $\{5, 2, 1, 2, 5\}$, i.e. the set $\{5, 2, 1\}$.

11. Each of the following formulas defines a function from $R^\#$ into $R^\#$. Find the range of each function.

$$(1)\ f(x) = x^3, \quad (2)\ g(x) = \sin x, \quad (3)\ h(x) = x^2 + 1$$

Solution:

(1) Every real number a has a real cube root $\sqrt[3]{a}$; hence

$$f(\sqrt[3]{a}) = (\sqrt[3]{a})^3 = a$$

In other words, the range of f is the entire set of real numbers.

(2) The sine of any real number will lie in the closed interval $[-1, 1]$. Also, all the numbers in this interval will be the sine of some real number. Consequently, the range of g is the interval $[-1, 1]$.

(3) If we add 1 to the square of each real number, we obtain the set of numbers which are greater than or equal to 1. In other words, the range of h is the infinite interval $[1, \infty)$.

EQUAL FUNCTIONS

12. Let the functions f_1, f_2, f_3, f_4 of $R^\#$ into $R^\#$ be defined by

(a) $f_1(x) = x^2$ (c) $f_3(z) = z^2$

(b) $f_2(y) = y^2$ (d) f_4 assigns to each real number its square

Which of these functions are equal?

Solution:

They are all equal to each other. The letters are merely dummy variables. Each function assigns the same number to every real number.

13. Let the functions f, g and h be defined by

(a) $f(x) = x^2$ where $0 \leq x \leq 1$

(b) $g(y) = y^2$ where $2 \leq y \leq 8$

(c) $h(z) = z^2$ where $z \in R^\#$

Which of these functions are equal?

Solution:

None of the functions are equal. Although the rules of correspondence are the same, the domains are different. Thus the functions are different.

ONE-ONE FUNCTIONS

14. Let $A = \{a, b, c, d, e\}$, and let B be the set of letters in the alphabet. Let the functions f, g and h of A into B be defined by:

(1) $f(a) = r,\ f(b) = a,\ f(c) = s,\ f(d) = r,\ f(e) = e$

(2) $g(a) = a,\ g(b) = c,\ g(c) = e,\ g(d) = r,\ g(e) = s$

(3) $h(a) = z,\ h(b) = y,\ h(c) = x,\ h(d) = y,\ h(e) = z.$

State whether or not each of these functions is one-one.

Solution:

Note that in order for a function to be one-one, it must assign different image points to different elements in the domain.

(1) f is not a one-one function since f assigns r to both a and d, i.e. $f(a) = f(d) = r$.

(2) g is a one-one function.

(3) h is not a one-one function since $h(a) = h(e)$.

15. State whether or not each of the following functions is one-one.

(1) To each person on the earth assign the number which corresponds to his age.

(2) To each country in the world assign the number of people which live in the country.

(3) To each book written by only one author assign the author.

(4) To each country in the world which has a prime minister assign its prime minister.

Solution:

(1) Many people in the world have the same age; hence this function is not one-one.

(2) Although it is possible for two countries to have the same number of people, statistics show that today this is not so; hence this function is one-one.

(3) It is possible for two different books to have the same author; hence this function is not one-one.

(4) No two different countries in the world have the same prime minister; thus this function is one-one.

16. Let $A = [-1, 1] = \{x \mid -1 \leqq x \leqq 1\}$, $B = [1, 3]$ and $C = [-3, -1]$. Let the functions $f_1 : A \to R^\#$, $f_2 : B \to R^\#$ and $f_3 : C \to R^\#$ be defined by the rule: To each number assign its square. Which of the functions are one-one?

Solution:

The function $f_1 : A \to R^\#$ is not one-one since $f_1(\frac{1}{2}) = f_1(-\frac{1}{2})$, that is, since two different numbers in the domain are assigned the same image.

The function $f_2 : B \to R^\#$ is one-one since the squares of unequal positive numbers are themselves unequal.

Also, $f_3 : C \to R^\#$ is one-one since the squares of unequal negative numbers are unequal.

Notice, once again, that a formula itself does not define a function. In fact, we have just seen that the same formula gives rise to different functions which have different properties.

17. Find the "largest" interval D on which the formula $f(x) = x^2$ defines a one-one function.

Solution:

As long as the interval D contains either positive or negative numbers, but not both, the function will be one-one. Thus D can be the infinite intervals $[0, \infty)$ or $(-\infty, 0]$. There can be other infinite intervals on which f will be one-one, but they will be subsets of one of these two.

18. Can a constant function be one-one?

Solution:

If the domain of a function contains a single element, the function will be a constant function and it will also be one-one.

19. On which sets A will the identity function $1_A : A \to A$ be one-one?

Solution:

A can be any set. The identity function is always one-one.

20. In Problem 7 we listed all possible functions of $A = \{a, b, c\}$ into $B = \{1, 0\}$. Which of the functions are one-one?

Solution:

None of the functions is one-one. In each function, at least two elements have the same image.

ONTO FUNCTIONS

21. Let $f : A \to B$. Find $f(A)$, i.e. the range of f, if f is an onto function.

Solution:

If f is an onto function then every element in the co-domain of f is in the range; hence $f(A) = B$.

22. Is the function $f : A \to A$ in Problem 8 onto?

Solution:

The numbers 1 and 4 in the co-domain are not the images of any elements in the domain; hence f is not an onto function. In other words, $f(A) = \{2, 3, 5\}$ is a proper subset of A.

23. Let $A = [-1, 1]$. Let functions f, g and h of A into A be defined by:

$$(1)\ f(x) = x^2, \qquad (2)\ g(x) = x^3, \qquad (3)\ h(x) = \sin x$$

Which function, if any, is onto?

Solution:

(1) No negative numbers appear in the range of f; hence f is not an onto function.

(2) The function g is onto, that is, $g(A) = A$.

(3) The function h is not onto. In fact, there is no number x in A such that $\sin x = 1$.

24. Can a constant function be an onto function?

Solution:

If the co-domain of a function f consists of a single element, then f is always a constant function and is onto.

25. On which sets A will the identity function $1_A : A \to A$ be onto?

Solution:

The identity function is always onto; hence A can be any set.

26. In Problem 7 all possible functions of $A = \{a, b, c\}$ into $B = \{1, 0\}$ are listed. Which of these, if any, are onto functions?

Solution:

All the functions are onto except f_1 and f_8.

PRODUCT FUNCTIONS

27. Let the functions $f : A \to B$ and $g : B \to C$ be defined by the diagram

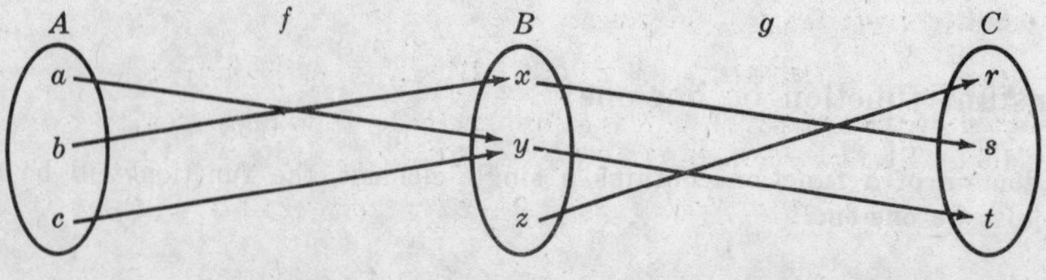

(a) Find the product function $(g \circ f): A \to C$.

(b) Find the ranges of f, g and $g \circ f$.

Solution:

(a) We use the definition of the product function and compute:

$$(g \circ f)(a) \equiv g(f(a)) = g(y) = t$$
$$(g \circ f)(b) \equiv g(f(b)) = g(x) = s$$
$$(g \circ f)(c) \equiv g(f(c)) = g(y) = t$$

Notice that we arrive at the same answer if we "follow the arrows":

$$a \to y \to t$$
$$b \to x \to s$$
$$c \to y \to t$$

(b) By the diagram, the range of f is $\{x, y\}$, and the range of g is $\{r, s, t\}$. By (a), the range of $g \circ f$ is $\{s, t\}$. Notice that the ranges of g and $g \circ f$ are different.

28. Let $A = \{1, 2, 3, 4, 5\}$ and let the functions $f: A \to A$ and $g: A \to A$ be defined by:

$$f(1) = 3, \quad f(2) = 5, \quad f(3) = 3, \quad f(4) = 1, \quad f(5) = 2$$
$$g(1) = 4, \quad g(2) = 1, \quad g(3) = 1, \quad g(4) = 2, \quad g(5) = 3$$

Find the composition functions $f \circ g$ and $g \circ f$.

Solution:

We use the definition of the product function and compute:

$$(f \circ g)(1) \equiv f(g(1)) = f(4) = 1$$
$$(f \circ g)(2) \equiv f(g(2)) = f(1) = 3$$
$$(f \circ g)(3) \equiv f(g(3)) = f(1) = 3$$
$$(f \circ g)(4) \equiv f(g(4)) = f(2) = 5$$
$$(f \circ g)(5) \equiv f(g(5)) = f(3) = 3$$

Also,

$$(g \circ f)(1) \equiv g(f(1)) = g(3) = 1$$
$$(g \circ f)(2) \equiv g(f(2)) = g(5) = 3$$
$$(g \circ f)(3) \equiv g(f(3)) = g(3) = 1$$
$$(g \circ f)(4) \equiv g(f(4)) = g(1) = 4$$
$$(g \circ f)(5) \equiv g(f(5)) = g(2) = 1$$

Notice that the functions $f \circ g$ and $g \circ f$ are not equal.

29. Let the functions $f: R^\# \to R^\#$ and $g: R^\# \to R^\#$ be defined by

$$f(x) = 2x + 1, \qquad g(x) = x^2 - 2$$

Find formulas which define the product functions $g \circ f$ and $f \circ g$.

Solution:

We first compute $g \circ f: R^\# \to R^\#$. Essentially we want to substitute the formula for f inside the formula for g. We use the definition of the product function as follows:

$$(g \circ f)(x) \equiv g(f(x)) = g(2x+1) = (2x+1)^2 - 2 = 4x^2 + 4x - 1$$

Perhaps the reader is more familiar with the process if the functions are defined as:

$$y = f(x) = 2x + 1, \qquad z = g(y) = y^2 - 2$$

Then y is eliminated from the two formulas:

$$z = y^2 - 2 = (2x - 1)^2 - 2 = 4x^2 + 4x - 1$$

The reader should become familiar with the first method. It is necessary to realize that x is merely a dummy variable. We now compute $f \circ g: R^\# \to R^\#$:

$$(f \circ g)(x) \equiv f(g(x)) = f(x^2 - 2) = 2(x^2 - 2) + 1 = 2x^2 - 3$$

30. Let the functions f and g on the real numbers $R^\#$ be defined by

$$f(x) = x^2 + 2x - 3, \qquad g(x) = 3x - 4$$

(1) Find formulas which define the product functions $g \circ f$ and $f \circ g$.

(2) Check the formulas by showing $(g \circ f)(2) = g(f(2))$ and $(f \circ g)(2) = f(g(2))$.

Solution:

(1) $(g \circ f)(x) \equiv g(f(x)) = g(x^2 + 2x - 3) = 3(x^2 + 2x - 3) - 4 = 3x^2 + 6x - 13$

$(f \circ g)(x) \equiv f(g(x)) = f(3x - 4) = (3x - 4)^2 + 2(3x - 4) - 3 = 9x^2 - 18x + 5$

(2) $(g \circ f)(2) = 3(2)^2 + 6(2) - 13 = 12 + 12 - 13 = 11$

$g(f(2)) = g(2^2 + 2(2) - 3) = g(5) = 3(5) - 4 = 11$

$(f \circ g)(2) = 9(2)^2 - 18(2) + 5 = 36 - 36 + 5 = 5$

$f(g(2)) = f(3(2) - 4) = f(2) = 2^2 + 2(2) - 3 = 5$

31. Prove: If $f : A \to B$ is onto and $g : B \to C$ is onto, then the product function $(g \circ f) : A \to C$ is onto.

Solution:

Let c be any element in C. Since g is onto, there exists an element $b \in B$ such that $g(b) = c$. Also, since f is onto, there exists an element $a \in A$ such that $f(a) = b$. Now $(g \circ f)(a) \equiv g(f(a)) = g(b) = c$. Thus for any $c \in C$, we have shown there is at least one element $a \in A$ such that $(g \circ f)(a) = c$. Therefore $g \circ f$ is an onto function.

32. Prove Theorem 4.1: Let $f : A \to B$, $g : B \to C$ and $h : C \to D$; then

$$(h \circ g) \circ f = h \circ (g \circ f)$$

Solution:

The two functions are equal if they assign the same image to each element in the domain, that is, if

$$((h \circ g) \circ f)(x) = (h \circ (g \circ f))(x)$$

for every $x \in A$. Computing,

$$((h \circ g) \circ f)(x) \equiv (h \circ g)(f(x)) \equiv h(g(f(x)))$$

and

$$(h \circ (g \circ f))(x) \equiv h((g \circ f)(x)) \equiv h(g(f(x)))$$

Hence

$$(h \circ g) \circ f = h \circ (g \circ f)$$

INVERSE OF A FUNCTION

33. Let $A = \{1, 2, 3, 4, 5\}$. Let the function $f : A \to A$ be defined by the diagram

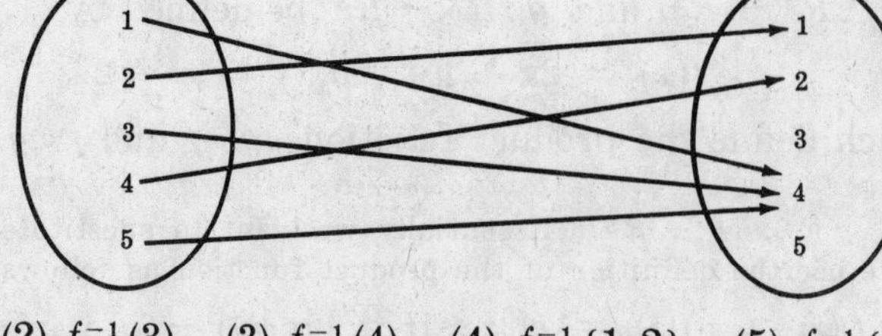

Find (1) $f^{-1}(2)$, (2) $f^{-1}(3)$, (3) $f^{-1}(4)$, (4) $f^{-1}\{1,2\}$, (5) $f^{-1}\{2,3,4\}$.

Solution:

(1) $f^{-1}(2)$ consists of those elements whose image is 2. Only 4 has the image 2; hence $f^{-1}(2) = \{4\}$.

(2) $f^{-1}(3) = \emptyset$ since 3 is not the image of any element in the domain.

(3) $f^{-1}(4) = \{1, 3, 5\}$ since $f(1) = 4$, $f(3) = 4$, $f(5) = 4$ and since 4 is not the image of any other element.

(4) $f^{-1}\{1, 2\}$ consists of those elements whose image is either 1 or 2; hence $f^{-1}\{1, 2\} = \{2, 4\}$.

(5) $f^{-1}\{2, 3, 4\} = \{4, 1, 3, 5\}$ since each of these numbers, and no others, has 2, 3 or 4 as an image point.

34. Let the function $f : R^\# \to R^\#$ be defined by $f(x) = x^2$. Find:
(1) $f^{-1}(25)$, (2) $f^{-1}(-9)$, (3) $f^{-1}([-1,1])$, (4) $f^{-1}((-\infty,0])$, (5) $f^{-1}([4,25])$.

Solution:

(1) $f^{-1}(25) = \{5,-5\}$ since $f(5) = 25$ and $f(-5) = 25$ and since the square of no other number is 25.

(2) $f^{-1}(-9) = \emptyset$ since there is no real number whose square is -9, i.e. the equation $x^2 = -9$ has no real root.

(3) $f^{-1}([-1,1]) = [-1,1]$ since $|x| \leqq 1$ implies $|x^2| \leqq 1$, i.e. if x belongs to $[-1,1]$ then $f(x) = x^2$ also belongs to $[-1,1]$.

(4) $f^{-1}((-\infty,0]) = \{0\}$ since $0^2 = 0 \in (-\infty,0]$ and since no other number squared belongs to $(-\infty,0]$.

(5) $f^{-1}([4,25])$ consists of those numbers whose squares belong to $[4,25]$, i.e. those numbers x such that $4 \leqq x^2 \leqq 25$. Hence

$$f^{-1}([4,25]) = \{x \mid 2 \leqq x \leqq 5 \text{ or } -5 \leqq x \leqq -2\}$$

35. Let $f : A \to B$. Find $f^{-1}(f(A))$, that is, find the inverse of the range of f.

Solution:

Since the image of every element in A is in the range of f,

$$f^{-1}(f(A)) = A$$

under all circumstances.

INVERSE FUNCTION

36. Let $f : A \to B$, and let f have an inverse function $f^{-1} : B \to A$. State two properties of the function f.

Solution:

The function f must be both one-one and onto.

37. Let $W = \{1,2,3,4,5\}$, and let the functions $f : W \to W$, $g : W \to W$ and $h : W \to W$ be defined by the following diagrams:

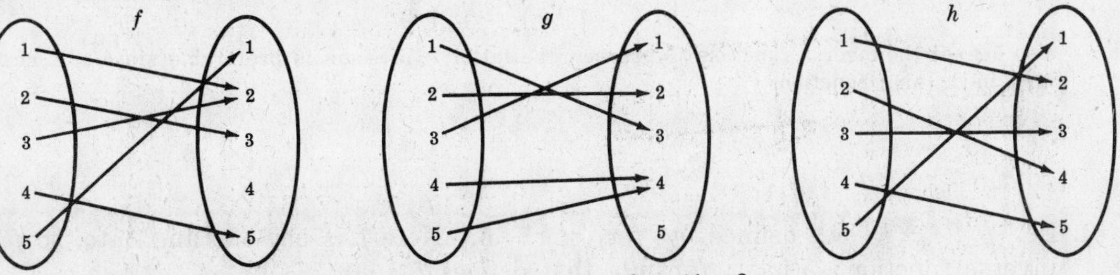

Which of these functions, if any, has an inverse function?

Solution:

In order for a function to have an inverse, the function must be both one-one and onto. Only h is one-one and onto; hence h, and only h, has an inverse function.

38. Let $A = [-1,1]$. Let the functions f_1, f_2, f_3 and f_4 of A into A be defined by:

(1) $f_1(x) = x^2$, (2) $f_2(x) = x^5$, (3) $f_3(x) = \sin x$, (4) $f_4(x) = \sin \frac{1}{2}\pi x$

State whether or not each of these functions has an inverse function.

Solution:

(1) f_1 is neither one-one or onto; hence f_1 has no inverse.

(2) f_2 is one-one since $x \neq y$ implies $x^5 \neq y^5$. Also, f_2 is onto. Hence f_2 has an inverse function.

(3) f_3 is a one-one function but not onto; hence f_3 has no inverse.

(4) f_4 has an inverse function since it is one-one and onto.

39. Prove: Let $f:A \to B$ and $g:B \to C$ have inverse functions $f^{-1}:B \to A$ and $g^{-1}:C \to B$. Then the composition function $g \circ f:A \to C$ has an inverse function which is $f^{-1} \circ g^{-1}:C \to A$.

Solution:

By Theorem 4.3, we must show that

$$(f^{-1} \circ g^{-1}) \circ (g \circ f) = 1 \qquad \text{and} \qquad (g \circ f) \circ (f^{-1} \circ g^{-1}) = 1$$

By repeated use of Theorem 4.1, the associative law for composition of functions, we compute

$$\begin{aligned}(f^{-1} \circ g^{-1}) \circ (g \circ f) &= f^{-1} \circ (g^{-1} \circ (g \circ f)) = f^{-1} \circ ((g^{-1} \circ g) \circ f) \\ &= f^{-1} \circ (1 \circ f) = f^{-1} \circ f = 1\end{aligned}$$

Notice that we use the property that $g^{-1} \circ g$ is the identity function, and that the product of 1, the identity function, and f is f. Similarly,

$$\begin{aligned}(g \circ f) \circ (f^{-1} \circ g^{-1}) &= g \circ (f \circ (f^{-1} \circ g^{-1})) = g \circ ((f \circ f^{-1}) \circ g^{-1}) \\ &= g \circ (1 \circ g^{-1}) = g \circ g^{-1} = 1\end{aligned}$$

40. Let $f:R^\# \to R^\#$ be defined by $f(x) = 2x - 3$. Note that f is one-one and onto, so f has an inverse function $f^{-1}:R^\# \to R^\#$. Find a formula that defines the inverse function f^{-1}.

Solution:

Let y be the image of x under the function f. Then

$$y = f(x) = 2x - 3$$

Consequently, x will be the image of y under the inverse function f^{-1}, i.e.,

$$x = f^{-1}(y)$$

Solving for x in terms of y in the above equation,

$$x = (y+3)/2$$

Then

$$f^{-1}(y) = (y+3)/2$$

is a formula defining the inverse function. Note y is merely a dummy variable; hence

$$f^{-1}(x) = (x+3)/2$$

also defines the inverse function. Moreover, the latter expression is preferable since x is customarily used in defining functions.

41. Let $f:R^\# \to R^\#$ be defined by $f(x) = x^3 + 5$. Note f is one-one and onto, so f has an inverse function. Find a formula that defines f^{-1}.

Solution:

Solve for x in terms of y: $y = x^3 + 5$, $y - 5 = x^3$, and $x = \sqrt[3]{y-5}$.
Then the inverse function is $f^{-1}(x) = \sqrt[3]{x-5}$.

42. Let $A = R^\# - \{3\}$ and $B = R^\# - \{1\}$. Let the function $f:A \to B$ be defined by

$$f(x) = \frac{x-2}{x-3}$$

Then f is one-one and onto. Find a formula that defines f^{-1}.

Solution:

Solving $y = \dfrac{x-2}{x-3}$ for x in terms of y, we obtain $x = \dfrac{2-3y}{1-y}$.
Hence the inverse function is $f^{-1}(x) = \dfrac{2-3x}{1-x}$.

MISCELLANEOUS PROBLEMS

43. Let the function $f: R^\# \to R^\#$ be defined by $f(x) = x^2 - 3x + 2$. Find:

(a) $f(-3)$ (e) $f(x^2)$ (i) $f(2x-3)$ (m) $f(f(x+1))$
(b) $f(2) - f(-4)$ (f) $f(y-z)$ (j) $f(2x-3) + f(x+3)$ (n) $f(x+h) - f(x)$
(c) $f(y)$ (g) $f(x+h)$ (k) $f(x^2 - 3x + 2)$ (o) $[f(x+h) - f(x)]/h$
(d) $f(a^2)$ (h) $f(x+3)$ (l) $f(f(x))$

Solution:

The function assigns to any element the square of the element minus 3 times the element plus 2.

(a) $f(-3) = (-3)^2 - 3(-3) + 2 = 9 + 9 + 2 = 20$

(b) $f(2) = (2)^2 - 3(2) + 2 = 0$, $f(-4) = (-4)^2 - 3(-4) + 2 = 30$. Then

$$f(2) - f(-4) = 0 - 30 = -30$$

(c) $f(y) = (y)^2 - 3(y) + 2 = y^2 - 3y + 2$

(d) $f(a^2) = (a^2)^2 - 3(a^2) + 2 = a^4 - 3a^2 + 2$

(e) $f(x^2) = (x^2)^2 - 3(x^2) + 2 = x^4 - 3x^2 + 2$

(f) $f(y-z) = (y-z)^2 - 3(y-z) + 2 = y^2 - 2yz + z^2 - 3y + 3z + 2$

(g) $f(x+h) = (x+h)^2 - 3(x+h) + 2 = x^2 + 2xh + h^2 - 3x - 3h + 2$

(h) $f(x+3) = (x+3)^2 - 3(x+3) + 2 = (x^2 + 6x + 9) - 3x - 9 + 2 = x^2 + 3x + 2$

(i) $f(2x-3) = (2x-3)^2 - 3(2x-3) + 2 = 4x^2 - 12x + 9 - 6x + 9 + 2 = 4x^2 - 18x + 20$

(j) Using (h) and (i), we have

$$f(2x-3) + f(x+3) = (4x^2 - 18x + 20) + (x^2 + 3x + 2) = 5x^2 - 15x + 22$$

(k) $f(x^2 - 3x + 2) = (x^2 - 3x + 2)^2 - 3(x^2 - 3x + 2) + 2 = x^4 - 6x^3 + 10x^2 - 3x$

(l) $f(f(x)) = f(x^2 - 3x + 2) = x^4 - 6x^3 + 10x^2 - 3x$

(m) $f(f(x+1)) = f([(x+1)^2 - 3(x+1) + 2]) = f([x^2 + 2x + 1 - 3x - 3 + 2])$
$= f(x^2 - x) = (x^2 - x)^2 - 3(x^2 - x) + 2 = x^4 - 2x^3 - 2x^2 + 3x + 2$

(n) By (g), $f(x+h) = x^2 + 2xh + h^2 - 3x - 3h + 2$. Hence

$$f(x+h) - f(x) = (x^2 + 2xh + h^2 - 3x - 3h + 2) - (x^2 - 3x + 2) = 2xh + h^2 - 3h$$

(o) Using (n), we have

$$[f(x+h) - f(x)]/h = (2xh + h^2 - 3h)/h = 2x + h - 3$$

44. Let the functions $f: R^\# \to R^\#$ and $g: R^\# \to R^\#$ be defined by $f(x) = 2x - 3$ and $g(x) = x^2 + 5$. Find (a) $f(5)$, (b) $g(-3)$, (c) $g(f(2))$, (d) $f(g(3))$, (e) $g(a-1)$, (f) $f(g(a-1))$, (g) $g(f(x))$, (h) $f(g(x+1))$, (i) $g(g(x))$.

Solution:

(a) $f(5) = 2(5) - 3 = 10 - 3 = 7$

(b) $g(-3) = (-3)^2 + 5 = 9 + 5 = 14$

(c) $g(f(2)) = g([2(2) - 3]) = g([4 - 3]) = g(1) = (1)^2 + 5 = 6$

(d) $f(g(3)) = f([3^2 + 5]) = f([9 + 5]) = f(14) = 2(14) - 3 = 25$

(e) $g(a-1) = (a-1)^2 + 5 = a^2 - 2a + 1 + 5 = a^2 - 2a + 6$

(f) Using (e), we have

$$f(g(a-1)) = f(a^2 - 2a + 6) = 2(a^2 - 2a + 6) - 3 = 2a^2 - 4a + 9$$

(g) $g(f(x)) = g(2x-3) = (2x-3)^2 + 5 = 4x^2 - 12x + 14$

(h) $f(g(x+1)) = f([(x+1)^2 + 5]) = f([x^2 + 2x + 1 + 5])$
$= f(x^2 + 2x + 6) = 2(x^2 + 2x + 6) - 3 = 2x^2 + 4x + 9$

(i) $g(g(x)) = g(x^2 + 5) = (x^2 + 5)^2 + 5 = x^4 + 10x^2 + 30$

Supplementary Problems

DEFINITION OF A FUNCTION

45. State whether or not each of the diagrams defines a function of $\{1, 2, 3\}$ into $\{4, 5, 6\}$.

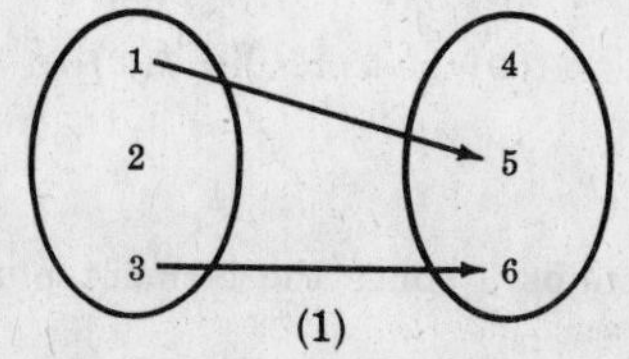

(1)

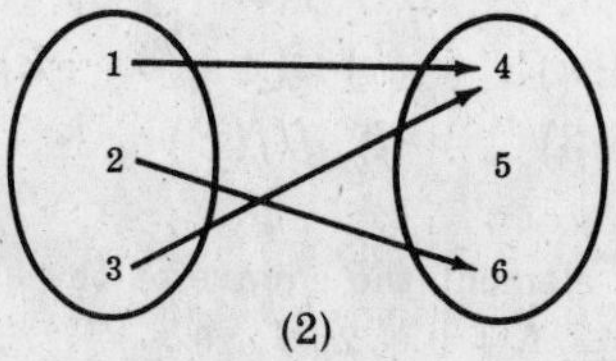

(2)

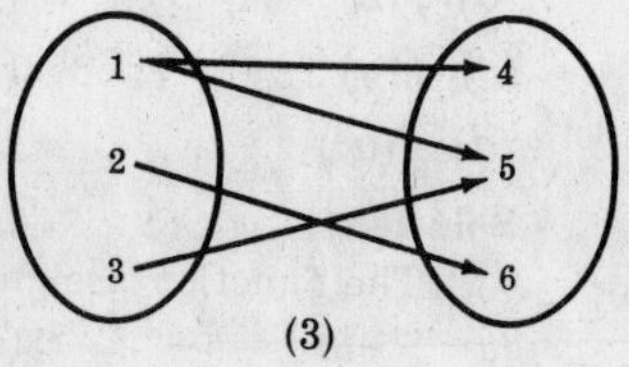

(3)

46. Redefine the following functions by use of a formula:

(1) To each real number let f assign its square plus 3.

(2) To each real number let g assign the number plus the absolute value of the number.

(3) To each real number greater than or equal to 3 let h assign the number cubed, and to each number less than 3 let h assign the number 4.

47. Let the function $f: R^\# \to R^\#$ be defined by $f(x) = x^2 - 4x + 3$. Find (1) $f(4)$, (2) $f(-3)$, (3) $f(y - 2z)$, (4) $f(x - 2)$.

48. Let the function $g: R^\# \to R^\#$ be defined by $g(x) = \begin{cases} x^2 - 3x & \text{if } x \geqq 2 \\ x + 2 & \text{if } x < 2 \end{cases}$.
Find (1) $g(5)$, (2) $g(0)$, (3) $g(-2)$.

49. Let $T = [-3, 5]$ and let the function $f: T \to R^\#$ be defined by $f(x) = 2x^2 - 7$. Find (*a*) $f(2)$, (*b*) $f(6)$, (*c*) $f(t - 2)$.

50. Let the function $h: R^\# \to R^\#$ be defined by $h(x) = \begin{cases} 2x + 5 & \text{if } x > 9 \\ x^2 - |x| & \text{if } x \in [-9, 9] \\ x - 4 & \text{if } x < -9 \end{cases}$.

Find (*a*) $h(3)$, (*b*) $h(12)$, (*c*) $h(-15)$, (*d*) $h(h(5))$, i.e. $h^2(5)$.

51. Let $X = \{2, 3\}$ and $Y = \{1, 3, 5\}$. How many different functions are there from X into Y?

RANGE OF A FUNCTION

52. The following diagrams define functions f, g and h which map the set $\{1, 2, 3, 4\}$ into itself.

f

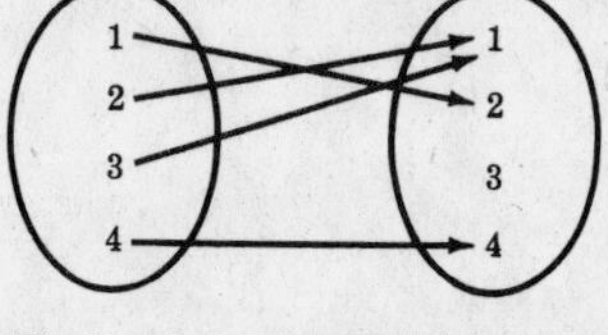

g

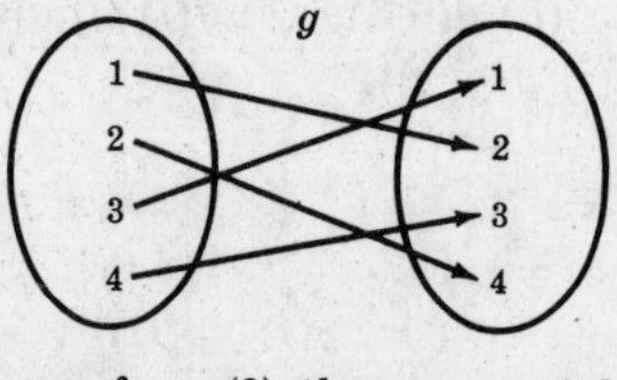

h

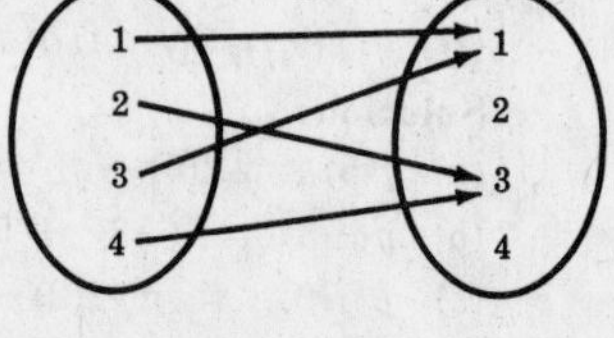

Find (1) the range of f, (2) the range of g, (3) the range of h.

53. Let $W = \{-1, 0, 2, 5, 11\}$. Let the function $f: W \to R^\#$ be defined by $f(x) = x^2 - x - 2$. Find the range of f.

54. Consider the following six functions:

$$f_1: [-2, 2] \to R^\# \qquad f_4: (-\infty, -5] \to R^\#$$
$$f_2: [0, 3] \to R^\# \qquad f_5: [-1, 4) \to R^\#$$
$$f_3: [-3, 0] \to R^\# \qquad f_6: [-5, 3) \to R^\#$$

If each function is defined by the same formula,

$$f(x) = x^2$$

i.e. if to each number x each function assigns x^2, find the range of (1) f_1, (2) f_2, (3) f_3, (4) f_4, (5) f_5, (6) f_6.

55. Consider the six functions in Problem 54. If each function is defined by the formula

$$f(x) = x^3$$

i.e. if to each number x each function assigns x^3, find the range of (1) f_1, (2) f_2, (3) f_3, (4) f_4, (5) f_5, (6) f_6.

56. Consider the functions in Problem 54. Suppose each function is defined by the formula

$$f(x) = x - 3$$

Find the range of (1) f_1, (2) f_2, (3) f_3, (4) f_4, (5) f_5, (6) f_6.

57. Consider the functions in Problem 54. Suppose each function is defined by the formula

$$f(x) = 2x + 4$$

Find the range of (1) f_1, (2) f_2, (3) f_3, (4) f_4, (5) f_5, (6) f_6.

58. Suppose $f: A \to B$. Which of the following is always true:
(1) $f(A) \subset B$, (2) $f(A) = B$, (3) $f(A) \supset B$.

ONE-ONE FUNCTIONS

59. Let $f: X \to Y$. State whether or not each of the following properties defines a one-one function:

(1) $f(a) = f(b)$ implies $a = b$ (3) $f(a) \neq f(b)$ implies $a \neq b$
(2) $a = b$ implies $f(a) = f(b)$ (4) $a \neq b$ implies $f(a) \neq f(b)$

60. State whether or not each of the functions in Problem 54 is one-one.

61. State whether or not each of the functions in Problem 55 is one-one.

62. State whether or not each of the functions in Problem 52 is one-one.

63. Prove: If $f: A \to B$ is one-one and if $g: B \to C$ is one-one, then the product function $g \circ f: A \to C$ is one-one.

PRODUCT FUNCTIONS

64. The functions $f: A \to B$, $g: B \to A$, $h: C \to B$, $F: B \to C$ and $G: A \to C$ are pictured in the diagram below.

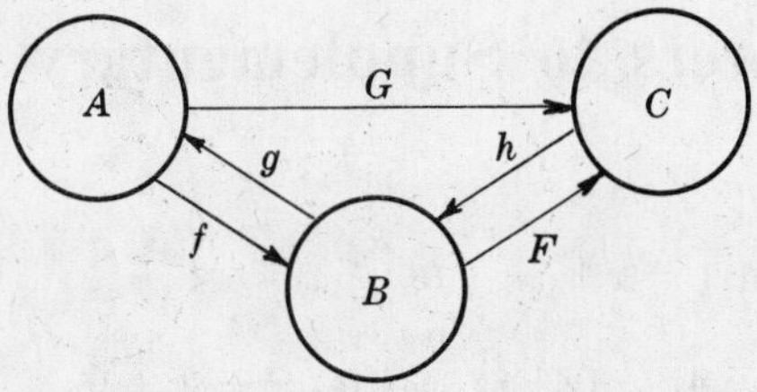

State whether or not each of the following defines a product function and if it does, determine its domain and co-domain:
(1) $g \circ f$, (2) $h \circ f$, (3) $F \circ f$, (4) $G \circ f$, (5) $g \circ h$, (6) $F \circ h$, (7) $h \circ G \circ g$, (8) $h \circ G$.

65. Consider the functions f, g and h in Problem 52. Find the product functions (1) $f \circ g$, (2) $h \circ f$, (3) $g \circ g$, i.e. g^2.

66. Let the functions $f: R\# \to R\#$ and $g: R\# \to R\#$ be defined by

$$f(x) = x^2 + 3x + 1, \quad g(x) = 2x - 3$$

Find formulas which define the product functions (1) $f \circ g$, (2) $g \circ f$, (3) $g \circ g$, (4) $f \circ f$.

67. Let the functions $f: R\# \to R\#$ and $g: R\# \to R\#$ be defined by

$$f(x) = x^2 - 2|x|, \quad g(x) = x^2 + 1$$

Find (a) $(g \circ f)(3)$, (b) $(f \circ g)(-2)$, (c) $(g \circ f)(-4)$, (d) $(f \circ g)(5)$.

INVERSE OF A FUNCTION

68. Let $f : R\# \to R\#$ be defined by $f(x) = x^2 + 1$. Find (1) $f^{-1}(5)$, (2) $f^{-1}(0)$, (3) $f^{-1}(10)$, (4) $f^{-1}(-5)$, (5) $f^{-1}([10, 26])$, (6) $f^{-1}([0, 5])$, (7) $f^{-1}([-5, 1])$, (8) $f^{-1}([-5, 5])$.

69. Let $g : R\# \to R\#$ be defined by $g(x) = \sin x$. Find (1) $g^{-1}(0)$, (2) $g^{-1}(1)$, (3) $g^{-1}(2)$, (4) $g^{-1}([-1, 1])$.

70. Let $f : A \to B$. Find $f^{-1}(B)$.

MISCELLANEOUS PROBLEMS

71. Let $f : R\# \to R\#$ be defined by $f(x) = 3x + 4$. Then f is one-one and onto. Give a formula that defines f^{-1}.

72. Let $A = R\# - \{-1/2\}$ and $B = R\# - \{1/2\}$. Let $f : A \to B$ be defined by

$$f(x) = (x - 3)/(2x + 1)$$

Then f is one-one and onto. Find a formula that defines f^{-1}.

73. Let $W = [0, \infty)$. Let the functions $f : W \to W$, $g : W \to W$ and $h : W \to W$ be defined by

$$f(x) = x^2, \quad g(x) = x^3 + 1, \quad h(x) = x + 2$$

Which of these functions, if any, is onto?

74. Let the function $f : R\# \to R\#$ be defined by $f(x) = x^2 + x - 2$. Find

(*a*) $f(3)$ (*b*) $f(-3) - f(2)$ (*c*) $f(x - 2)$ (*d*) $f(f(-2))$ (*e*) $f(y)$ (*f*) $f(x + h)$ (*g*) $f(x + h) - f(x)$ (*h*) $f(f(x))$ (*i*) $f^{-1}(10)$ (*j*) $f^{-1}(4)$ (*k*) $f^{-1}(-5)$

75. Let $f : A \to B$, $g : B \to A$, and let $g \circ f = 1_A$, the identity function on A. State whether each of the following is true or false.

(1) $g = f^{-1}$.
(2) f is an onto function.
(3) f is a one-one function.
(4) g is an onto function.
(5) g is a one-one function.

Answers to Supplementary Problems

45. (1) No, (2) Yes, (3) No

46. (1) $f(x) = x^2 + 3$, (2) $g(x) = x + |x|$, (3) $h(x) = \begin{cases} x^3 & \text{if } x \geqq 3 \\ 4 & \text{if } x < 3 \end{cases}$

47. (1) 3, (2) 24, (3) $y^2 - 4yz + 4z^2 - 4y + 8z + 3$, (4) $x^2 - 8x + 15$

48. (1) 10, (2) 2, (3) 0

49. (*a*) 1, (*b*) Undefined since 6 is not in the domain, (*c*) $2t^2 - 8t + 1$ when $-1 \leqq t \leqq 7$

50. (*a*) 6, (*b*) 29, (*c*) −19, (*d*) 45

51. Nine

52. (1) {1, 2, 4}, (2) {1, 2, 3, 4}, (3) {1, 3}

53. {0, −2, 18, 108}

54. (1) $[0, 4]$, (2) $[0, 9]$, (3) $[0, 9]$, (4) $[25, \infty)$, (5) $[0, 16)$, (6) $[0, 25]$

55. (1) $[-8,8]$, (2) $[0,27]$, (3) $[-27,0]$, (4) $(-\infty,-125]$, (5) $[-1,64)$, (6) $[-125,27)$

56. (1) $[-5,-1]$, (2) $[-3,0]$, (3) $[-6,-3]$, (4) $(-\infty,-8]$, (5) $[-4,1)$, (6) $[-8,0)$

57. (1) $[0,8]$, (2) $[4,10]$, (3) $[-2,4]$, (4) $(-\infty,-6]$, (5) $[2,12)$, (6) $[-6,10)$

58. $f(A) \subset B$

59. (1) Yes, (2) No, (3) No, (4) Yes

60. (1) No, (2) Yes, (3) Yes, (4) Yes, (5) No, (6) No

61. They are all one-one.

62. Only g is one-one.

63. We must show that $(g \circ f)(a) = (g \circ f)(b)$ implies $a = b$. Let $(g \circ f)(a) = (g \circ f)(b)$. Then, by definition of product function, $g(f(a)) \equiv (g \circ f)(a) = (g \circ f)(b) \equiv g(f(b))$. Since g is one-one, $f(a) = f(b)$; and since f is one-one, $a = b$. Hence $g \circ f$ is one-one.

64. (1) $g \circ f : A \to A$, (2) Not defined, (3) $F \circ f : A \to C$, (4) Not defined, (5) $g \circ h : C \to A$, (6) $F \circ h : C \to C$, (7) $h \circ G \circ g : B \to B$, (8) $h \circ G : A \to B$.

65. (1) $f \circ g$ (2) $h \circ f$ (3) $g \circ g$

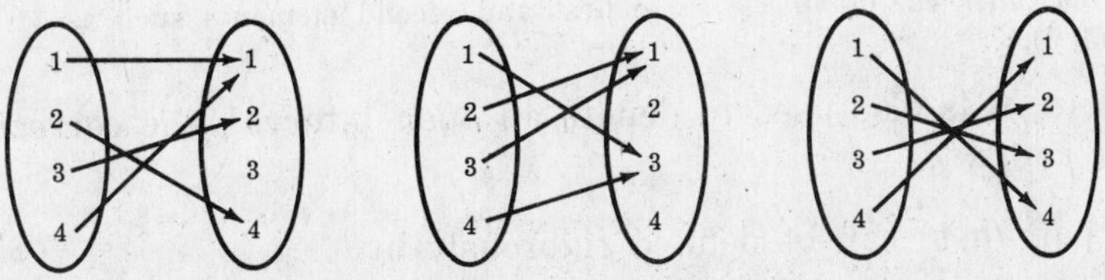

66. (1) $(f \circ g)(x) = 4x^2 - 6x + 1$
(2) $(g \circ f)(x) = 2x^2 + 6x - 1$
(3) $(g \circ g)(x) = 4x - 9$
(4) $(f \circ f)(x) = x^4 + 6x^3 + 14x^2 + 15x + 5$

67. (a) 10, (b) 15, (c) 65, (d) 624

68. (1) $\{-2, 2\}$ (2) $\emptyset$ (3) $\{3, -3\}$ (4) $\emptyset$ (5) $\{x \mid -5 \leqq x \leqq -3 \text{ or } 3 \leqq x \leqq 5\}$ (6) $\{x \mid -2 \leqq x \leqq 2\}$ (7) $\{0\}$ (8) $\{x \mid -2 \leqq x \leqq 2\}$

69. (1) $\{\ldots, -2\pi, -\pi, 0, \pi, 2\pi, \ldots\} = \{x \mid x = n\pi \text{ where } n \in Z\}$
(2) $\{x \mid x = (\pi/2) + 2\pi n \text{ where } n \in Z\}$
(3) $\emptyset$, (4) $R^{\#}$, the set of all real numbers.

70. $f^{-1}(B) = A$

71. $f^{-1}(x) = (x-4)/3$

72. $f^{-1}(x) = (3+x)/(1-2x)$

73. Only f is onto.

74. (a) 10 (b) 0 (c) $x^2 - 3x$ (d) -2 (e) $y^2 + y - 2$ (f) $x^2 + 2xh + h^2 + x + h - 2$ (g) $2xh + h^2 + h$ (h) $x^4 + 2x^3 - 2x^2 - 3x$ (i) $\{3, -4\}$ (j) $\{2, -3\}$ (k) $\emptyset$

75. (1) False, (2) False, (3) True, (4) True, (5) False

Chapter 5

Product Sets and Graphs of Functions

ORDERED PAIRS

Intuitively, an *ordered pair* consists of two elements, say a and b, in which one of them, say a, is designated as the first element and the other as the second element. An ordered pair is denoted by

$$(a, b)$$

Two ordered pairs (a, b) and (c, d) are equal if and only if $a = c$ and $b = d$.

Example 1.1: The ordered pairs $(2, 3)$ and $(3, 2)$ are different.

Example 1.2: The points in the Cartesian plane shown in Fig. 5-1 below represent ordered pairs of real numbers.

Example 1.3: The set $\{2, 3\}$ is not an ordered pair since the elements 2 and 3 are not distinguished.

Example 1.4: Ordered pairs can have the same first and second elements such as $(1, 1)$, $(4, 4)$ and $(5, 5)$.

Although the notation (a, b) is also used to denote an open interval, the correct meaning will be clear from the context.

Remark 5.1: An ordered pair (a, b) can be defined rigorously by

$$(a, b) \equiv \{\{a\}, \{a, b\}\}$$

From this definition, the fundamental property of ordered pairs can be proven:

$$(a, b) = (c, d) \quad \text{implies} \quad a = c \text{ and } b = d$$

PRODUCT SET

Let A and B be two sets. The *product set* of A and B consists of all ordered pairs (a, b) where $a \in A$ and $b \in B$. It is denoted by

$$A \times B$$

which reads "A cross B". More concisely,

$$A \times B = \{(a, b) \mid a \in A, b \in B\}$$

Example 2.1: Let $A = \{1, 2, 3\}$ and $B = \{a, b\}$. Then the product set

$$A \times B = \{(1, a), (1, b), (2, a), (2, b), (3, a), (3, b)\}$$

Example 2.2: Let $W = \{s, t\}$. Then

$$W \times W = \{(s, s), (s, t), (t, s), (t, t)\}$$

Example 2.3: The Cartesian plane shown in Fig. 5.1 is the product set of the real numbers with itself, i.e. $R^{\#} \times R^{\#}$.

The product set $A \times B$ is also called the *Cartesian product* of A and B. It is named after the mathematician Descartes who, in the seventeenth century, first investigated the set $R^{\#} \times R^{\#}$. It is also for this reason that $R^{\#} \times R^{\#}$, as pictured in Fig. 5-1, is called the Cartesian plane.

Remark 5.2: If set A has n elements and set B has m elements then the product set $A \times B$ has n times m elements, i.e. nm elements. If either A or B is the null set then $A \times B$ is also the null set. Lastly, if either A or B is infinite and the other is not empty, then $A \times B$ is infinite.

Remark 5.3: The Cartesian product of two sets is not commutative; more specifically,

$$A \times B \neq B \times A$$

unless $A = B$ or one of the factors is empty.

COORDINATE DIAGRAMS

The reader is familiar with the Cartesian plane $R^\# \times R^\#$, as shown in Fig. 5-1 below. Each point P represents an ordered pair (a, b) of real numbers. A vertical line through P meets the horizontal axis at a and a horizontal line through P meets the vertical axis at b, as in Fig. 5-1.

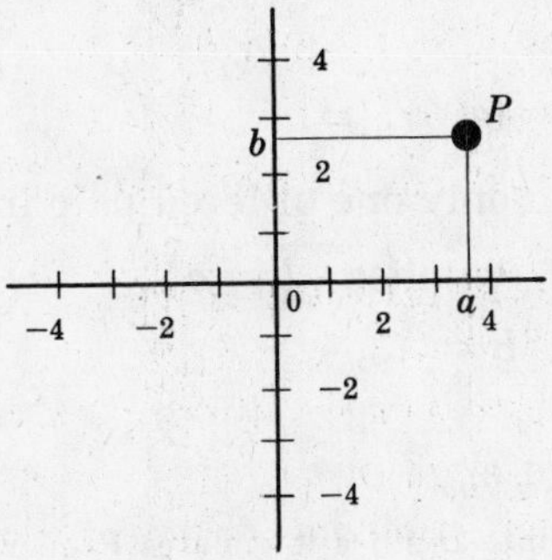

Fig. 5-1

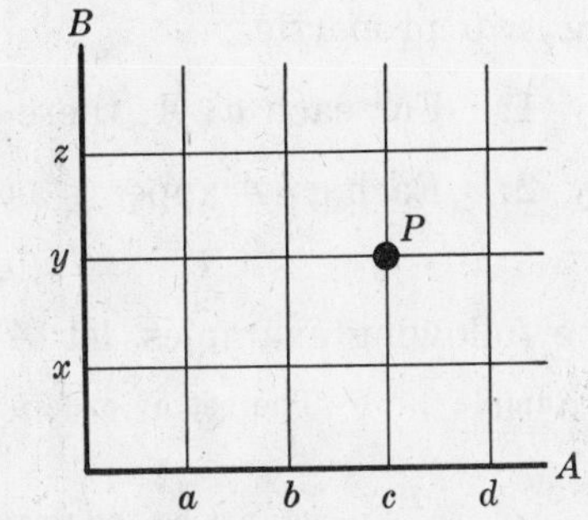

Fig. 5-2

The Cartesian product of any two sets, if they do not contain too many elements, can be displayed on a coordinate diagram in a similar manner. For example, if $A = \{a, b, c, d\}$ and $B = \{x, y, z\}$, then the coordinate diagram of $A \times B$ is as shown in Fig. 5-2 above. Here the elements of A are displayed on the horizontal axis and the elements of B are displayed on the vertical axis. Notice that the vertical lines through the elements of A and the horizontal lines through the elements of B meet in 12 points. These points represent $A \times B$ in the obvious way. The point P is the ordered pair (c, y).

GRAPH OF A FUNCTION

Let f be a function of A into B, that is, let $f: A \rightarrow B$. The graph f^* of the function f consists of all ordered pairs in which $a \in A$ appears as a first element and its image appears as its second element. In other words,

$$f^* = \{(a, b) \mid a \in A, \; b = f(a)\}$$

Notice that f^*, the graph of $f: A \rightarrow B$, is a subset of $A \times B$.

Example 3.1: Let the function $f: A \rightarrow B$ be defined by the diagram

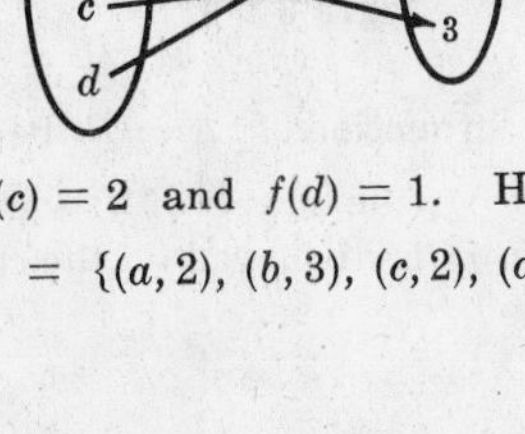

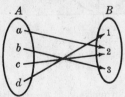

Then $f(a) = 2$, $f(b) = 3$, $f(c) = 2$ and $f(d) = 1$. Hence the graph of f is

$$f^* = \{(a, 2), (b, 3), (c, 2), (d, 1)\}$$

Example 3.2: Let $W = \{1, 2, 3, 4\}$. Let the function $f: W \to R^{\#}$ be defined by

$$f(x) = x + 3$$

Then the graph of f is

$$f^* = \{(1,4), (2,5), (3,6), (4,7)\}$$

Example 3.3: Let N be the natural numbers $1, 2, 3, \ldots$. Let the function $g: N \to N$ be defined by

$$g(x) = x^3$$

Then the graph of g is

$$g^* = \{(1,1), (2,8), (3,27), (4,64), \ldots\}$$

PROPERTIES OF THE GRAPH OF A FUNCTION

Let $f: A \to B$. We recall two properties of the function f. First, for each element $a \in A$ there is assigned an element in B. Secondly, there is only one element in B which is assigned to each $a \in A$. In view of these properties of f, the graph f^* of f has the following two properties:

Property 1: For each $a \in A$, there is an ordered pair $(a, b) \in f^*$.

Property 2: Each $a \in A$ appears as the first element in only one ordered pair in f^*, that is,

$$(a, b) \in f^*, \ (a, c) \in f^* \quad \text{implies} \quad b = c$$

In the following examples, let $A = \{1, 2, 3, 4\}$ and $B = \{3, 4, 5, 6\}$.

Example 4.1: The set of ordered pairs

$$\{(1,5), (2,3), (4,6)\}$$

cannot be the graph of a function of A into B since it violates Property 1. Specifically, $3 \in A$ and there is no ordered pair in which 3 is a first element.

Example 4.2: The set of ordered pairs

$$\{(1,5), (2,3), (3,6), (4,6), (2,4)\}$$

cannot be the graph of a function of A into B since it violates Property 2, that is, the element $2 \in A$ appears as the first element in two different ordered pairs $(2,3)$ and $(2,4)$.

GRAPHS AND COORDINATE DIAGRAMS

Let f^* be the graph of a function $f: A \to B$. As f^* is a subset of $A \times B$, it can be displayed, i.e. graphed, on the coordinate diagram of $A \times B$.

Example 5.1: Let $f(x) = x^2$ define a function on the interval $-2 \leq x \leq 4$. Then the graph of f is displayed in Fig. 5-3 below in the usual way:

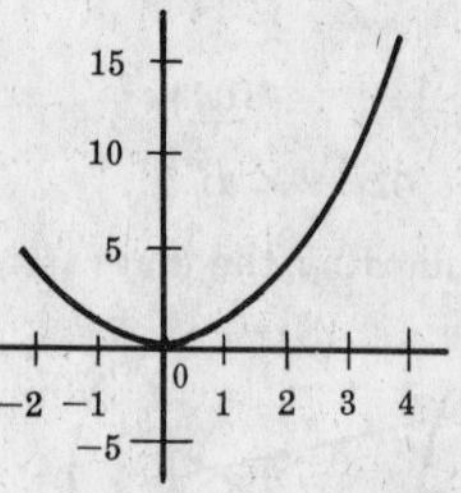

Fig. 5-3

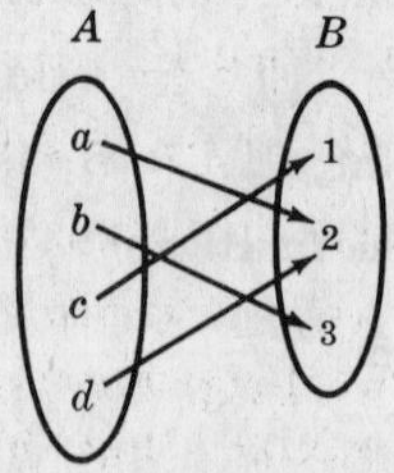

Fig. 5-4

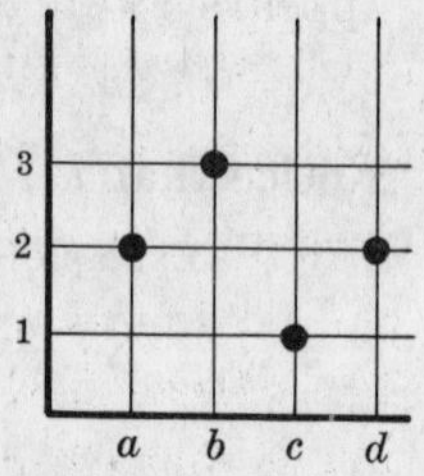

Fig. 5-5

Example 5.2: Let a function $f: A \to B$ be defined by the diagram shown in Fig. 5-4 above.

Here f^*, the graph of f, consists of the ordered pairs $(a, 2)$, $(b, 3)$, $(c, 1)$ and $(d, 2)$. Then f^* is displayed on the coordinate diagram of $A \times B$ as shown in Fig. 5-5 above.

PROPERTIES OF GRAPHS OF FUNCTIONS ON COORDINATE DIAGRAMS

Let $f: A \to B$. Then f^*, the graph of f, has the two properties listed previously:

Property 1: For each $a \varepsilon A$, there is an ordered pair $(a, b) \varepsilon f^*$.

Property 2: If $(a, b) \varepsilon f^*$ and $(a, c) \varepsilon f^*$, then $b = c$.

Therefore, if f^* is displayed on the coordinate diagram of $A \times B$, it has the following two properties:

Property 1: Each vertical line will contain at least one point of f^*.

Property 2: Each vertical line will contain only one point of f^*.

Example 6.1: Let $A = \{a, b, c\}$ and $B = \{1, 2, 3\}$. Consider the sets of points in the two coordinate diagrams of $A \times B$ below.

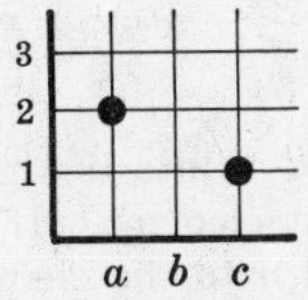

(1)

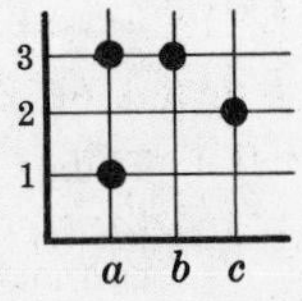

(2)

In (1), the vertical line through b does not contain a point of the set; hence the set of points cannot be the graph of a function of A into B.

In (2), the vertical line through a contains two points of the set; hence this set of points cannot be the graph of a function of A into B.

Example 6.2: The circle $x^2 + y^2 = 9$, pictured below, cannot be the graph of a function since there are vertical lines which contain more than one point of the circle.

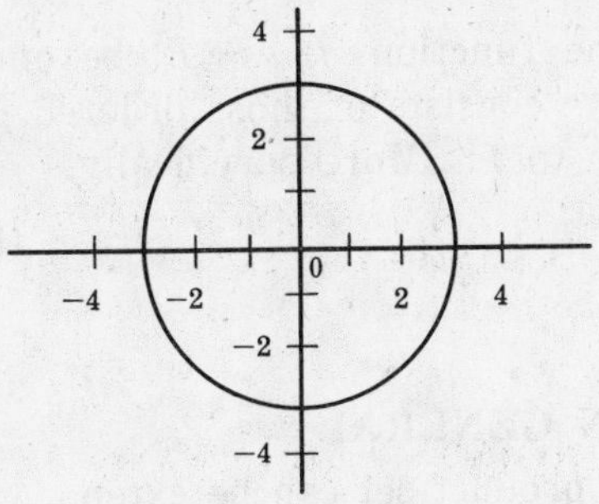

$x^2 + y^2 = 9$ is plotted

FUNCTIONS AS SETS OF ORDERED PAIRS

Let f^* be a subset of $A \times B$, the Cartesian product of sets A and B; and let f^* have the two properties discussed previously:

Property 1: For each $a \varepsilon A$, there is an ordered pair $(a, b) \varepsilon f^*$.

Property 2: No two different ordered pairs in f^* have the same first element.

Thus we have a rule that assigns to each element $a \varepsilon A$ the element $b \varepsilon B$ that appears in the ordered pair $(a, b) \varepsilon f^*$. Property 1 guarantees that each element in A will have an image, and Property 2 guarantees that the image is unique. Accordingly, f^* is a function of A into B.

In view of the correspondence between functions $f: A \to B$ and subsets of $A \times B$ with Property 1 and Property 2 above, we redefine a function by the

Definition 5.1: A function f of A into B is a subset of $A \times B$ in which each $a \varepsilon A$ appears as the first element in one and only one ordered pair belonging to f.

Although this definition of a function may seem artificial, it has the advantage that it does not use such undefined terms as "assigns", "rule", "correspondence".

Example 7.1: Let $A = \{a, b, c\}$ and $B = \{1, 2, 3\}$. Furthermore, let

$$f = \{(a, 2), (c, 1), (b, 2)\}$$

Then f has Property 1 and Property 2. Hence f is a function of A into B, which is also illustrated in the following diagram:

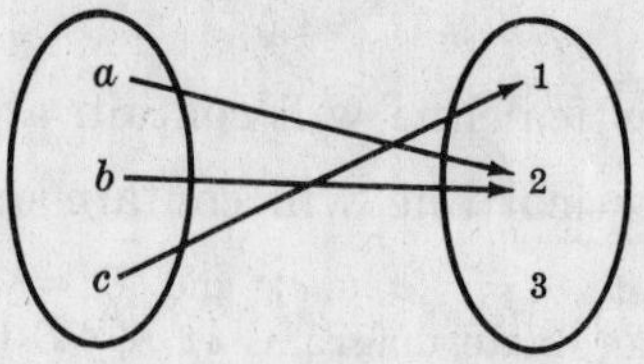

Example 7.2: Let $V = \{1, 2, 3\}$ and $W = \{a, e, i, o, u\}$. Also, let

$$f = \{(1, a), (2, e), (3, i), (2, u)\}$$

Then f is not a function of V into W since two different ordered pairs in f, $(2, e)$ and $(2, u)$, have the same first element. If f is to be a function of V into W, then it cannot assign both e and u to the element $2 \in V$.

Example 7.3: Let $S = \{1, 2, 3, 4\}$ and $T = \{1, 3, 5\}$. Let

$$f = \{(1, 1), (2, 5), (4, 3)\}$$

Then f is not a function of S into T since $3 \in S$ does not appear as the first element in any ordered pair belonging to f.

The geometrical implication of Definition 5.1 is stated in

Remark 5.4: Let f be a set of points in the coordinate diagram of $A \times B$. If every vertical line contains one and only one point of f, then f is a function of A into B.

Remark 5.5: Let the function $f: A \to B$ be one-one and onto. Then the inverse function f^{-1} consists of those ordered pairs which when reversed, i.e. permuted, belong to f. More specifically,

$$f^{-1} = \{(b, a) \mid (a, b) \in f\}$$

PRODUCT SETS IN GENERAL

The concept of a product set can be extended to more than two sets in a natural way. The Cartesian product of sets A, B and C, denoted by

$$A \times B \times C$$

consists of all ordered triplets (a, b, c) where $a \in A$, $b \in B$ and $c \in C$. Analogously, the Cartesian product of n sets $A_1, A_2, \ldots, A_n$, denoted by

$$A_1 \times A_2 \times \cdots \times A_n$$

consists of all ordered n-tuples $(a_1, a_2, \ldots, a_n)$ where $a_1 \in A_1, \ldots, a_n \in A_n$. Here an ordered n-tuple has the obvious intuitive meaning, that is, it consists of n elements, not necessarily distinct, in which one of them is designated as the first element, another as the second element, etc.

Example 8.1: In three dimensional Euclidean geometry each point represents an ordered triplet, i.e. its x-component, its y-component and its z-component.

Example 8.2: Let $A = \{a, b\}$, $B = \{1, 2, 3\}$ and $C = \{x, y\}$. Then

$$\begin{aligned} A \times B \times C = \{&(a, 1, x), (a, 1, y), (a, 2, x), \\ &(a, 2, y), (a, 3, x), (a, 3, y), \\ &(b, 1, x), (b, 1, y), (b, 2, x), \\ &(b, 2, y), (b, 3, x), (b, 3, y)\} \end{aligned}$$

Solved Problems

ORDERED PAIRS AND PRODUCT SETS

1. Let W = {John, Jim, Tom} and let V = {Betty, Mary}. Find $W \times V$.

Solution:

$W \times V$ consists of all ordered pairs (a, b) where $a \in W$ and $b \in V$. Consequently,

$W \times V$ = {(John, Betty), (John, Mary), (Jim, Betty), (Jim, Mary), (Tom, Betty), (Tom, Mary)}

2. Suppose the ordered pairs $(x+y, 1)$ and $(3, x-y)$ are equal. Find x and y.

Solution:

If $(x+y, 1) = (3, x-y)$ then, by the fundamental property of ordered pairs,

$$x+y = 3 \quad \text{and} \quad 1 = x-y$$

The solution of these simultaneous equations is $x=2$, $y=1$.

3. Find the ordered pairs corresponding to the points P_1, P_2, P_3 and P_4 which appear in the coordinate diagram of $A \times B$ in Fig. 5-6. Here, $A = \{a, b, c, d, e\}$ and $B = \{a, e, i, o, u\}$.

Solution:

The vertical line through P_1 crosses b on the A axis, and the horizontal line through P_1 crosses i on the B axis; thus P_1 corresponds to the ordered pair (b, i). Similarly, $P_2 = (a, a)$, $P_3 = (d, u)$ and $P_4 = (e, e)$.

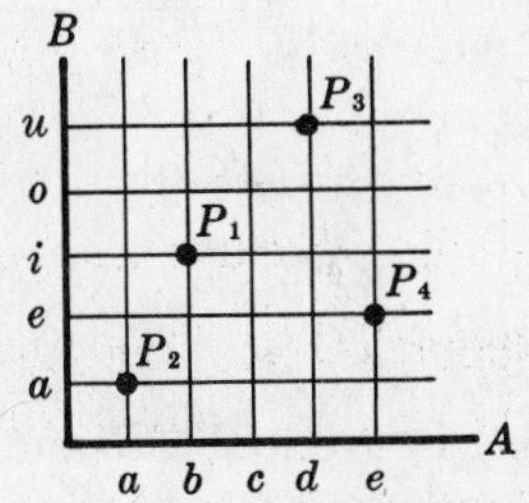

Fig. 5-6

4. Let $A = \{a, b\}$, $B = \{2, 3\}$ and $C = \{3, 4\}$. Find

(1) $A \times (B \cup C)$, (2) $(A \times B) \cup (A \times C)$, (3) $A \times (B \cap C)$, (4) $(A \times B) \cap (A \times C)$

Solution:

(1) First compute $B \cup C = \{2, 3, 4\}$. Then

$$A \times (B \cup C) = \{(a, 2), (a, 3), (a, 4), (b, 2), (b, 3), (b, 4)\}$$

(2) First find $A \times B$ and $A \times C$:

$$A \times B = \{(a, 2), (a, 3), (b, 2), (b, 3)\}$$
$$A \times C = \{(a, 3), (a, 4), (b, 3), (b, 4)\}$$

Then compute the union of the two sets:

$$(A \times B) \cup (A \times C) = \{(a, 2), (a, 3), (b, 2), (b, 3), (a, 4), (b, 4)\}$$

Notice, from (1) and (2), that

$$A \times (B \cup C) = (A \times B) \cup (A \times C)$$

(3) First compute $B \cap C = \{3\}$. Then

$$A \times (B \cap C) = \{(a, 3), (b, 3)\}$$

(4) In (2), $A \times B$ and $A \times C$ were computed. The intersection of $A \times B$ and $A \times C$ consists of the ordered pairs which belong to both sets, i.e.,

$$(A \times B) \cap (A \times C) = \{(a, 3), (b, 3)\}$$

Notice, from (3) and (4), that

$$A \times (B \cap C) = (A \times B) \cap (A \times C)$$

5. Sketch, by shading the appropriate area, the product set

$$\{x \mid 1 \leq x < 4\} \times \{x \mid -2 \leq x \leq 3\}$$

on the coordinate diagram of $R^{\#} \times R^{\#}$.

Solution:

Draw light vertical lines through 1 and 4 on the horizontal axis, and horizontal lines through -2 and 3 on the vertical axis as shown in Fig. 5-7 below.

The rectangular area bordered by the four lines, together with three of its sides, represents the product set. Shade the diagram as shown in Fig. 5-8 below.

Notice that the side of the rectangle which does not belong to the product set is shaded by a dotted line.

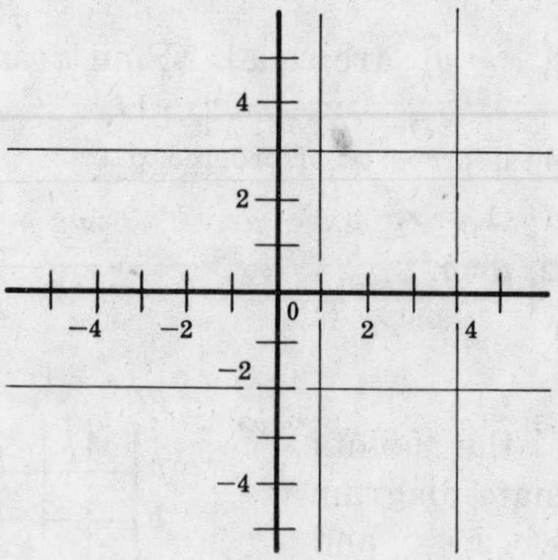

Fig. 5-7

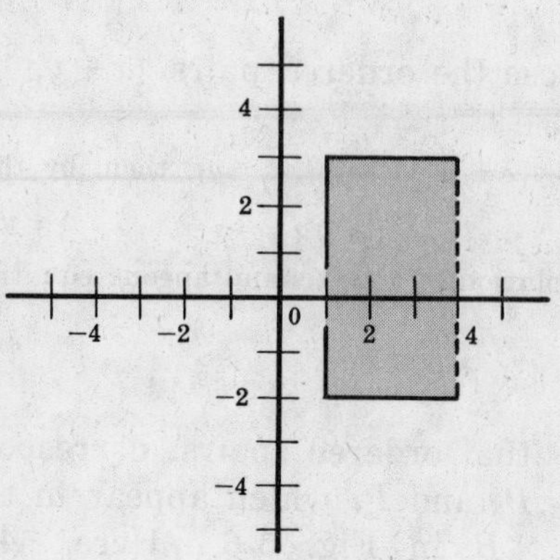

Fig. 5-8

6. Prove: $A \subset B$ and $C \subset D$ implies $(A \times C) \subset (B \times D)$.

Solution:

Let (x, y) be any element in $A \times C$; then $x \,\varepsilon\, A$ and $y \,\varepsilon\, C$. By hypothesis, A is a subset of B and C is a subset of D; then $x \,\varepsilon\, B$ and $y \,\varepsilon\, D$. Thus the ordered pair (x, y) belongs to $B \times D$. We have shown that $(x, y) \,\varepsilon\, A \times C$ implies $(x, y) \,\varepsilon\, B \times D$; hence $A \times C$ is a subset of $B \times D$.

7. Let $A = \{1, 2, 3\}$, $B = \{2, 4\}$ and $C = \{3, 4, 5\}$. Find $A \times B \times C$.

Solution:

A convenient method of finding $A \times B \times C$ is through the so-called "tree diagram" shown below:

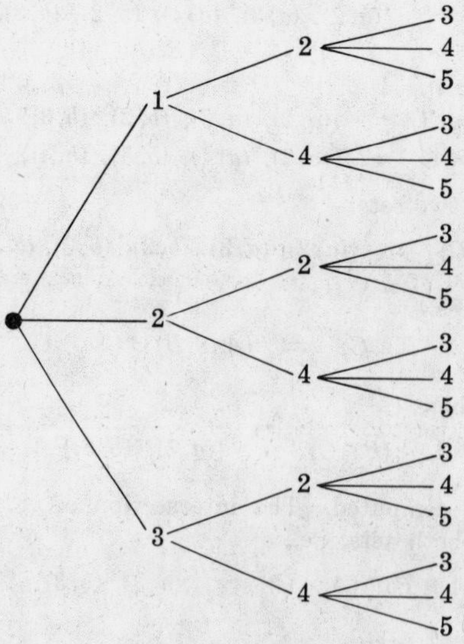

(1, 2, 3)
(1, 2, 4)
(1, 2, 5)
(1, 4, 3)
(1, 4, 4)
(1, 4, 5)
(2, 2, 3)
(2, 2, 4)
(2, 2, 5)
(2, 4, 3)
(2, 4, 4)
(2, 4, 5)
(3, 2, 3)
(3, 2, 4)
(3, 2, 5)
(3, 4, 3)
(3, 4, 4)
(3, 4, 5)

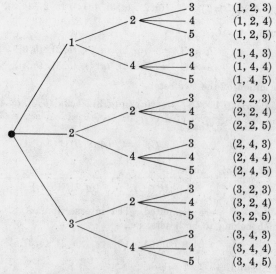

$A \times B \times C$ consists of the ordered triplets on the right of the "tree".

8. Prove: $A \times (B \cap C) = (A \times B) \cap (A \times C)$.

Solution:

We first show $A \times (B \cap C)$ is a subset of $(A \times B) \cap (A \times C)$. Let (x, y) be any element of $A \times (B \cap C)$; then $x \in A$ and $y \in B \cap C$. By definition of intersection, y belongs to both B and to C. Since $x \in A$ and $y \in B$, then $(x, y) \in A \times B$. Also, since $x \in A$ and $y \in C$, then $(x, y) \in A \times C$. Accordingly, (x, y) belongs to the intersection of $A \times B$ and $A \times C$. We have therefore proven $A \times (B \cap C) \subset (A \times B) \cap (A \times C)$.

Next we want to show that $(A \times B) \cap (A \times C)$ is a subset of $A \times (B \cap C)$. Now let (z, w) be any element of $(A \times B) \cap (A \times C)$; then (z, w) belongs to $A \times B$ and (z, w) belongs to $A \times C$. It now follows that $z \in A$ and $w \in B$, and $z \in A$ and $w \in C$. Since w belongs to both B and C, then $w \in B \cap C$. We have $z \in A$ and $w \in B \times C$; then $(z, w) \in A \times (B \cap C)$. We have just proven that $(A \times C) \cap (A \times C)$ is a subset of $A \times (B \cap C)$. By Definition 1.1, the sets are equal.

9. Let $S = \{a, b\}$, $W = \{1, 2, 3, 4, 5\}$ and $V = \{3, 5, 7, 9\}$. Find $(S \times W) \cap (S \times V)$.

Solution:

The product set $(S \times W) \cap (S \times V)$ can be found by first computing $S \times W$ and $S \times V$, and then computing the intersection of these sets. But, by Problem 8,

$$(S \times W) \cap (S \times V) = S \times (W \cap V)$$

Then $W \cap V = \{3, 5\}$, and

$$(S \times W) \cap (S \times V) = S \times (W \cap V) = \{(a, 3), (a, 5), (b, 3), (b, 5)\}$$

GRAPHS OF FUNCTIONS

10. Let $W = \{1, 2, 3, 4\}$, and let the function $f: W \to R^{\#}$ be defined by the formula $f(x) = x^2$. Find the graph f^* of the function f.

Solution:

First compute $f(1) = 1^2 = 1$, $f(2) = 2^2 = 4$, $f(3) = 3^2 = 9$, $f(4) = 4^2 = 16$. The graph f^* of f consists of the ordered pairs $(x, f(x))$, i.e. (x, x^2) where $x \in W$. Accordingly, $f^* = \{(1, 1), (2, 4), (3, 9), (4, 16)\}$.

11. Let $V = \{a, b, c, d\}$, and let the function $g: V \to V$ be defined by the diagram shown in Fig. 5-9 below. Find the graph g^* of the function g, and sketch g^* on the coordinate diagram of $V \times V$.

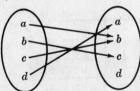

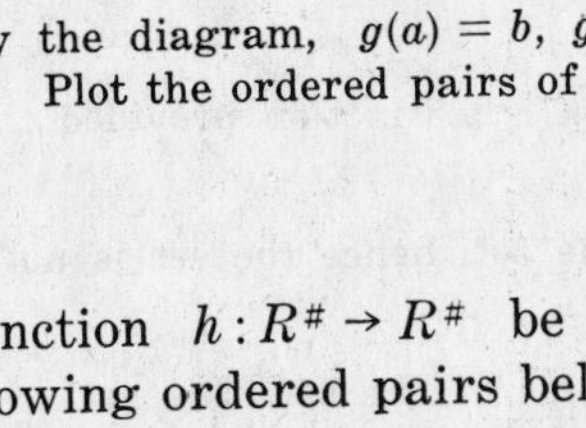

Fig. 5-9

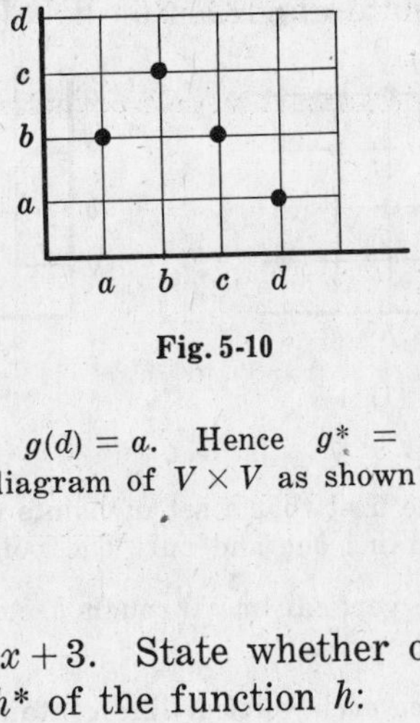

Fig. 5-10

Solution:

Note, by the diagram, $g(a) = b$, $g(b) = c$, $g(c) = b$ and $g(d) = a$. Hence $g^* = \{(a, b), (b, c), (c, b), (d, a)\}$. Plot the ordered pairs of g^* on the coordinate diagram of $V \times V$ as shown in Fig. 5-10 above.

12. Let the function $h: R^{\#} \to R^{\#}$ be defined by $h(x) = x + 3$. State whether or not each of the following ordered pairs belongs to the graph h^* of the function h:

(a) $(2, 6)$, (b) $(8, 11)$, (c) $(10, 12)$, (d) $(4, 7)$, (e) $(-6, -9)$, (f) $(-1, 2)$.

Solution:

(a) $h(2) = 2 + 3 = 5$; so $(2, 6)$ does not belong to h^*.
(b) $h(8) = 8 + 3 = 11$; so $(8, 11)$ does belong to h^*.
(c) $h(10) = 10 + 3 = 13$; so $(10, 12) \notin h^*$.
(d) $h(4) = 4 + 3 = 7$; so $(4, 7) \in h^*$.
(e) $h(-6) = -6 + 3 = -3$; so $(-6, -9) \notin h^*$.
(f) $h(-1) = -1 + 3 = 2$; so $(-1, 2) \in h^*$.

13. Let $S = \{a, e, i, o, u\}$. To each letter in S let the function g assign the letter which follows it in the alphabet. Find the graph g^* of the function g.

Solution:

First find $g(a) = b$, $g(e) = f$, $g(i) = j$, $g(o) = p$ and $g(u) = v$. Thus

$$g^* = \{(a, b), (e, f), (i, j), (o, p), (u, v)\}$$

FUNCTIONS AS ORDERED PAIRS

14. Let $V = \{1, 2, 3, 4\}$. State whether or not each of the following sets of ordered pairs is a function of V into V.

(1) $f_1 = \{(2, 3), (1, 4), (2, 1), (3, 2), (4, 4)\}$

(2) $f_2 = \{(3, 1), (4, 2), (1, 1)\}$

(3) $f_3 = \{(2, 1), (3, 4), (1, 4), (2, 1), (4, 4)\}$

(4) $f_4 = \{(2, 3), (1, 6), (4, 2), (3, 4)\}$

Solution:

Note first, by Definition 5.1, that a subset f of $V \times V$ is a function $f: V \to V$ if each $x \varepsilon V$ appears as the first element in one and only one ordered pair in f.

(1) Since two different ordered pairs $(2, 3) \varepsilon f_1$ and $(2, 1) \varepsilon f_1$ have the same first element, f_1 is not a function of V into V.

(2) The element $2 \varepsilon V$ does not appear as the first element in any ordered pair belonging to f_2. Hence f_2 is not a function of V into V.

(3) The set f_3 is a function of V into V. Although 2 appears as the first element in two ordered pairs, these two ordered pairs are equal.

(4) Although each element in V appears as the first element in one and only one ordered pair belonging to f_4, the set f_4 is not a function of V into V since f_4 is not a subset of $V \times V$. Specifically, $(1, 6) \varepsilon f_4$ but $(1, 6) \not\varepsilon V \times V$.

15. Let $W = \{a, b, c, d\}$. State whether or not the set of points in each of the following coordinate diagrams of $W \times W$ is a function from W into W.

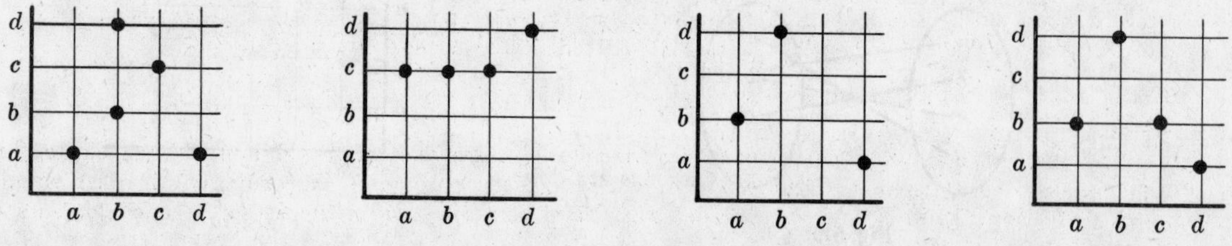

(1) (2) (3) (4)

Solution:

Note first that a set of points on a coordinate diagram is a function provided that every vertical line contains one and only one point of the set.

(1) The vertical line through b contains two points of the set; hence the set is not a function from W into W.

(2) Since each vertical line contains one and only one point of the set, this set is a function of W into W. The fact that the horizontal line through c contains three points does not violate the properties of a function.

(3) The vertical line through c contains no point of the set; hence this set is not a function of W into W.

(4) For the same reason as in (2), this set is a function of W into W.

16. Let $R = \{1, 2, 3, 4, 5, 6\}$ and let $S = \{1, 2, 3, 4\}$. The set g of points in the coordinate diagram of $R \times S$ in the figure at right is a function of R into S.

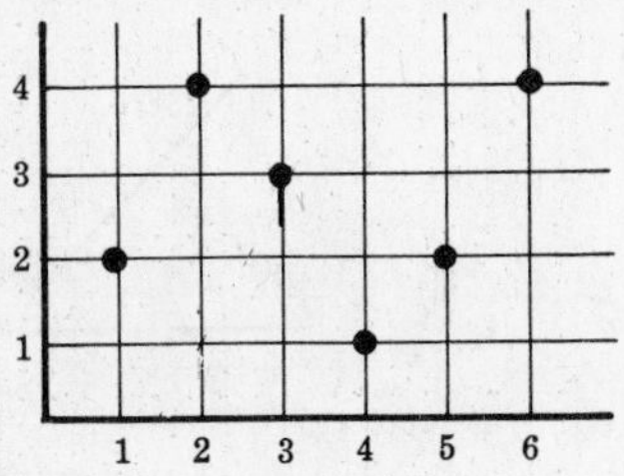

(a) Find $g(2)$, $g(4)$, $g(6)$. (b) Find $g^{-1}(2)$, $g^{-1}(3)$, $g^{-1}(4)$.
(c) Find $\{x \mid x \in R,\ g(x) < 3\}$.

Solution:

(a) To find $g(2)$ we look at the point in g which lies on the vertical line through 2; the point is $(2, 4)$. Hence $g(2) = 4$, the second element in the ordered pair.

The vertical line through 4 contains the point $(4, 1)$ of g; hence $g(4) = 1$.

The vertical line through 6 contains the point $(6, 4)$; hence $g(6) = 4$.

(b) To find $g^{-1}(2)$ we look at the points that lie on the horizontal line through 2. The points are $(1, 2)$ and $(5, 2)$. Then $g^{-1}(2)$ consists of the first elements of these ordered pairs, i.e. $g^{-1}(2) = \{1, 5\}$. Note that the ordered pairs $(1, 2)$ and $(5, 2)$ in g mean that $g(1) = 2$ and $g(5) = 2$.

The horizontal line through 3 contains only the point $(3, 3)$ of g; then $g^{-1}(3) = \{3\}$.

The horizontal line through 4 contains the points $(2, 4)$ and $(6, 4)$ of g; then $g^{-1}(4) = \{2, 6\}$.

(c) First note that $g(1) = 2$, $g(2) = 4$, $g(3) = 3$, $g(4) = 1$, $g(5) = 2$, $g(6) = 4$. The set $\{x \mid x \in R,\ g(x) < 3\}$ consists of those elements in R whose image is less than 3, i.e. whose image is 1 or 2. The set is $\{1, 4, 5\}$. Geometrically, this set consists of the first elements of the points in g which lie below the horizontal line through 3.

17. Let h be a set of points in the coordinate diagram of $E \times F$ which is a function of E into F.

(a) If each horizontal line contains at most one point of h, what type of function is h?

(b) If each horizontal line contains at least one point of h, what type of function is h?

Solution:

(a) If each horizontal line contains at most one point of h, then, for every $x \in F$, $h^{-1}(x)$ is empty or consists of one element in E. Thus h is a one-one function.

(b) If each horizontal line contains at least one point of h, then, for every $x \in F$, $h^{-1}(x)$ is not empty. Hence h is an onto function.

18. Under what conditions will the set of ordered pairs

$$f = \{(1, 5), (3, 1), (4, 7), (-2, -3)\}$$

define a function of A into B?

Solution:

The set f will define a function of A into B if f is a subset of $A \times B$ and if each element in A appears as the first element in one and only one ordered pair in f. Accordingly, A must equal the set of first elements of f, i.e. $A = \{1, 3, 4, -2\}$; and B must contain the set of second elements of f, i.e. $\{5, 1, 7, -3\} \subset B$.

19. Let $W = [-4, 4]$. State whether or not each of the following sets of points displayed on a coordinate diagram of $W \times W$ is a function of W into W.

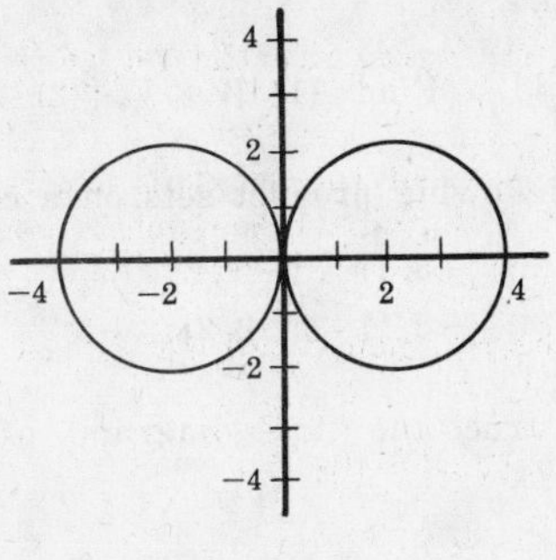

(1)

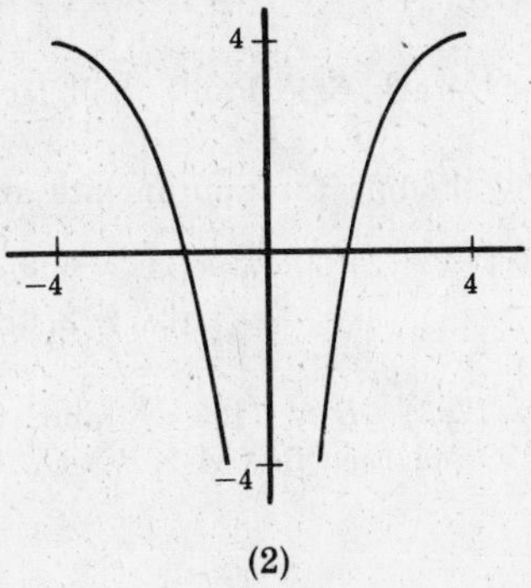

(2)

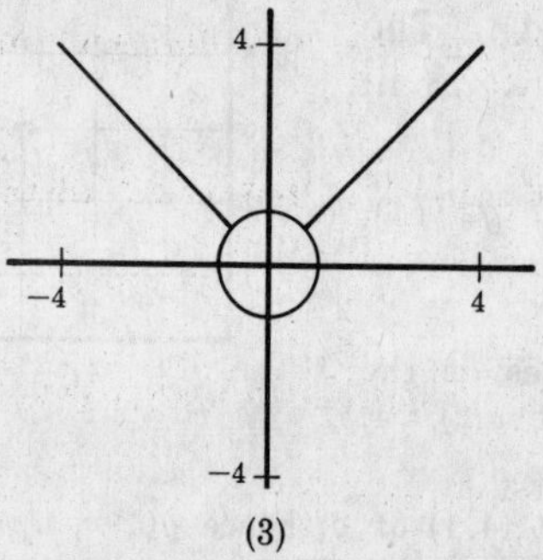

(3)

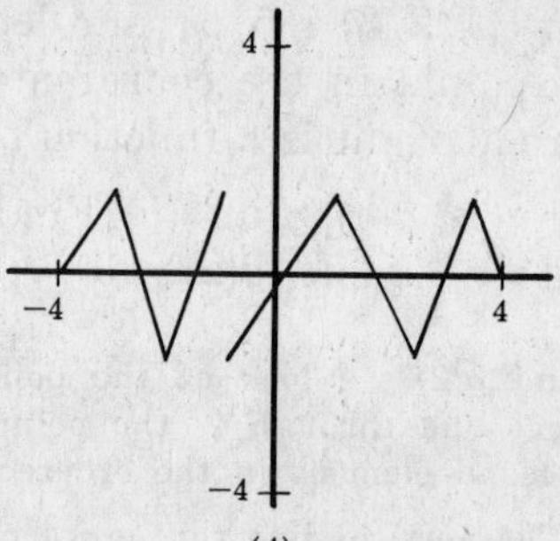

(4)

Solution:

Note first, by Remark 5.4, that a set of points on a coordinate diagram is a function if every vertical line contains one and only one point of the set.

(1) Since vertical lines contain two points of the set, the set of points is not a function of W into W.

(2) Since the vertical lines close to the vertical axis contain no point of the set, the set of points is not a function of W into W.

(3) The vertical lines through the circle will contain two points of the set; hence the set of points is not a function of W into W.

(4) The set of points is a function of W into W since each vertical line contains one and only one point of the set.

20. Let $A = \{a, b, c, d\}$. The set

$$\{(a, b),\ (b, d),\ (c, a),\ (d, c)\}$$

is a one-one, onto function of A into A. Find the inverse function.

Solution:

To find the inverse function, permute, i.e. reverse, the elements in each ordered pair. Then the inverse function is

$$\{(b, a),\ (d, b),\ (a, c),\ (c, d)\}$$

Supplementary Problems

ORDERED PAIRS AND PRODUCT SETS

21. Suppose $(y-2, 2x+1) = (x-1, y+2)$. Find x and y.

22. Find the ordered pairs corresponding to the points P_1, P_2, P_3 and P_4 which appear below in the coordinate diagram of $\{1, 2, 3, 4\} \times \{2, 4, 6, 8\}$.

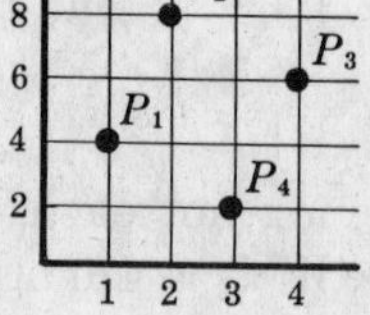

23. Let $W = \{\text{Mark, Eric, Paul}\}$ and let $V = \{\text{Eric, David}\}$. Find (1) $W \times V$, (2) $V \times W$, (3) $V \times V$.

24. Sketch, by shading the appropriate area, each of the following product sets on a coordinate diagram of $R^\# \times R^\#$.

(1) $[-3, 3] \times [-1, 2]$ (3) $[-3, 1) \times (-\infty, 2]$

(2) $(-2, 3] \times [-3, \infty)$ (4) $[-3, 1) \times (-2, 2]$

25. Let $A = \{2, 3\}$, $B = \{1, 3, 5\}$ and $C = \{3, 4\}$. Construct the "tree diagram" of $A \times B \times C$, as in Problem 7, and then find $A \times B \times C$.

26. Let $S = \{a, b, c\}$, $T = \{b, c, d\}$ and $W = \{a, d\}$. Construct the "tree diagram" of $S \times T \times W$ and then find $S \times T \times W$.

27. Suppose sets V, W and Z have 3, 4 and 5 elements respectively. How many elements are there in (1) $V \times W \times Z$, (2) $Z \times V \times W$, (3) $W \times Z \times V$?

28. Let $A = B \cap C$. Which, if any, of the following is true?

$$(1)\quad A \times A = (B \times B) \cap (C \times C) \qquad (2)\quad A \times A = (B \times C) \cap (C \times B)$$

GRAPHS OF FUNCTIONS

29. Let $M = \{1, 2, 3, 4, 5\}$ and let the function $f: M \to R^{\#}$ be defined by

$$f(x) = x^2 + 2x - 1$$

Find the graph of f.

30. Let $W = \{1, 2, 3, 4\}$ and let a function $g: W \to W$ be defined by the diagram

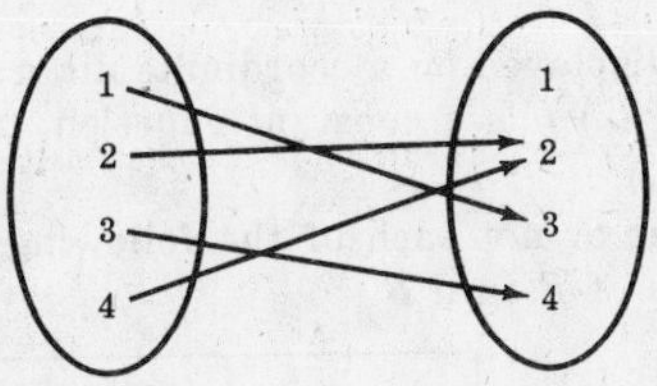

(1) Find the graph of g. (2) Display the graph of g on the coordinate diagram of $W \times W$.

31. Let the function $h: R^{\#} \to R^{\#}$ be defined by the formula

$$h(x) = 2x - 1$$

State whether or not each of the following ordered pairs belongs to the graph of h: (a) $(3, 5)$, (b) $(-2, -5)$, (c) $(-4, -7)$, (d) $(8, 17)$, (e) $(-3, -5)$, (f) $(4, 7)$.

32. Let the function g assign to each name in the set

{Betty, Martin, David, Alan, Rebecca}

the number of different letters needed to spell the name. Find the graph of g.

33. Each of the following formulas defines a function of $R^{\#}$ into $R^{\#}$. Plot the graph of each of these functions on a coordinate diagram of $R^{\#} \times R^{\#}$, the Cartesian plane.

(1) $f(x) = 2x - 1$ (3) $f(x) = |x|$

(2) $f(x) = x^2 - 2x - 1$ (4) $f(x) = x - 2|x|$

FUNCTIONS AS ORDERED PAIRS

34. Let $W = \{a, b, c, d\}$. State whether or not each of the following sets of ordered pairs is a function of W into W.

(1) $\{(b, a), (c, d), (d, a), (c, d), (a, d)\}$ (3) $\{(a, b), (b, b), (c, b), (d, b)\}$

(2) $\{(d, d), (c, a), (a, b), (d, b)\}$ (4) $\{(a, a), (b, a), (a, b), (c, d)\}$

35. Let $V = \{1, 2, 3, 4\}$. State whether or not the set of points in each of the following coordinate diagrams of $V \times V$ is a function of V into V.

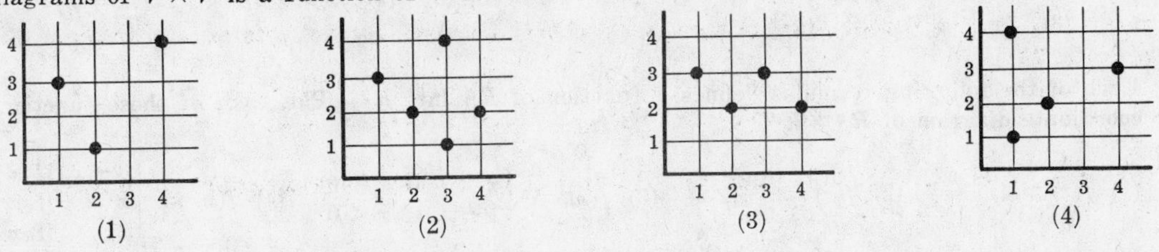

36. Let $A = \{1, 2, 3, 4, 5\}$. Let f be the set of points displayed in the first coordinate diagram of $A \times A$, and let g be the set of points displayed in the second diagram.

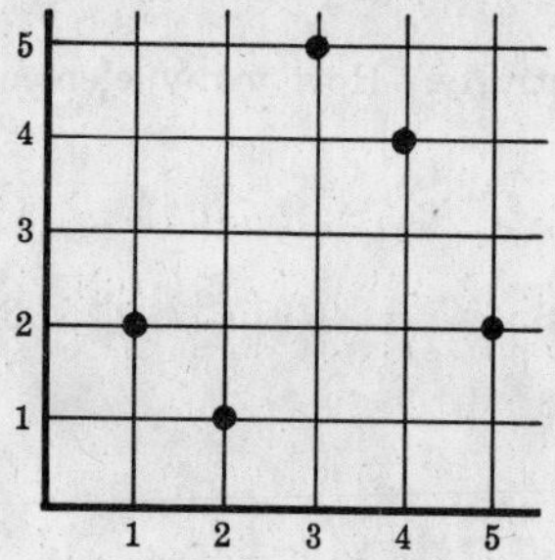

f is displayed

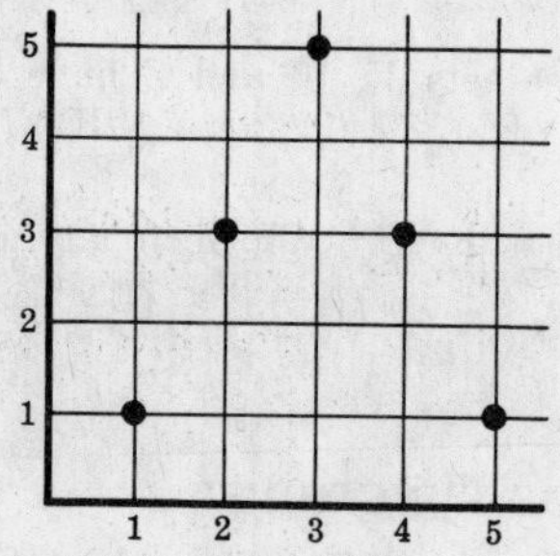

g is displayed

Thus f and g are functions of A into A. Find

(1) $f(3)$ (3) $f^{-1}(2)$ (5) $f^{-1}(4)$ (7) product function $f \circ g$ (9) $\{x \mid f(x) \leqq 4\}$
(2) $g(5)$ (4) $g^{-1}(1)$ (6) product function $g \circ f$ (8) $f^{-1}(\{1, 2\})$ (10) $\{x \mid g(x) > 2\}$

37. Let the function $f: A \to B$ be displayed on a coordinate diagram of $A \times B$. What geometrical property does f have if (1) f is one-one, (2) f is a constant function, (3) f is onto, (4) f has an inverse?

38. Let $B = [-4, 4]$. State whether or not each of the following sets of points displayed on a coordinate diagram of $B \times B$ is a function of B into B.

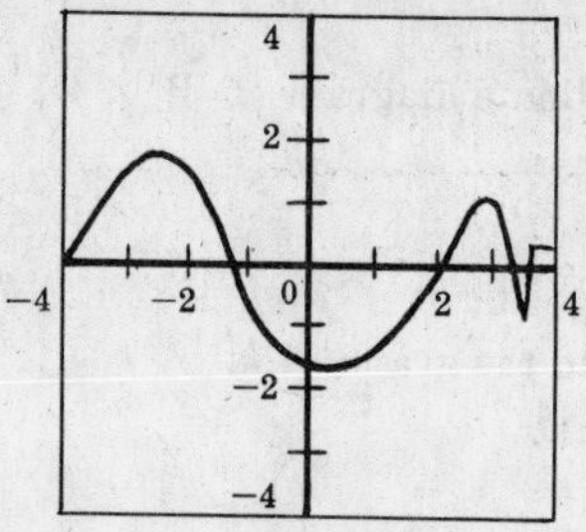

(1)

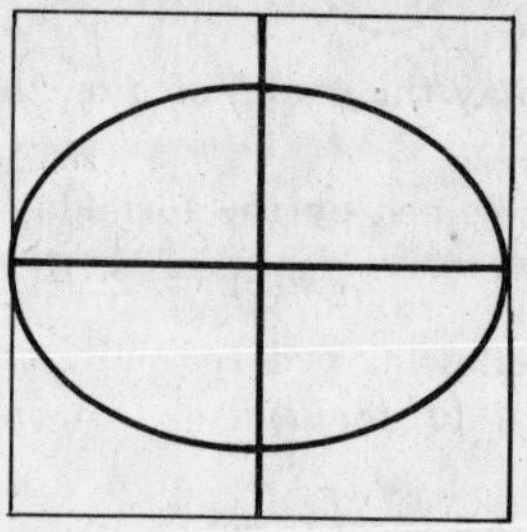
(2)

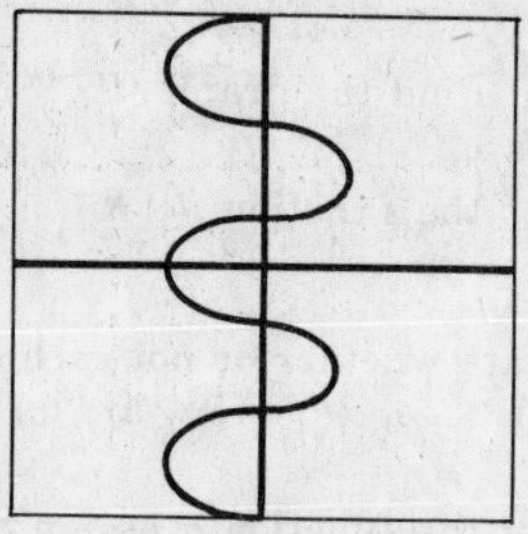
(3)

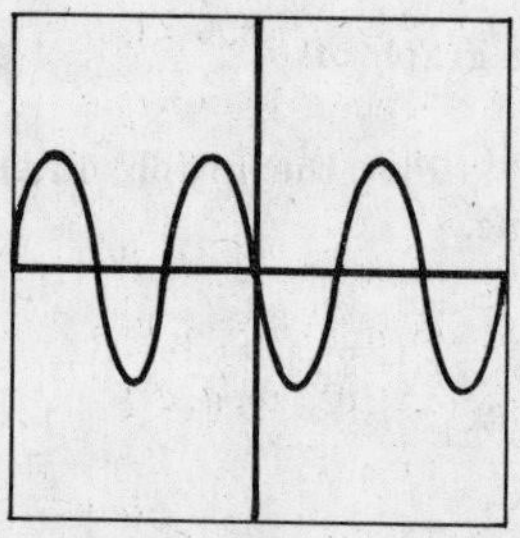
(4)

(5)

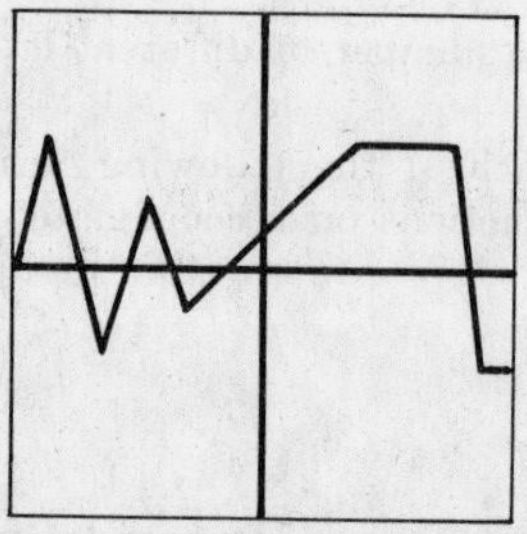
(6)

MISCELLANEOUS PROBLEMS

39. Sketch, by shading the appropriate area, each of the following product sets on a coordinate diagram of $R^\# \times R^\#$.

(1) $\{x \mid -3 < x \leqq 2\} \times \{x \mid -2 < x < 4\}$
(2) $\{x \mid |x| < 3\} \times \{x \mid |x| \leqq 1\}$
(3) $\{x \mid |x| \leqq 2\} \times \{x \mid x > -3\}$
(4) $\{x \mid x < 1\} \times \{x \mid -2 \leqq x < 3\}$
(5) $\{x \mid x > -2\} \times \{x \mid x \leqq 3\}$

40. Each of the following formulas defines a function of $R^\#$ into $R^\#$. Plot each of these functions on a coordinate diagram of $R^\# \times R^\#$.

(1) $f(x) = 4x - x^2$ (2) $f(x) = x + 2|x|$ (3) $f(x) = \begin{cases} x^2 & \text{if } x \geqq 0 \\ 1 - x & \text{if } x < 0 \end{cases}$ (4) $f(x) = \begin{cases} 3 - x & \text{if } x > 2 \\ x & \text{if } |x| \leqq 2 \\ 2 & \text{if } x < -2 \end{cases}$

Answers to Supplementary Problems

21. $x = 2,\ y = 3.$

22. $P_1 = (1, 4),\ P_2 = (2, 8),\ P_3 = (4, 6),\ P_4 = (3, 2).$

23. (1) $W \times V$ = {(Mark, Eric), (Mark, David), (Eric, Eric), (Eric, David), (Paul, Eric), (Paul, David)}.
(2) $V \times W$ = {(Eric, Mark), (David, Mark), (Eric, Eric), (David, Eric), (Eric, Paul), (David, Paul)}.
(3) $V \times V$ = {(Eric, Eric), (Eric, David), (David, Eric), (David, David)}.

24. (1)

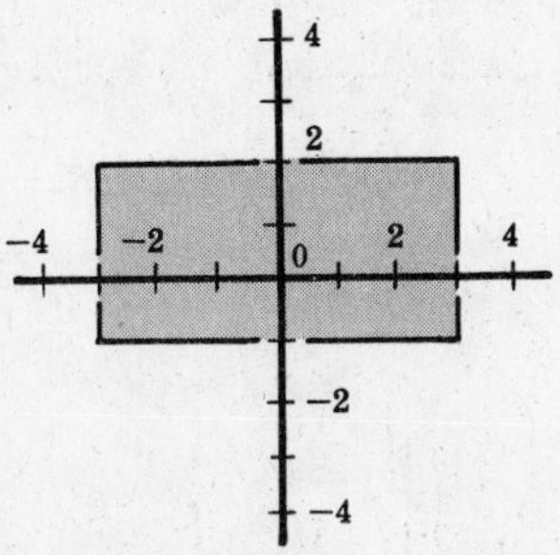

(2)

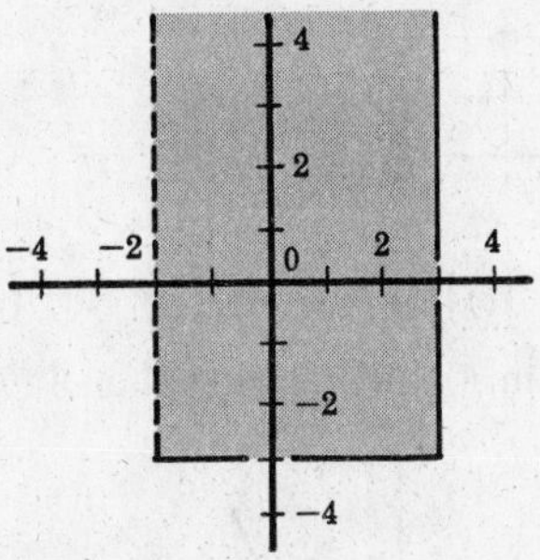

(3)

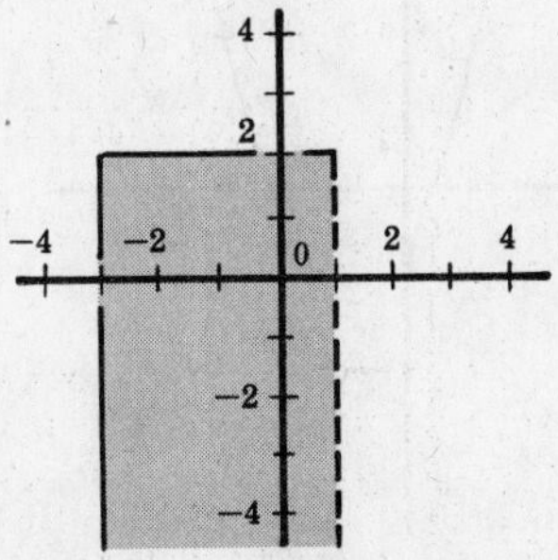

(4)

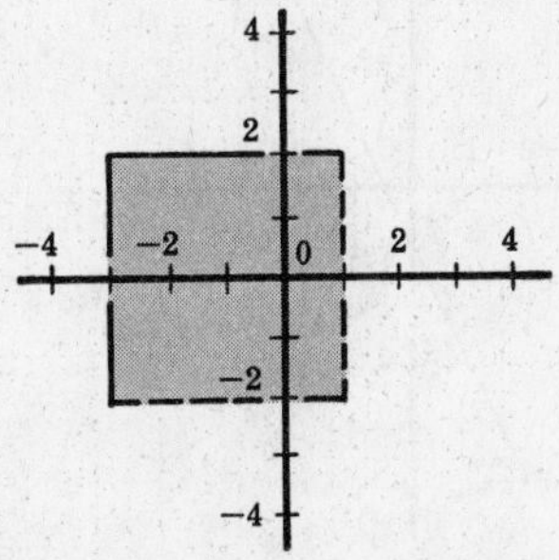

25. See Fig. 5-11 below.

$$A \times B \times C = \{(2,1,3),\ (2,1,4),\ (2,3,3),\ (2,3,4),\ (2,5,3),\ (2,5,4),\ (3,1,3),\ (3,1,4),\ (3,3,3),\ (3,3,4),\ (3,5,3),\ (3,5,4)\}$$

26. See Fig. 5-12 below.

$$S \times T \times W = \{(a,b,a),\ (a,b,d),\ (a,c,a),\ (a,c,d),\ (a,d,a),\ (a,d,d),\ (b,b,a),\ (b,b,d),\ (b,c,a),\ (b,c,d),\ (b,d,a),\ (b,d,d),\ (c,b,a),\ (c,b,d),\ (c,c,a),\ (c,c,d),\ (c,d,a),\ (c,d,d)\}$$

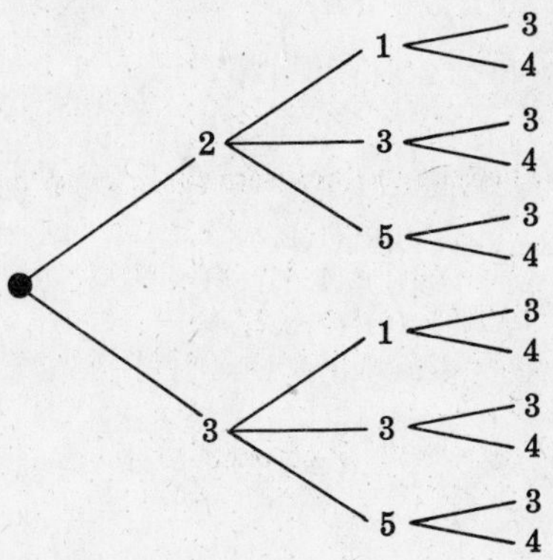

Fig. 5-11

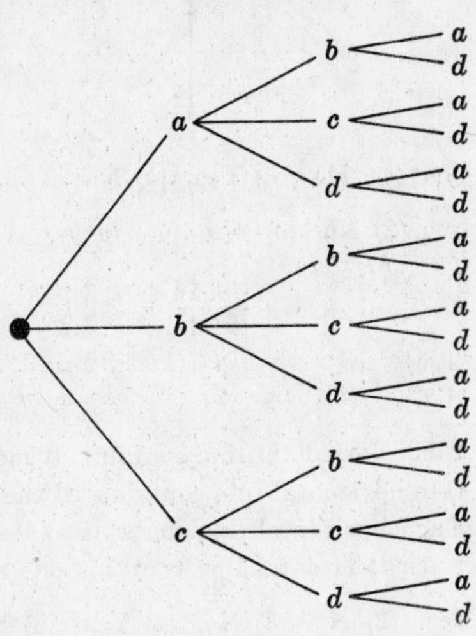

Fig. 5-12

27. Each has 60 elements.

28. Both are true.

$$A \times A = (B \times B) \cap (C \times C) = (B \times C) \cap (C \times B)$$

29. $\{(1,2), (2,7), (3,14), (4,23), (5,34)\}$

30. (1) $\{(1,3), (2,2), (3,4), (4,2)\}$

(2)

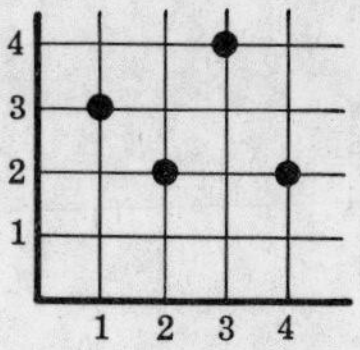

31. (*a*) Yes, (*b*) Yes, (*c*) No, (*d*) No, (*e*) No, (*f*) Yes

32. {(Betty, 4), (Martin, 6), (David, 4), (Alan, 3), (Rebecca, 5)}

33. (1)

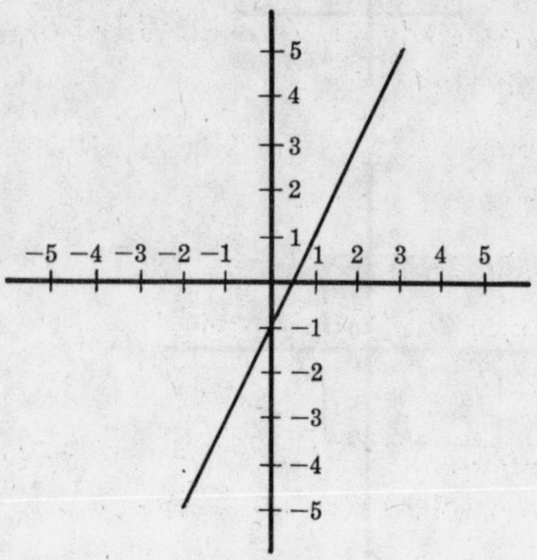

(2)

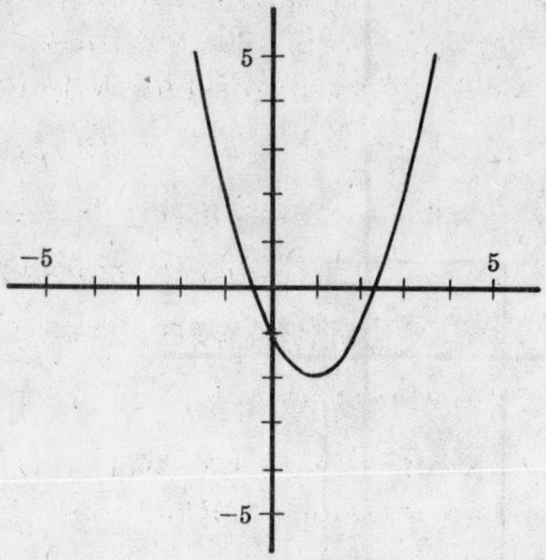

(3)

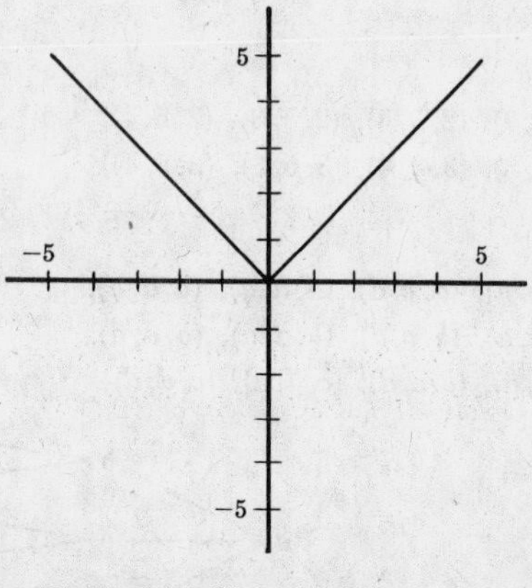

(4)

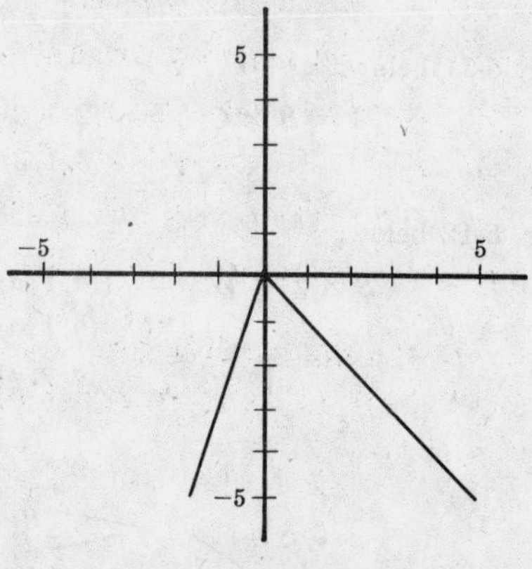

34. (1) Yes, (2) No, (3) Yes, (4) No

35. (1) No, (2) No, (3) Yes, (4) No

36. (1) 5
(2) 1
(3) $\{1,5\}$
(4) $\{1,5\}$
(5) $\{4\}$
(6) $\{(1,3), (2,1), (3,1), (4,3), (5,3)\}$
(7) $\{(1,2), (2,5), (3,2), (4,5), (5,2)\}$
(8) $\{1,2,5\}$
(9) $\{1,2,4,5\}$
(10) $\{2,3,4\}$

37. (1) Each horizontal line contains at most one point.
(2) One horizontal line contains all the points.
(3) Each horizontal line contains at least one point.
(4) Each horizontal line contains one and only one point.

38. (1) Yes, (2) No, (3) No, (4) Yes, (5) No, (6) Yes

Chapter 6

Relations

PROPOSITIONAL FUNCTIONS, OPEN SENTENCES

A *propositional function* defined on the Cartesian product $A \times B$ of two sets A and B is an expression denoted by

$$P(x, y)$$

which has the property that $P(a, b)$, where a and b are substituted for the variables x and y respectively in $P(x, y)$, is true or false for any ordered pair $(a, b) \in A \times B$. For example, if A is the set of playwrights and B is the set of plays, then

$$P(x, y) = \text{"}x \text{ wrote } y\text{"}$$

is a propositional function on $A \times B$. In particular,

$$P(\text{Shakespeare, Hamlet}) = \text{"Shakespeare wrote Hamlet"}$$
$$P(\text{Shakespeare, Faust}) = \text{"Shakespeare wrote Faust"}$$

are true and false respectively.

The expression $P(x, y)$ by itself shall be called an *open sentence in two variables* or, simply, an *open sentence.* Other examples of open sentences are as follows:

Example 1.1: "x is less than y."

Example 1.2: "x weighs y pounds."

Example 1.3: "x divides y."

Example 1.4: "x is the wife of y."

Example 1.5: "The square of x plus the square of y is sixteen", i.e. "$x^2 + y^2 = 16$".

Example 1.6: "Triangle x is similar to triangle y."

In all of our examples there are two variables. It is also possible to have open sentences in one variable such as "x is in the United Nations", or in more than two variables such as "x times y equals z".

RELATIONS

A *relation* R consists of the following:

(1) a set A

(2) a set B

(3) an open sentence $P(x, y)$ in which $P(a, b)$ is either true or false for any ordered pair (a, b) belonging to $A \times B$.

We then call R a *relation from A to B* and denote it by

$$R = (A,\ B,\ P(x, y))$$

Furthermore, if $P(a, b)$ is true we write

$$a\,R\,b$$

which reads "a is related to b". On the other hand, if $P(a, b)$ is not true we write

$$a \not R\, b$$

which reads "a is not related to b".

Example 2.1: Let $R_1 = (R^\#, R^\#, P(x, y))$ where $P(x, y)$ reads "x is less than y". Then R_1 is a relation since $P(a, b)$, i.e. "$a < b$", is either true or false for any ordered pair (a, b) of real numbers. Moreover, since $P(2, \pi)$ is true we can write

$$2\,R_1\,\pi$$

and since $P(5, \sqrt{2})$ is false we can write

$$5 \not R_1 \sqrt{2}$$

Example 2.2: Let $R_2 = (A, B, P(x, y))$ where A is the set of men, B is the set of women, and $P(x, y)$ reads "x is the husband of y". Then R_2 is a relation.

Example 2.3: Let $R_3 = (N, N, P(x, y))$ where N is the natural numbers and $P(x, y)$ reads "x divides y". Then R_3 is a relation. Furthermore,

$$3\,R_3\,12, \quad 2 \not R_3\, 7, \quad 5\,R_3\,15, \quad 6 \not R_3\, 13$$

Example 2.4: Let $R_4 = (A, B, P(x, y))$ where A is the set of men, B is the set of women and $P(x, y)$ reads "x divides y". Then R_4 is not a relation since $P(a, b)$ has no meaning if a is a man and b is a woman.

Example 2.5: Let $R_5 = (N, N, P(x, y))$ where N is the natural numbers and $P(x, y)$ reads "x is less than y". Then R_5 is a relation.

Notice that R_1 and R_5 are not the same relation even though the same open sentence is used to define each relation.

Let $R = (A, B, P(x, y))$ be a relation. We then say that the open sentence $P(x, y)$ *defines a relation* from A to B. Furthermore, if $A = B$, then we say that $P(x, y)$ defines a relation in A, or that R is a relation in A.

Example 2.6: The open sentence $P(x, y)$, which reads "x is less than y", defines a relation in the rational numbers.

Example 2.7: The open sentence "x is the husband of y" defines a relation from the set of men to the set of women.

Terminology: Some authors call the expression $P(x, y)$ a relation. They then assume implicitly that the variables x and y range, respectively, over some sets A and B, i.e. that $P(x, y)$ is a propositional function defined on some product set $A \times B$. We shall adhere to the previous terminology where $P(x, y)$ is simply an open sentence and, hence, a relation consists of $P(x, y)$ and two given sets A and B.

SOLUTION SETS AND GRAPHS OF RELATIONS

Let $R = (A, B, P(x, y))$ be a relation. The *solution set* R^* of the relation R consists of the elements (a, b) in $A \times B$ for which $P(a, b)$ is true. In other words,

$$R^* = \{(a, b) \mid a \in A,\ b \in B,\ P(a, b) \text{ is true}\}$$

Notice that R^*, the solution set of a relation R from A to B, is a subset of $A \times B$. Hence R^* can be displayed, i.e. plotted or sketched, on the coordinate diagram of $A \times B$.

The *graph* of a relation R from A to B consists of those points on the coordinate diagram of $A \times B$ which belong to the solution set of R.

Example 3.1: Let $R = (A, B, P(x,y))$ where $A = \{2, 3, 4\}$, $B = \{3, 4, 5, 6\}$, and $P(x,y)$ reads "x divides y". Then the solution set of R is

$$R^* = \{(2,4), (2,6), (3,3), (3,6), (4,4)\}$$

The solution set of R is displayed on the coordinate diagram of $A \times B$ as shown in Fig. 6-1 below.

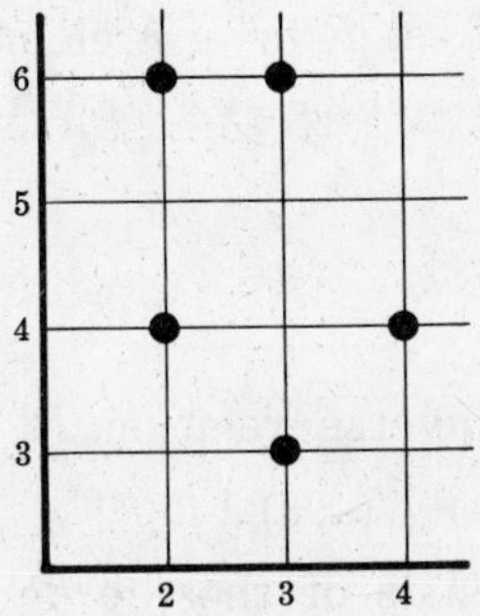

Fig. 6-1

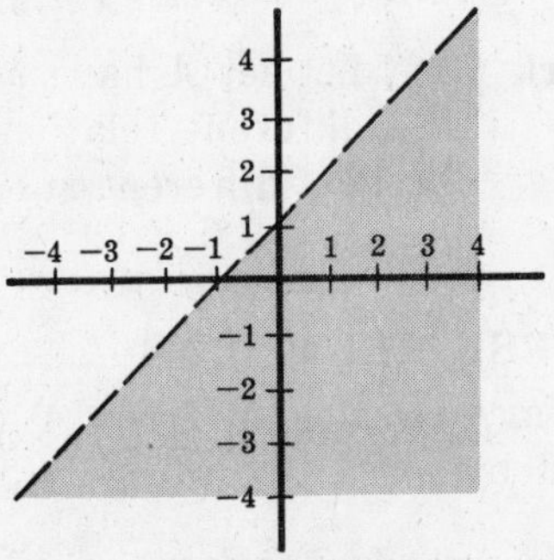

R^* is shaded

Fig. 6-2

Example 3.2: Let R be the relation in the real numbers defined by

$$y < x + 1$$

The shaded area in the coordinate diagram of $R^\# \times R^\#$ shown in Fig. 6-2 above consists of the points which belong to R^*, the solution set of R, that is, is the graph of R.

Notice that R^* consists of the points below the line $y = x + 1$. The line $y = x + 1$ is dashed in order to show that the points on the line do not belong to R^*.

RELATIONS AS SETS OF ORDERED PAIRS

Let R^* be any subset of $A \times B$. We can define a relation $R = (A, B, P(x,y))$ where $P(x,y)$ reads

"The ordered pair (x, y) belongs to R^*"

The solution set of this relation R is the original set R^*. Thus to every relation $R = (A, B, P(x,y))$ there corresponds a unique solution set R^* which is a subset of $A \times B$, and to every subset R^* of $A \times B$ there corresponds a relation $R = (A, B, P(x,y))$ for which R^* is its solution set. In view of this one-one correspondence between relations $R = (A, B, P(x,y))$ and subsets R^* of $A \times B$, we redefine a relation by the

Definition 6.1: A relation R from A to B is a subset of $A \times B$.

Although Definition 6.1 of a relation may seem artificial it has the advantage that we do not use in this definition of a relation the undefined concepts "open sentence" and "variable".

Example 4.1: Let $A = \{1, 2, 3\}$ and $B = \{a, b\}$. Then

$$R = \{(1,a), (1,b), (3,a)\}$$

is a relation from A to B. Furthermore,

$$1\,R\,a, \quad 2 \not R\, b, \quad 3\,R\,a, \quad 3 \not R\, b$$

Example 4.2: Let $W = \{a, b, c\}$. Then

$$R = \{(a,b), (a,c), (c,c), (c,b)\}$$

is a relation in W. Moreover,

$$a \not R\, a, \quad b \not R\, a, \quad c\,R\,c, \quad a\,R\,b$$

Example 4.3: Let

$$R = \{(x,y) \mid x \in R^\#, y \in R^\#, y < x^2\}$$

Then R is a set of ordered pairs of real numbers, i.e. a subset of $R^\# \times R^\#$. Hence R is a relation in the real numbers which could also be defined by

$$R = (R^\#, R^\#, P(x,y))$$

where $P(x,y)$ reads "y is less than x^2".

Remark 6.1: Let set A have m elements and set B have n elements. Then there are 2^{mn} different relations from A to B, since $A \times B$, which has mn elements, has 2^{mn} different subsets.

INVERSE RELATIONS

Every relation R from A to B has an inverse relation R^{-1} from B to A which is defined by

$$R^{-1} = \{(b,a) \mid (a,b) \in R\}$$

In other words, the inverse relation R^{-1} consists of those ordered pairs which when reversed, i.e. permuted, belong to R.

Example 5.1: Let $A = \{1, 2, 3\}$ and $B = \{a, b\}$. Then

$$R = \{(1,a), (1,b), (3,a)\}$$

is a relation from A to B. The inverse relation of R is

$$R^{-1} = \{(a,1), (b,1), (a,3)\}$$

Example 5.2: Let $W = \{a, b, c\}$. Then

$$R = \{(a,b), (a,c), (c,c), (c,b)\}$$

is a relation in W. The inverse relation of R is

$$R^{-1} = \{(b,a), (c,a), (c,c), (b,c)\}$$

REFLEXIVE RELATIONS

Let $R = (A, A, P(x,y))$ be a relation in a set A, i.e. let R be a subset of $A \times A$. Then R is called a *reflexive relation* if, for every $a \in A$,

$$(a,a) \in R$$

In other words, R is reflexive if every element in A is related to itself.

Example 6.1: Let $V = \{1, 2, 3, 4\}$ and

$$R = \{(1,1), (2,4), (3,3), (4,1), (4,4)\}$$

Then R is not a reflexive relation since $(2,2)$ does not belong to R. Notice that all ordered pairs (a,a) must belong to R in order for R to be reflexive.

Example 6.2: Let A be the set of triangles in the Euclidean plane. The relation R in A defined by the open sentence "x is similar to y" is a reflexive relation since every triangle is similar to itself.

Example 6.3: Let R be the relation in the real numbers defined by the open sentence "x is less than y", i.e. "$x < y$". Then R is not reflexive since $a \nless a$ for any real number a.

Example 6.4: Let $\mathcal{A}$ be a family of sets, and let R be the relation in $\mathcal{A}$ defined by "x is a subset of y". Then R is a reflexive relation since every set is a subset of itself.

SYMMETRIC RELATIONS

Let R be a subset of $A \times A$, i.e. let R be a relation in A. Then R is called a *symmetric relation* if

$$(a,b) \in R \quad \text{implies} \quad (b,a) \in R$$

that is, if a is related to b then b is also related to a.

Example 7.1: Let $S = \{1, 2, 3, 4\}$, and let

$$R = \{(1,3), (4,2), (2,4), (2,3), (3,1)\}$$

Then R is not a symmetric relation since

$$(2,3) \in R \quad \text{but} \quad (3,2) \notin R$$

Example 7.2: Let A be the set of triangles in the Euclidean plane, and let R be the relation in A which is defined by the open sentence "x is similar to y". Then R is symmetric, since if triangle a is similar to triangle b then b is also similar to a.

Example 7.3: Let R be the relation in the natural numbers N which is defined by "x divides y". Then R is not symmetric since 2 divides 4 but 4 does not divide 2. In other words,

$$(2,4) \in R \quad \text{but} \quad (4,2) \notin R$$

Remark 6.2: Since $(a, b) \in R$ implies (b, a) belongs to the inverse relation R^{-1}, R is a symmetric relation if and only if

$$R = R^{-1}$$

ANTI-SYMMETRIC RELATIONS

A relation R in a set A, i.e. a subset of $A \times A$, is called an *anti-symmetric relation* if

$$(a, b) \in R \text{ and } (b, a) \in R \quad \text{implies} \quad a = b$$

In other words, if $a \neq b$ then possibly a is related to b or possibly b is related to a, but never both.

Example 8.1: Let N be the natural numbers and let R be the relation in N defined by "x divides y". Then R is anti-symmetric since

$$a \text{ divides } b \text{ and } b \text{ divides } a \quad \text{implies} \quad a = b$$

Example 8.2: Let $W = \{1, 2, 3, 4\}$, and let

$$R = \{(1,3), (4,2), (4,4), (2,4)\}$$

Then R is not an anti-symmetric relation in W since

$$(4,2) \in R \quad \text{and} \quad (2,4) \in R$$

Example 8.3: Let $\mathcal{A}$ be a family of sets, and let R be the relation in $\mathcal{A}$ defined by "x is a subset of y". Then R is anti-symmetric since

$$A \subset B \text{ and } B \subset A \quad \text{implies} \quad A = B$$

Remark 6.3: Let D denote the *diagonal line* of $A \times A$, i.e. the set of all ordered pairs $(a, a) \in A \times A$. Then a relation R in A is anti-symmetric if and only if

$$R \cap R^{-1} \subset D$$

TRANSITIVE RELATIONS

A relation R in a set A is called a *transitive relation* if

$$(a, b) \in R \text{ and } (b, c) \in R \quad \text{implies} \quad (a, c) \in R$$

In other words, if a is related to b and b is related to c, then a is related to c.

Example 9.1: Let A be the set of people on earth. Let R be the relation in A defined by the open sentence "x loves y". If a loves b and b loves c, it does not necessarily follow that a loves c. Accordingly, R is not a transitive relation.

Example 9.2: Let R be the relation in the real numbers defined by "x is less than y". Then, as previously shown,

$$a < b \text{ and } b < c \quad \text{implies} \quad a < c$$

Thus R is a transitive relation.

Example 9.3: Let $W = \{a, b, c\}$, and let

$$R = \{(a, b), (c, b), (b, a), (a, c)\}$$

Then R is not a transitive relation since

$$(c, b) \in R \text{ and } (b, a) \in R \quad \text{but} \quad (c, a) \notin R$$

Example 9.4: Let $\mathcal{A}$ be a family of sets, and let R be the relation in $\mathcal{A}$ defined by "x is a subset of y". Then R is a transitive relation since

$$A \subset B \text{ and } B \subset C \quad \text{implies} \quad A \subset C$$

EQUIVALENCE RELATIONS

A relation R in a set A is an *equivalence relation* if

(1) R is reflexive, that is, for every $a \in A$, $(a, a) \in R$,

(2) R is symmetric, that is, $(a, b) \in R$ implies $(b, a) \in R$,

(3) R is transitive, that is, $(a, b) \in R$, and $(b, c) \in R$ implies $(a, c) \in R$.

In a later chapter we will more fully study equivalence relations in sets. Now we just give two examples of equivalence relations.

Example 10.1: Let A be the set of triangles in the Euclidean plane. Let R be the relation on A defined by "x is similar to y". Then, as proven in geometry, R is reflexive, symmetric and transitive. Thus R is an equivalence relation.

Example 10.2: The most important example of an equivalence relation is that of "equality". For any elements in any set:

(1) $a = a$,

(2) $a = b$ implies $b = a$,

(3) $a = b$ and $b = c$ implies $a = c$.

DOMAIN AND RANGE OF A RELATION

Let R be a relation from A to B, that is, let R be a subset of $A \times B$. The *domain* D of the relation R is the set of all first elements of the ordered pairs which belong to R, that is,

$$D = \{a \mid a \in A, (a, b) \in R\}$$

The *range* E of the relation R consists of all the second elements which appear in the ordered pairs in R, that is,

$$E = \{b \mid b \in B, (a, b) \in R\}$$

Notice that the domain of a relation from A to B is a subset of A, and its range is a subset of B.

Example 11.1: Let $A = \{1, 2, 3, 4\}$, $B = \{a, b, c\}$, and

$$R = \{(2, a), (4, a), (4, c)\}$$

Then the domain of R is the set $\{2, 4\}$, and the range of R is the set $\{a, c\}$.

Example 11.2: Let the relation R in the real numbers be defined by the open sentence "$4x^2 + 9y^2 = 36$". R is displayed on the coordinate diagram of $R^\# \times R^\#$ as shown in the figure at right. The domain of R is the closed interval $[-3, 3]$, and the range of R is the closed interval $[-2, 2]$.

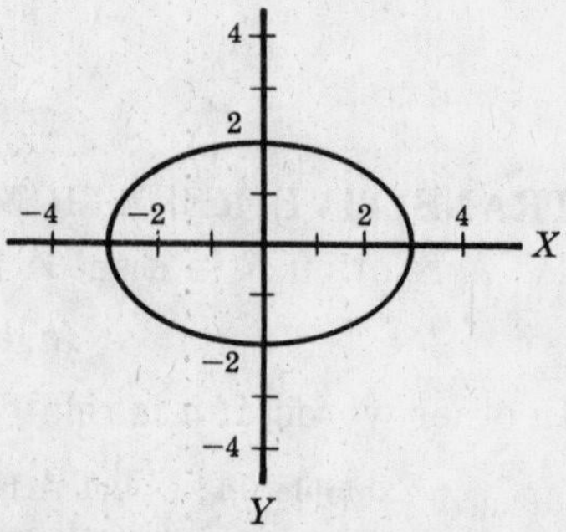

Remark 6.4: Let a relation R from A to B be displayed on the coordinate diagram of $A \times B$. Then $a \in A$ is in the domain of R if, and only if, the vertical line through a contains a point of the graph of R. Also, $b \in B$ is in the range of R if, and only if, the horizontal line through b contains a point of the graph of R.

RELATIONS AND FUNCTIONS

We repeat

Definition 5.1: A function f of A into B is a subset of $A \times B$ in which each $a \varepsilon A$ appears in one and only one ordered pair belonging to f.

Since every subset of $A \times B$ is a relation, a function is a special type of a relation. In fact, the terms "domain" and "range" appeared both in the discussion of functions and in the discussion of relations.

An important problem in mathematics is to determine whether or not a relation R in the real numbers defined by an equation of the form

$$F(x, y) = 0$$

is, in fact, a function. In other words, does or does not the relation defined by

$$F(x, y) = 0$$

define a function $y = f(x)$?

In general, this problem is extremely difficult. Here, we are only able to answer this question for very simple equations.

Example 12.1: Let R be the relation in the real numbers defined by

$$x^2 + y^2 = 25$$

R is displayed on the coordinate diagram of $R^\# \times R^\#$ as shown in Fig. 6-3.

Notice that R is a circle of radius 5 with center at the origin. Notice, further, that many vertical lines contain more than one point of R. In particular, $(3, 4) \varepsilon R$ and $(3, -4) \varepsilon R$. Thus the relation R is not a function.

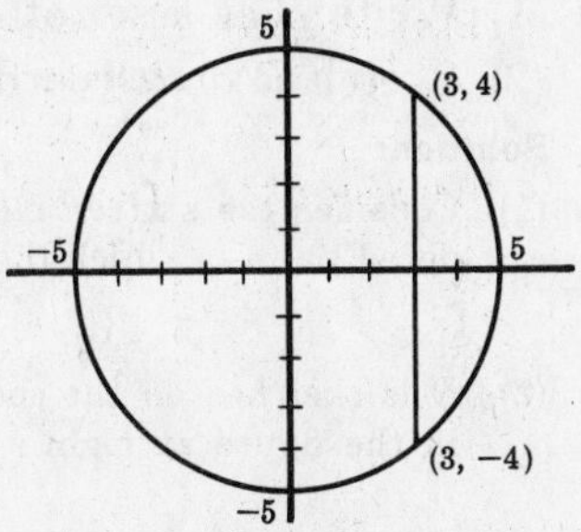

R is displayed

Fig. 6-3

Example 12.2: Let $A = [-5, 5]$, $B = [0, \infty)$, and let R be the relation from A to B defined by

$$x^2 + y^2 = 25$$

R is displayed on the coordinate diagram of $A \times B$ shown in Fig. 6-4 below.

Notice that R is the upper half of a circle. Notice further that each vertical line contains one and only one point of R; hence R is a function.

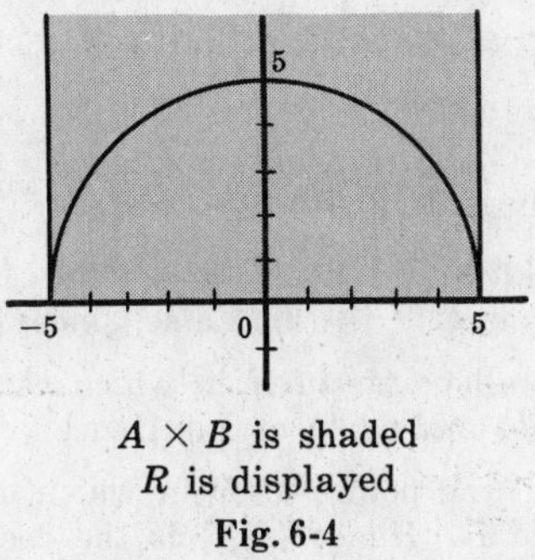

$A \times B$ is shaded

R is displayed

Fig. 6-4

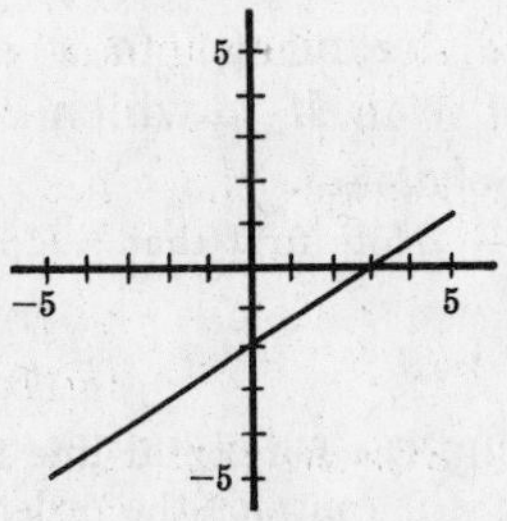

R is displayed

Fig. 6-5

Example 12.3: Let R be the relation in the real numbers defined by

$$2x - 3y = 6$$

R is displayed on the coordinate diagram of $R^\# \times R^\#$ shown in Fig. 6-5 above. Notice that R is a straight line and that every vertical line contains one and only one point of R; thus R is a function. Furthermore, by solving for y in terms of x in the equation above, we obtain a formula that defines the function R:

$$y = f(x) = \frac{2x - 6}{3}$$

Solved Problems

BASIC PROPERTIES OF RELATIONS

1. Let R be the relation from $A = \{1,2,3,4\}$ to $B = \{1,3,5\}$ which is defined by the open sentence "x is less than y".

(1) Find the solution set of R, that is, write R as a set of ordered pairs.

(2) Plot R on a coordinate diagram of $A \times B$.

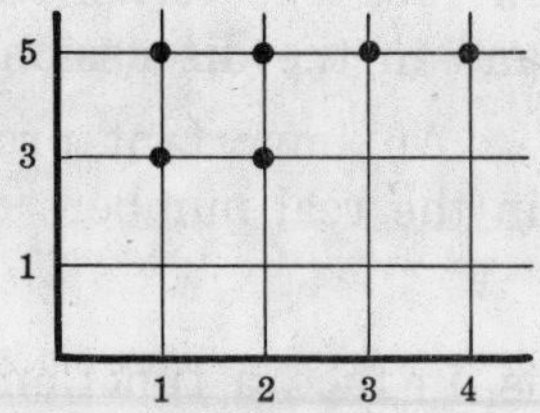

Solution:

(1) R consists of those ordered pairs $(a, b) \in A \times B$ for which $a < b$; hence

$$R = \{(1,3),\ (1,5),\ (2,3),\ (2,5),\ (3,5),\ (4,5)\}$$

(2) R is sketched on the coordinate diagram of $A \times B$ as shown in the figure at right.

2. Let R be the relation from $E = \{2,3,4,5\}$ to $F = \{3,6,7,10\}$ which is defined by the open sentence "x divides y".

(1) Write R as a set of ordered pairs, i.e. find the solution set of R.

(2) Sketch R on a coordinate diagram of $E \times F$.

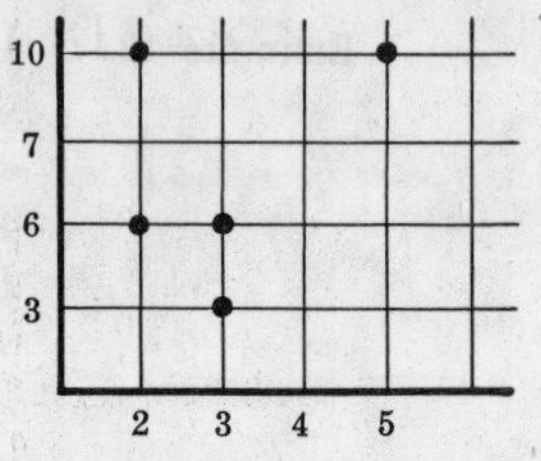

Solution:

(1) Consider the sixteen elements in $E \times F$ and choose those ordered pairs in which the first element divides the second; then

$$R = \{(2,6),\ (2,10),\ (3,3),\ (3,6),\ (5,10)\}$$

(2) R is sketched on the coordinate diagram of $E \times F$ as shown in the figure at right.

3. Let $M = \{a, b, c, d\}$, and let a relation R in M be the set of points displayed in the coordinate diagram of $M \times M$ shown in the figure at right.

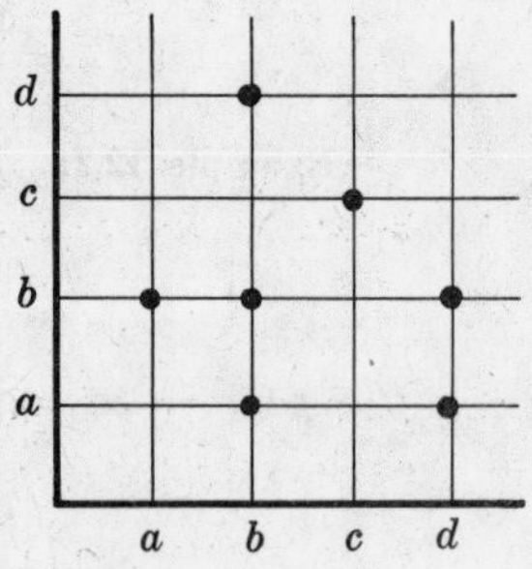

(1) State whether each of the following is true or false:

$(a)\ c\,R\,b,\quad (b)\ d \not{R}\, a,\quad (c)\ a \not{R}\, c,\quad (d)\ b \not{R}\, b$

(2) Find $\{x \mid (x, b) \in R\}$, that is, find all the elements in M which are related to b.

(3) Find $\{x \mid (d, x) \in R\}$, that is, find all those elements in M to which d is related.

Solution:

(1) Note first that $x\,R\,y$ is true if and only if (x, y) belongs to R.

(a) False, since $(c, b) \notin R$. (c) True, since $(a, c) \notin R$.
(b) False, since $(d, a) \in R$. (d) False, since $(b, b) \in R$.

(2) The horizontal line through b contains all points of R in which b appears as the second element; it contains the ordered pairs (a, b), (b, b) and (d, b) of R. Hence $\{a, b, d\}$ is the desired set.

(3) The vertical line through d contains all the points of R in which d appears as the first element; it contains the points (d, a) and (d, b) of R. Hence $\{a, b\}$ is the desired set.

4. Each of the following open sentences defines a relation in the real numbers. Sketch each relation on a coordinate diagram of $R^\# \times R^\#$.

(1) $y = x^2$ (3) $y < 3 - x$ (5) $y \geqq x^3$
(2) $y \leqq x^2$ (4) $y \geqq \sin x$ (6) $y > x^3$

Solution:

In order to sketch a relation on the real numbers which is defined by an open sentence of the form

(a) $y = f(x)$
(b) $y > f(x)$
(c) $y \geqq f(x)$
(d) $y < f(x)$
(e) $y \leqq f(x)$

first plot $y = f(x)$ in the usual manner. Then the relation, i.e. the desired set, will consist of the points

(a) on $y = f(x)$
(b) above $y = f(x)$
(c) above and on $y = f(x)$
(d) below $y = f(x)$
(e) below and on $y = f(x)$

Thus the following are the sketches of the above relations:

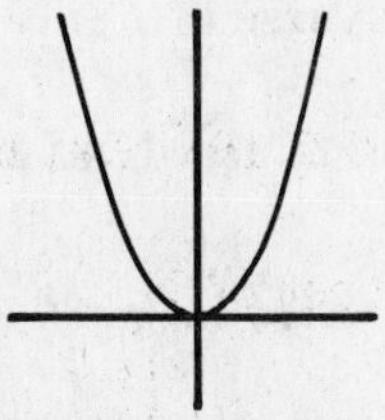

(1) $y = x^2$

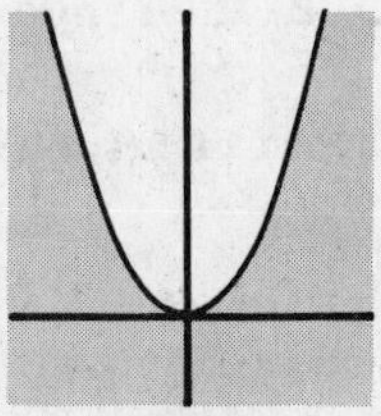

(2) $y \leqq x^2$

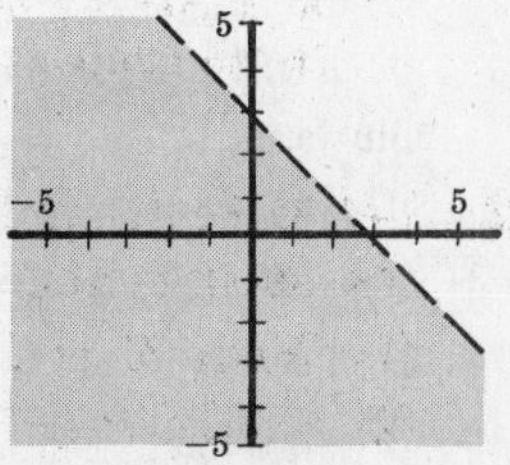

(3) $y < 3 - x$

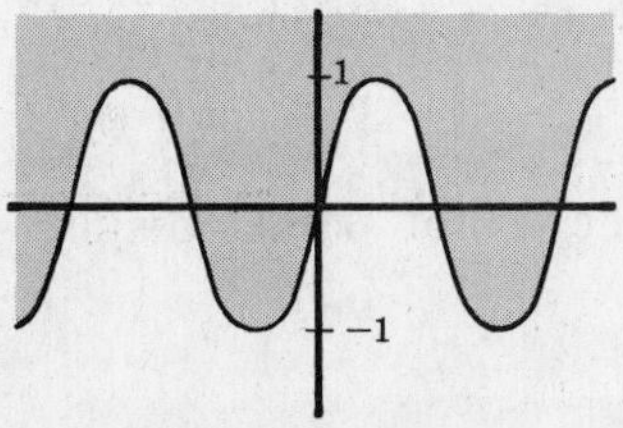

(4) $y \geqq \sin x$

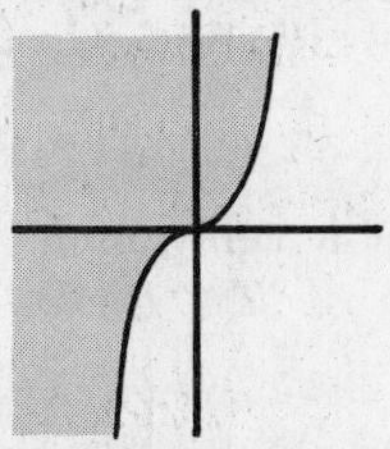

(5) $y \geqq x^3$

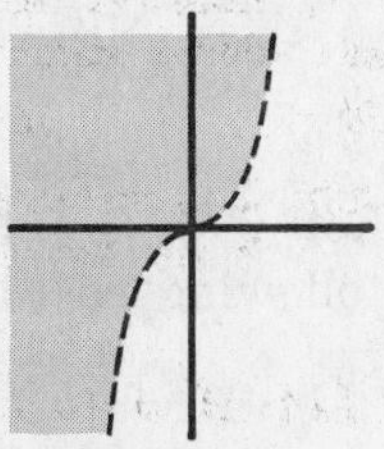

(6) $y > x^3$

Notice, once again, that the curve $y = f(x)$ is drawn with dashes if the points on $y = f(x)$ do not belong to the relation.

5. Each of the following open sentences defines a relation in the real numbers. Sketch each relation on a coordinate diagram of $R \times R$.

(1) $x^2 + y^2 < 16$ (or $x^2 + y^2 - 16 < 0$)
(2) $x^2 - 4y^2 \geqq 9$ (or $x^2 - 4y^2 - 9 \geqq 0$)
(3) $x^2 + y^2 \geqq 16$
(4) $x^2 - 4y^2 < 9$

Solution:

In order to sketch a relation in the real numbers which is defined by an open sentence of the form $f(x, y) < 0$ (or $\leqq, >, \geqq$), first plot $f(x, y) = 0$. The curve $f(x, y) = 0$ will, in simple situations, partition the plane into various regions. The relation will consist of all the points in possibly one or more regions. Test one or more points in each region in order to determine whether or not *all* the points in the region belong to the relation.

The sketch of each of the above relations is as follows:

(1)

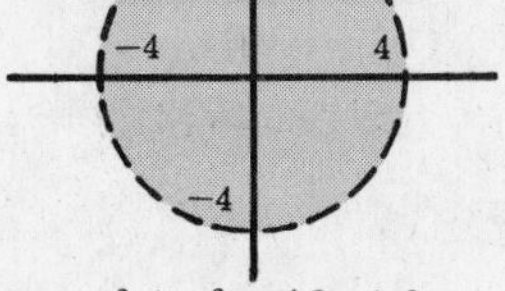

$x^2 + y^2 - 16 < 0$

(2)

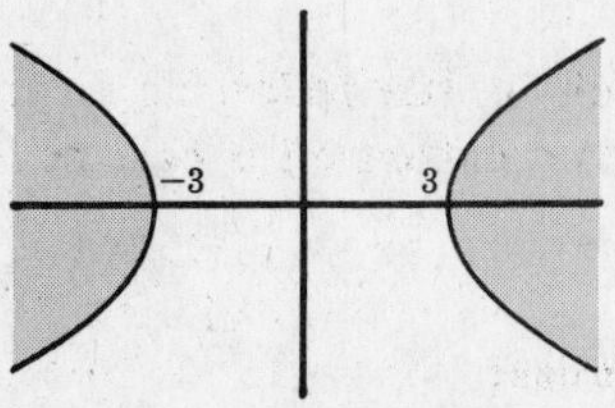

$x^2 - 4y^2 - 9 \geqq 0$

(3)

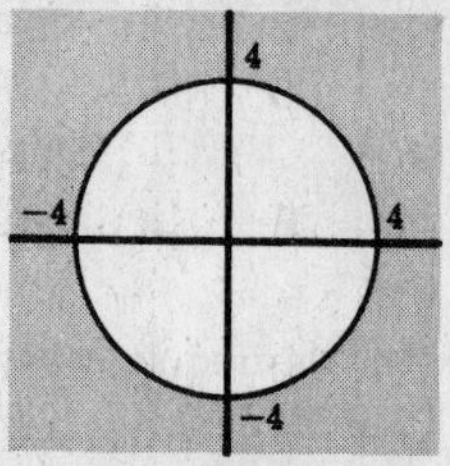

$x^2 + y^2 \geq 16$

(4)

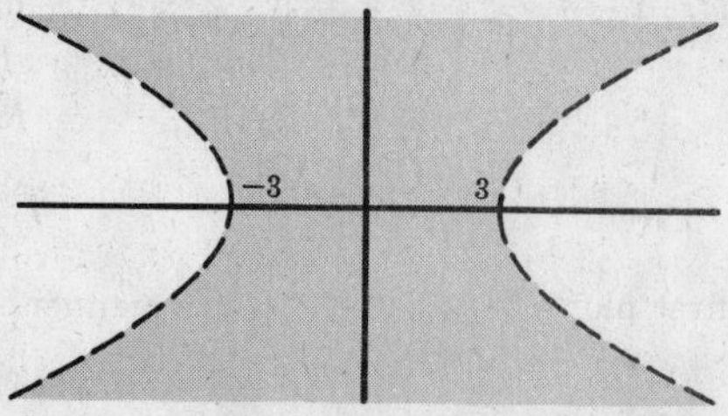

$x^2 - 4y^2 < 9$

DOMAINS, RANGES AND INVERSES

6. Consider the relation $R = \{(1,5), (4,5), (1,4), (4,6), (3,7), (7,6)\}$.

Find (1) the domain of R, (2) the range of R, (3) the inverse of R.

Solution:

(1) The domain of R consists of the set of first elements in R; hence the domain of R is

$$\{1, 4, 3, 7\}$$

(2) The range of R consists of the set of second elements in R; hence the range of R is

$$\{5, 4, 6, 7\}$$

(3) The inverse of R consists of the same pairs as are in R but in the reverse order; hence

$$R^{-1} = \{(5,1), (5,4), (4,1), (6,4), (7,3), (6,7)\}$$

7. Let $T = \{1,2,3,4,5\}$, and let a relation R in T be the set of points displayed in the following coordinate diagram of $T \times T$

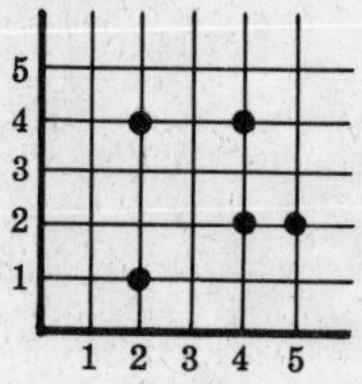

Find (1) the domain of R, (2) the range of R, (3) the inverse of R. (4) Sketch R^{-1} on a coordinate diagram of $T \times T$.

Solution:

(1) The element $x \in T$ is in the domain of R if and only if the vertical line through x contains a point of R. Thus the domain of R is the set $\{2,4,5\}$, since the vertical line through each of these elements, and only these elements, contains points of R.

(2) The element $x \in T$ is in the range of R if and only if the horizontal line through x contains a point of R. Thus the range of R is the set $\{1,2,4\}$, since the horizontal line through each of these elements, and only these elements, contains at least one point of R.

(3) Since

$$R = \{(2,1), (2,4), (4,2), (4,4), (5,2)\}$$
$$R^{-1} = \{(1,2), (4,2), (2,4), (4,4), (2,5)\}$$

(4) R^{-1} is displayed on a coordinate diagram of $T \times T$ as follows:

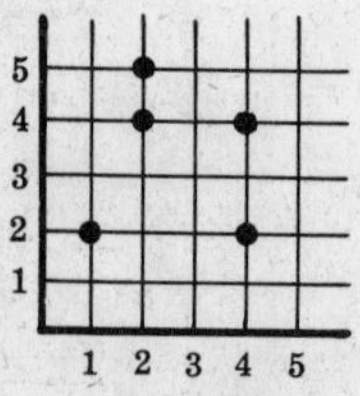

R^{-1} is sketched

8. Let $R = \{(x, y) \mid x \in R^\#,\ y \in R^\#,\ 4x^2 + 9y^2 = 36\}$. The sketch of R on the coordinate diagram of $R^\# \times R^\#$ follows:

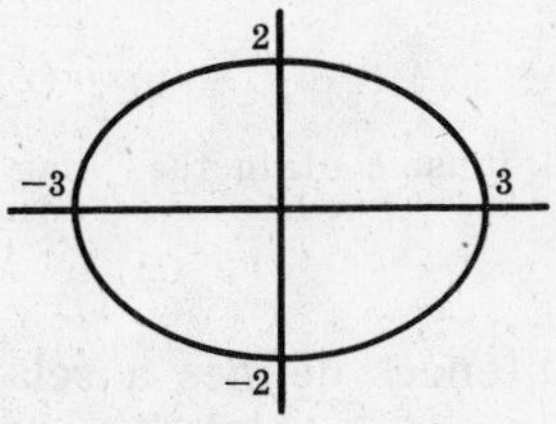

Find (1) the domain of R, (2) the range of R, (3) R^{-1}.

Solution:

(1) The domain of R is the interval $[-3, 3]$, since the vertical line through each of these numbers, and only these numbers, contains at least one point of R.

(2) The range of R is the interval $[-2, 2]$, since the horizontal line through each of these elements, and only these elements, contains at least one point of R.

(3) R^{-1} is found by interchanging x and y in the open sentence that defines R; hence

$$R^{-1} = \{(x, y) \mid x \in R^\#,\ y \in R^\#,\ 9x^2 + 4y^2 = 36\}$$

9. What relationships, if any, exist between the domain and range of a relation R and the domain and range of R^{-1}?

Solution:

Since R^{-1} consists of the same pairs as are in R except in the reverse order, each first element in R will be a second element in R^{-1} and each second element in R will be a first element in R^{-1}. Consequently, the domain of R is the range of R^{-1} and the range of R is the domain of R^{-1}.

10. Let R be the relation in the natural numbers $N = \{1, 2, 3, \ldots\}$ defined by the open sentence "$2x + y = 10$", that is, let

$$R = \{(x, y) \mid x \in N,\ y \in N,\ 2x + y = 10\}$$

Find (1) the domain of R, (2) the range of R, (3) R^{-1}.

Solution:

Note first that the solution set of $2x + y = 10$ is

$$R = \{(1, 8),\ (2, 6),\ (3, 4),\ (4, 2)\}$$

even though there are an infinite number of elements in N.

(1) The domain of R, which consists of the first elements of R, is $\{1, 2, 3, 4\}$.

(2) The range of R, which consists of the second elements of R, is $\{8, 6, 4, 2\}$.

(3) R^{-1} is found by interchanging x and y in the open sentence that defines R; hence

$$R^{-1} = \{(x, y) \mid x \in N,\ y \in N,\ x + 2y = 10\}$$

Also, as R^{-1} consists of the same pairs as are in R except in the reverse order, R^{-1} can be defined by

$$R^{-1} = \{(8, 1),\ (6, 2),\ (4, 3),\ (2, 4)\}$$

REFLEXIVE RELATIONS

11. When is a relation R in a set A *not* reflexive?

Solution:

R is not reflexive if there is at least one element $a \in A$ such that $(a, a) \notin R$.

12. Let $W = \{1, 2, 3, 4\}$ and $R = \{(1, 1), (1, 3), (2, 2), (3, 1), (4, 4)\}$. Is R reflexive?

Solution:

R is not reflexive since $3 \in W$ and $(3, 3) \notin R$.

13. Let A be any set and let D be the "diagonal line" of $A \times A$, that is, D is the set of all (a, a) where $a \in A$. What relationship is there, if any, between any reflexive relation R in A and D?

Solution:

Every reflexive relation R in A must contain the "diagonal line". In other words, D is a subset of R if R is reflexive.

14. Each of the following open sentences defines a relation R in the natural numbers N. State whether or not each is a reflexive relation.

(1) "x is less than or equal to y". (3) "$x + y = 10$"
(2) "x divides y". (4) "x and y are relatively prime".

Solution:

(1) Since $a \leq a$ for every $a \in N$, $(a, a) \in R$. Hence R is reflexive.
(2) Since every number divides itself, the relation is reflexive.
(3) Since $3 + 3 \neq 10$, 3 is not related to itself. Hence R is not reflexive.
(4) The greatest common divisor of 5 and 5 is 5; thus $(5, 5) \notin R$. Hence R is not reflexive.

15. Let $E = \{1, 2, 3\}$. Consider the following relations in E:

$R_1 = \{(1,2), (3,2), (2,2), (2,3)\}$ $R_4 = \{(1,2)\}$
$R_2 = \{(1,2), (2,3), (1,3)\}$ $R_5 = E \times E$
$R_3 = \{(1,1), (2,2), (2,3), (3,2), (3,3)\}$

State whether or not each of these relations is reflexive.

Solution:

If a relation in E is reflexive, then $(1, 1)$, $(2, 2)$ and $(3, 3)$ must belong to the relation. Therefore only R_3 and R_5 are reflexive.

SYMMETRIC RELATIONS

16. When is a relation R in a set A *not* symmetric?

Solution:

R is not symmetric if there are elements $a \in A$, $b \in A$ such that

$$(a, b) \in R, \quad (b, a) \notin R$$

Note $a \neq b$, otherwise $(a, b) \in R$ implies $(b, a) \in R$.

17. Let $V = \{1, 2, 3, 4\}$ and $R = \{(1,2), (3,4), (2,1), (3,3)\}$. Is R symmetric?

Solution:

R is not symmetric, since $3 \in V$, $4 \in V$, $(3, 4) \in R$ and $(4, 3) \notin R$.

18. Is there any set A in which every relation in A is symmetric?

Solution:

If A is the null set or if A contains only one element, then every relation in A is symmetric.

19. Each of the following open sentences defines a relation R in the natural numbers N. State whether or not each relation is symmetric.

(1) "x is less than or equal to y" (3) "$x + y = 10$"
(2) "x divides y" (4) "$x + 2y = 10$"

Solution:

(1) Since $3 \leq 5$ but $5 \nleq 3$, $(3, 5) \in R$ and $(5, 3) \notin R$. Thus R is not symmetric.
(2) Since 2 divides 4 but 4 does not divide 2, $(2, 4) \in R$ and $(4, 2) \notin R$. Hence R is not symmetric.

(3) If $a+b=10$ then $b+a=10$; in other words, if $(a,b) \in R$ then $(b,a) \in R$. Hence R is symmetric.

(4) Note that $(2,4) \in R$ but $(4,2) \notin R$, i.e. $2+2(4) = 10$ but $4+2(2) \neq 10$. Thus R is not symmetric.

20. Let $E = \{1,2,3\}$. Consider the following relations in E:

$$R_1 = \{(1,1),(2,1),(2,2),(3,2),(2,3)\} \qquad R_4 = \{(1,1),(3,2),(2,3)\}$$
$$R_2 = \{(1,1)\} \qquad R_5 = E \times E$$
$$R_3 = \{(1,2)\}$$

State whether or not each of these relations is symmetric.

Solution:

(1) R_1 is not symmetric since $(2,1) \in R_1$ but $(1,2) \notin R_1$.

(2) R_2 is symmetric.

(3) R_3 is not symmetric since $(1,2) \in R_3$ but $(2,1) \notin R_3$.

(4) R_4 is symmetric.

(5) R_5 is symmetric.

21. Prove: Let R and R' be symmetric relations in a set A; then $R \cap R'$ is a symmetric relation in A.

Solution:

Note first that R and R' are subsets of $A \times A$; hence $R \cap R'$ is also a subset of $A \times A$ and is, therefore, a relation in A.

Let (a,b) belong to $R \cap R'$. Then $(a,b) \in R$ and $(a,b) \in R'$. Since R and R' are symmetric, (b,a) also belongs to R and (b,a) also belongs to R'; hence $(b,a) \in R \cap R'$.

We have shown that $(a,b) \in R \cap R'$ implies $(b,a) \in R \cap R'$; hence $R \cap R'$ is symmetric.

ANTI-SYMMETRIC RELATIONS

22. When is a relation R in a set A not anti-symmetric?

Solution:

R is not anti-symmetric if there exists elements $a \in A$, $b \in A$, $a \neq b$ such that $(a,b) \in R$ and $(b,a) \in R$.

23. Let $W = \{1,2,3,4\}$ and $R = \{(1,2),(3,4),(2,2),(3,3),(2,1)\}$. Is R anti-symmetric?

Solution:

R is not anti-symmetric, since $1 \in W$, $2 \in W$, $1 \neq 2$, $(1,2) \in R$ and $(2,1) \in R$.

24. Can a relation R in a set A be both symmetric and anti-symmetric?

Solution:

Any subset of the "diagonal line" of $A \times A$, that is, any relation R in A in which

$$(a,b) \in R \text{ implies } a = b$$

is both symmetric and anti-symmetric.

25. Let $E = \{1,2,3\}$. Consider the following relations in E:

$$R_1 = \{(1,1),(2,1),(2,2),(3,2),(2,3)\} \qquad R_4 = \{(1,1),(2,3),(3,2)\}$$
$$R_2 = \{(1,1)\} \qquad R_5 = E \times E$$
$$R_3 = \{(1,2)\}$$

State whether or not each of these relations is anti-symmetric.

Solution:

(1) R_1 is not anti-symmetric since $(3,2) \in R_1$ and $(2,3) \in R_1$.

(2) R_2 is anti-symmetric.

(3) R_3 is anti-symmetric.

(4) R_4 is not anti-symmetric since $(2,3) \in R_4$ and $(3,2) \in R_4$.

(5) R_5 is not anti-symmetric for the same reasons as for R_4.

26. Let $E = \{1,2,3\}$. Give an example of a relation R in E such that R is neither symmetric nor anti-symmetric.

Solution:

The relation $R = \{(1,2), (2,1), (2,3)\}$ is not symmetric since $(2,3) \in R$ but $(3,2) \notin R$. R is also not anti-symmetric since $(1,2) \in R$ and $(2,1) \in R$.

27. Each of the following open sentences defines a relation R in the natural numbers N. State whether or not each of the relations is anti-symmetric.

(1) "x is less than or equal to y" (3) "$x + 2y = 10$"
(2) "x is less than y" (4) "x divides y"

Solution:

(1) Since $a \leq b$ and $b \leq a$ implies $a = b$, R is anti-symmetric.

(2) If $a \neq b$, then either $a < b$ or $b < a$; hence R is anti-symmetric.

(3) The solution set is $R = \{(2,4), (4,3), (6,2), (8,1)\}$. Note that $R \cap R^{-1} = \emptyset$, which is a subset of the "diagonal line" of $N \times N$. Hence R is anti-symmetric.

(4) Since a divides b and b divides a implies $a = b$, R is anti-symmetric.

TRANSITIVE RELATIONS

28. When is a relation in a set A *not* transitive?

Solution:

R is not transitive if there exist elements a, b and c belonging to A, not necessarily distinct, such that

$$(a,b) \in R, \ (b,c) \in R \quad \text{but} \quad (a,c) \notin R$$

29. Let $W = \{1,2,3,4\}$ and $R = \{(1,2), (4,3), (2,2), (2,1), (3,1)\}$. Is R transitive?

Solution:

R is not transitive since $(4,3) \in R$, $(3,1) \in R$ but $(4,1) \notin R$.

30. Let $W = \{1,2,3,4\}$ and $R = \{(2,2), (2,3), (1,4), (3,2)\}$. Is R transitive?

Solution:

R is not transitive since $(3,2) \in R$, $(2,3) \in R$ but $(3,3) \notin R$.

31. Each of the following open sentences defines a relation R in the natural numbers N. State whether or not each relation is transitive.

(1) "x is less than or equal to y" (3) "$x + y = 10$"
(2) "x divides y" (4) "$x + 2y = 5$"

Solution:

(1) Since $a \leq b$ and $b \leq c$ implies $a \leq c$, the relation is transitive.

(2) If x divides y and y divides z, then x divides z; that is,

$$(x,y) \in R, \ (y,z) \in R \quad \text{implies} \quad (x,z) \in R$$

Hence R is transitive.

(3) Note that $2 + 8 = 10$, $8 + 2 = 10$ and $2 + 2 \neq 10$; that is,

$$(2,8) \in R, \ (8,2) \in R \quad \text{but} \quad (2,2) \notin R$$

Hence R is not transitive.

(4) R is not transitive, since $(3,1) \in R$, $(1,2) \in R$ but $(3,2) \notin R$; that is,

$$3 + 2(1) = 5, \ 1 + 2(2) = 5 \quad \text{but} \quad 3 + 2(2) \neq 5$$

32. Prove: If a relation R is transitive, then its inverse relation R^{-1} is also transitive.

Solution:

Let (a, b) and (b, c) belong to R^{-1}; then $(c, b) \in R$ and $(b, a) \in R$. Since R is transitive, (c, a) also belongs to R; hence $(a, c) \in R^{-1}$.

We have shown that $(a, b) \in R^{-1}$, $(b, c) \in R^{-1}$ implies $(a, c) \in R^{-1}$; hence R^{-1} is transitive.

33. Let $E = \{1, 2, 3\}$. Consider the following relations in E:

$$R_1 = \{(1,2), (2,2)\} \qquad R_4 = \{(1,1)\}$$
$$R_2 = \{(1,2), (2,3), (1,3), (2,1), (1,1)\} \qquad R_5 = E \times E$$
$$R_3 = \{(1,2)\}$$

State whether or not each of these relations is transitive.

Solution:

Each of the relations is transitive except R_2. R_2 is not transitive, since

$$(2,1) \in R_2, \ (1,2) \in R_2 \quad \text{but} \quad (2,2) \notin R_2$$

RELATIONS AND FUNCTIONS

34. Let $W = \{1, 2, 3, 4\}$. Consider the following relations in W:

$$R_1 = \{(1,2), (2,3), (3,4), (4,1)\} \qquad R_4 = \{(1,2), (3,4), (4,1)\}$$
$$R_2 = \{(1,1), (1,2), (1,3), (1,4)\} \qquad R_5 = \{(2,1), (4,4), (3,1), (2,3)\}$$
$$R_3 = \{(1,1), (2,1), (3,1), (4,1)\}$$

State whether or not each of these relations is a function of W into W.

Solution:

Note first that a relation R in W is a function of W into W if and only if each $a \in W$ appears as the first element in one and only one ordered pair in R.

(1) R_1 is a function.

(2) R_2 is not a function since $1 \in W$, and 1 appears as the first element in $(1,1) \in R_2$ and $(1,2) \in R_2$.

(3) R_3 is a function.

(4) R_4 is not a function since $2 \in W$, but 2 does not appear as the first element in any ordered pair in R_4.

(5) R_5 is not a function since the two different ordered pairs $(2,1)$ and $(2,3)$, which belong to R_5, have the same first element.

35. Let the relation R from A to B be sketched on the coordinate diagram of $A \times B$. How could one determine geometrically whether or not R is, in fact, a function of A into B?

Solution:

If every vertical line contains exactly one point of R, then R is a function of A into B.

36. Let $A = [-4,4]$, $B = [0,4]$, $C = [-2,0]$ and $D = [-4,0]$, and let the open sentence $P(x,y)$ read "$x^2 + y^2 = 16$". Consider the following relations:

$$(1) \ R_1 = (A, B, P(x,y)) \qquad (4) \ R_4 = (A, C, P(x,y))$$
$$(2) \ R_2 = (B, A, P(x,y)) \qquad (5) \ R_5 = (A, D, P(x,y))$$
$$(3) \ R_3 = (B, B, P(x,y))$$

State whether or not each of these relations is a function.

Solution:

Note first that the relation R in the real numbers defined by $x^2 + y^2 = 16$ consists of the points on a circle of radius 4 with center at the origin, which is shown in Fig. 6-6 below.

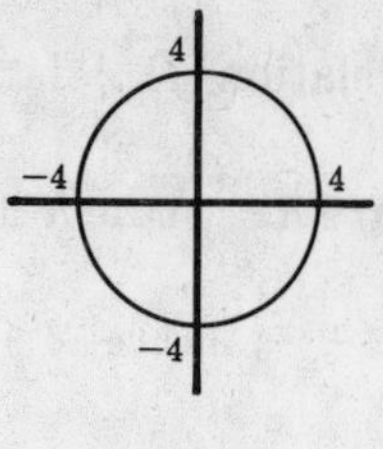

R is plotted
Fig. 6-6

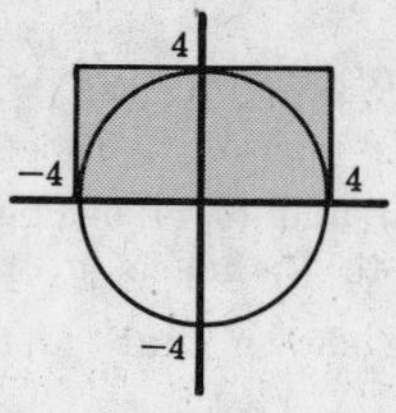

$A \times B$ is shaded
R is plotted
Fig. 6-7

(1) Sketch $A \times B$ on the coordinate diagram of $R^\# \times R^\#$, in which R is displayed by shading the appropriate area as shown in Fig. 6-7 above.

The intersection of the circle R and $A \times B$ is R_1. Note that each vertical line through A contains exactly one point of R_1. Hence R_1 is a function.

(2) Sketch $B \times A$ and R on a coordinate diagram of $R^\# \times R^\#$ as shown in Fig. 6-8 below.

$R_2 = R \cap (B \times A)$. Note that there are vertical lines through elements of B which contain two points of R_2; hence R_2 is not a function.

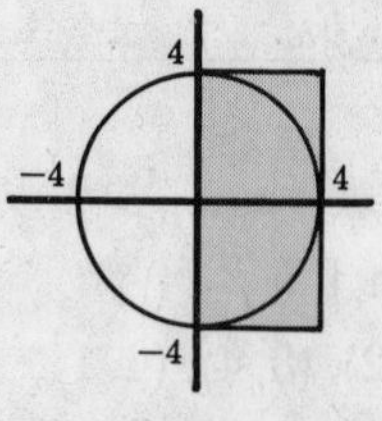

$B \times A$ is shaded
R is plotted
Fig. 6-8

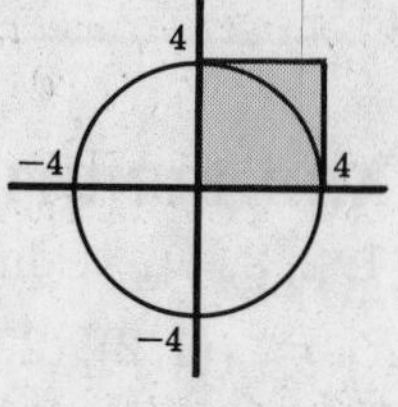

$B \times B$ is shaded
R is plotted
Fig. 6-9

(3) Sketch $B \times B$ and R on a coordinate diagram of $R^\# \times R^\#$ as shown in Fig. 6-9 above.

Each vertical line through any element in B contains exactly one point of $R_3 = R \cap (B \times B)$; hence R_3 is a function.

(4) Sketch $A \times C$ and R on a coordinate diagram of $R^\# \times R^\#$ as shown in Fig. 6-10 below.

The vertical line through $0 \in A$ contains no point of $R_4 = R \cap (A \times C)$; hence R_4 is not a function.

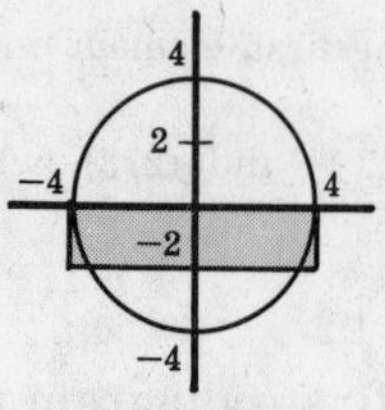

$A \times C$ is shaded
R is plotted
Fig. 6-10

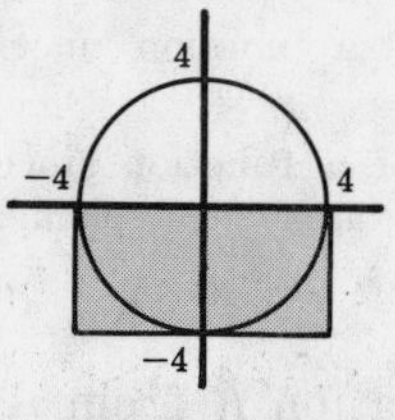

$A \times D$ is shaded
R is plotted
Fig. 6-11

(5) Sketch $A \times D$ and R on a coordinate diagram of $R^\# \times R^\#$ as shown in Fig. 6-11 above.

Notice that each vertical line through any element in A contains exactly one point of $R_5 = R \cap (A \times D)$; hence R_5 is a function of A into D.

MISCELLANEOUS PROBLEMS

37. Let R be the relation in the natural numbers N defined by the open sentence "$(x - y)$ is divisible by 5"; that is, let

$$R = \{(x, y) \mid x \in N,\ y \in N,\ (x - y) \text{ is divisible by } 5\}$$

Prove that R is an equivalence relation.

Solution:

Let $a \in N$; then $(a - a) = 0$ is divisible by 5, and hence $(a, a) \in R$. Thus R is reflexive.

Let $(a, b) \in R$; then $(a-b)$ is divisible by 5, and hence $(b-a) = -(a-b)$ is also divisible by 5. Thus (b, a) belongs to R. Since

$$(a, b) \in R \quad \text{implies} \quad (b, a) \in R$$

R is symmetric.

Let $(a, b) \in R$ and $(b, c) \in R$; then $(a-b)$ and $(b-c)$ are each divisible by 5. Hence $(a-c) = (a-b)+(b-c)$ is also divisible by 5, i.e. (a, c) belongs to R. Since

$$(a, b) \in R \text{ and } (b, c) \in R \quad \text{implies} \quad (a, c) \in R$$

R is transitive.

Since R is reflexive, symmetric and transitive, R is, by definition, an equivalence relation.

38. Let R and R' be relations in a set A. Prove each of the following two statements:

(1) If R is symmetric and R' is symmetric, then $R \cup R'$ is symmetric.

(2) If R is reflexive and R' is any relation, then $R \cup R'$ is reflexive.

Solution:

(1) If $(a, b) \in R \cup R'$, then (a, b) belongs to R or R', which are symmetric. Hence (b, a) also belongs to R or R'. Then $(b, a) \in R \cup R'$ and $R \cup R'$ is symmetric.

(2) R is reflexive if and only if R contains the "diagonal line" D of $A \times A$. But $D \subset R$ and $R \subset R \cup R'$ implies $D \subset R \cup R'$. Therefore $R \cup R'$ is reflexive.

39. Let R and R' be relations in a set A. Show that each of the following statements is false by giving a counter example, that is, an example for which it is not true.

(1) If R is anti-symmetric and R' is anti-symmetric, then $R \cup R'$ is anti-symmetric.

(2) If R is transitive and R' is transitive, then $R \cup R'$ is transitive.

Solution:

(1) $R = \{(1,2)\}$ and $R' = \{(2,1)\}$ are each anti-symmetric; but $R \cup R' = \{(1,2), (2,1)\}$ is not anti-symmetric.

(2) $R = \{(1,2)\}$ and $R' = \{(2,3)\}$ are each transitive; but $R \cup R' = \{(1,2), (2,3)\}$ is not transitive.

40. Let

$$R = \{(x, y) \mid x \in R^\#,\ y \in R^\#,\ y \geqq x^2\}$$
$$R' = \{(x, y) \mid x \in R^\#,\ y \in R^\#,\ y \leqq x+2\}$$

Note that R and R' are both relations in the real numbers.

(1) Sketch the relation $R \cap R'$ on a coordinate diagram of $R^\# \times R^\#$.

(2) Find the domain of $R \cap R'$.

(3) Find the range of $R \cap R'$.

Solution:

(1) Sketch R on a coordinate diagram of $R^\# \times R^\#$, as in Problem 4, and shade R with strokes slanting upward to the right (////); and on the same coordinate diagram sketch R' with strokes slanting downward to the right (\\\\), as shown in Fig. 6-12. The cross-hatched area is $R \cap R'$. Thus $R \cap R'$ is displayed in Fig. 6-13.

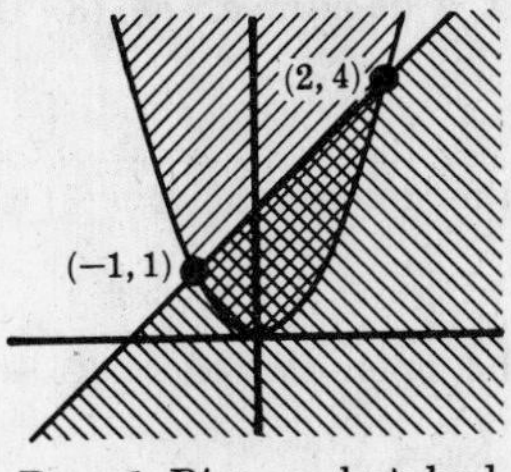

R and R' are sketched

Fig. 6-12

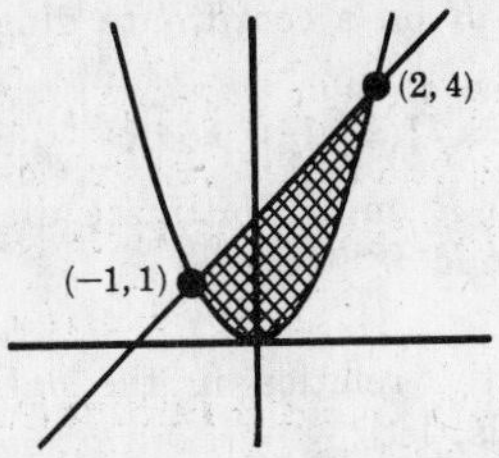

Fig. 6-13

(2) The domain of $R \cap R'$ is $[-1, 2]$, since a vertical line through each point in this interval, and only these points, will contain a point of $R \cap R'$.

(3) The range of $R \cap R'$ is $[0, 4]$, since a horizontal line through each point in this interval, and only these points, will contain at least one point of $R \cap R'$.

Supplementary Problems

BASIC PROPERTIES OF RELATIONS

41. Let R be the relation in $A = \{2, 3, 4, 5\}$ which is defined by the open sentence "x and y are relatively prime", i.e. "the only common divisor of x and y is 1".

(1) Find the solution set of R, that is, write R as a set of ordered pairs.
(2) Sketch R on a coordinate diagram of $A \times A$.

42. Let R be the relation in $B = \{2, 3, 4, 5, 6\}$ defined by the open sentence "$|x - y|$ is divisible by 3". Write R as a set of ordered pairs.

43. Let $C = \{1, 2, 3, 4, 5\}$, and let the relation R in C be the set of points displayed in the following coordinate diagram of $C \times C$.

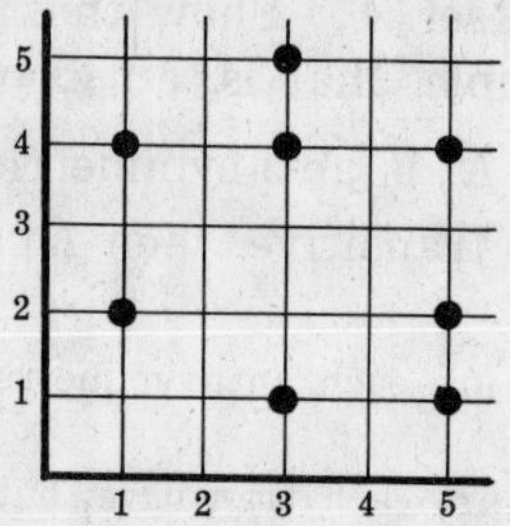

(1) State whether each is true or false: (a) $1\,R\,4$, (b) $2\,R\,5$, (c) $3 \not R\, 1$, (d) $5 \not R\, 3$.

(2) Write each of the following subsets of C in tabular form:

(a) $\{x \mid 3\,R\,x\}$ (c) $\{x \mid (x, 2) \notin R\}$
(b) $\{x \mid (4, x) \in R\}$ (d) $\{x \mid x\,R\,5\}$

Find (3) the domain of R, (4) the range of R, (5) R^{-1}.

44. Each of the following open sentences defines a relation in the real numbers. Sketch each relation on a coordinate diagram of $R^\# \times R^\#$.

(1) $y < x^2 - 4x + 2$ (3) $x < y^2$
(2) $y \geqq \frac{x}{2} + 2$ (4) $x \geqq \sin y$

45. Let $R = \{(x, y) \mid x \in R^\#, y \in R^\#, x^2 + 4y^2 \leqq 16\}$.

(1) Sketch R on a coordinate diagram of $R^\# \times R^\#$. Find (2) the domain of R, (3) the range of R.

46. Let $R = \{(x, y) \mid x \in R^\#, y \in R^\#, x^2 - y^2 \geqq 4\}$.

(1) Sketch R on a coordinate diagram of $R^\# \times R^\#$. Find (2) the domain of R, (3) the range of R.
(4) Define R^{-1}.

47. Let R be the relation in the natural numbers N defined by the open sentence "$x + 3y = 12$". In other words, let

$$R = \{(x, y) \mid x \in N, y \in N, x + 3y = 12\}$$

(1) Write R as a set of ordered pairs. Find (2) the domain of R, (3) the range of R, (4) R^{-1}.

48. Let R be the relation in the natural numbers N defined by $2x + 4y = 15$.
(1) Write R as a set of ordered pairs. Find (2) the domain of R, (3) the range of R, (4) R^{-1}.

REFLEXIVE, SYMMETRIC, ANTI-SYMMETRIC AND TRANSITIVE RELATIONS

49. Let $W = \{1, 2, 3, 4\}$. Consider the following relations in W:

$$R_1 = \{(1,1), (1,2)\} \qquad R_4 = \{(1,1), (2,2), (3,3)\}$$
$$R_2 = \{(1,1), (2,3), (4,1)\} \qquad R_5 = W \times W$$
$$R_3 = \{(1,3), (2,4)\}$$

State whether or not each of the relations is (1) symmetric, (2) anti-symmetric, (3) transitive, (4) reflexive.

50. State whether each of the following statements is true or false. Assume R and R' are relations in a set A.
(1) If R is symmetric, then R^{-1} is symmetric.
(2) If R is anti-symmetric, then R^{-1} is anti-symmetric.
(3) If R is reflexive, then $R \cap R^{-1} \neq \emptyset$.
(4) If R is symmetric, then $R \cap R^{-1} \neq \emptyset$.
(5) If R is transitive and R' is transitive, then $R \cup R'$ is transitive.
(6) If R is transitive and R' is transitive, then $R \cap R'$ is transitive.
(7) If R is anti-symmetric and R' is anti-symmetric, then $R \cup R'$ is anti-symmetric.
(8) If R is anti-symmetric and R' is anti-symmetric, then $R \cap R'$ is anti-symmetric.
(9) If R is reflexive and R' is reflexive, then $R \cup R'$ is reflexive.
(10) If R is reflexive and R' is reflexive, then $R \cap R'$ is reflexive.

51. Let L be the set of lines in the Euclidean plane and let R be the relation in L defined by "x is parallel to y". State whether or not R is (1) reflexive, (2) symmetric, (3) anti-symmetric, (4) transitive. (Assume a line is parallel to itself.)

52. Let L be the set of lines in the Euclidean plane and let R be the relation in L defined by "x is perpendicular to y". State whether or not R is (1) reflexive, (2) symmetric, (3) anti-symmetric, (4) transitive.

53. Let $\mathcal{A}$ be a family of sets and let R be the relation in $\mathcal{A}$ defined by "x is disjoint from y". State whether or not R is (1) reflexive, (2) symmetric, (3) anti-symmetric, (4) transitive.

54. What type of relation is R if (1) $R \cap R^{-1} = \emptyset$, (2) $R = R^{-1}$?

55. Each of the following open sentences defines a relation in the natural numbers N.

(1) "x is greater than y" (3) "x times y is the square of a number"
(2) "x is a multiple of y" (4) "$x + 3y = 12$"

State whether or not each of the relations is (*a*) reflexive, (*b*) symmetric, (*c*) anti-symmetric, (*d*) transitive.

RELATIONS AND FUNCTIONS

56. Let $T = \{a, b, c, d\}$. Consider the following relations in T:

(1) $R_1 = \{(a, b), (b, c), (c, d), (d, a)\}$ (4) $R_4 = \{(a, a), (b, a), (c, a), (d, d)\}$
(2) $R_2 = \{(b, a), (c, d), (b, a), (a, b), (d, b)\}$ (5) $R_5 = \{(b, a), (a, c), (d, d)\}$
(3) $R_3 = \{(d, c), (c, b), (a, b), (d, d)\}$

State whether or not each of the relations is a function.

57. Let $A = [-4, 4]$, $B = [0, 4]$, $C = [-4, 0]$, and let the open sentence $P(x, y)$ read "$x^2 + 4y^2 = 16$". Consider the following relations:

(1) $R_1 = (A, B, P(x, y))$ (3) $R_3 = (B, A, P(x, y))$
(2) $R_2 = (A, C, P(x, y))$ (4) $R_4 = (B, C, P(x, y))$

Sketch each relation on a Cartesian plane as in Problem 36, and then state whether or not the relation is a function.

58. Let $A = [0, \infty)$, $B = (-\infty, 0]$, $C = [2, \infty)$, $D = (-\infty, -2]$, and let the open sentence $P(x, y)$ read "$x^2 - y^2 = 4$". Consider the relation

$$R = (X, Y, P(x, y))$$

where X and Y are unknown sets. If X and Y are each permitted to be any of the above four sets, which of the sixteen relations are functions? (*Hint*: First plot $P(x, y)$ on a Cartesian plane.)

59. Let A be any set.

(1) Is there more than one reflexive relation in A which is a function?

(2) Is there any reflexive relation in A which is a function?

60. Prove: Let A be non-empty and let R be a transitive relation in A which does not contain any of the "diagonal elements" $(x, x) \in A \times A$; then R is not a function in A.

MISCELLANEOUS PROBLEMS

61. Consider the following relations in the real numbers:

$$R = \{(x, y) \mid x \in R\#,\ y \in R\#,\ x^2 + y^2 \leqq 25\}$$
$$R' = \{(x, y) \mid x \in R\#,\ y \in R\#,\ y \geqq 4x^2/9\}$$

(1) Sketch the relation $R \cap R'$ on a coordinate diagram of $R\# \times R\#$.

(2) Find the domain of $R \cap R'$. (3) Find the range of $R \cap R'$.

62. Consider each of the following sets of ordered pairs of real numbers, i.e. relations in $R\#$:

(1) $\{(x, y) \mid x^2 + y^2 \leqq 25\} \cap \{(x, y) \mid y \geqq 3x/4\}$
(2) $\{(x, y) \mid x^2 + y^2 \geqq 25\} \cap \{(x, y) \mid y \geqq 4x^2/9\}$
(3) $\{(x, y) \mid x^2 + y^2 \leqq 25\} \cup \{(x, y) \mid y \geqq 4x^2/9\}$
(4) $\{(x, y) \mid x^2 + y^2 < 25\} \cap \{(x, y) \mid y < 3x/4\}$

Sketch each relation on a coordinate diagram of $R\# \times R\#$ and state its domain and range.

63. Let A be the set of people. Each open sentence below defines a relation in A. For each of these relations find an open sentence, sometimes called the "inverse sentence", that defined the inverse relation.

(1) "x is the husband of y"
(2) "x is older than y"
(3) "x is taller than y"
(4) "x is richer than y"
(5) "x is more intelligent than y"

64. Let N be the natural numbers. Each open sentence below defines a relation in N. For each of the relations, find an open sentence which defines the inverse relation.

(1) "x is greater than y"
(2) "x is greater than or equal to y"
(3) "x is a multiple of y"
(4) "$2x + 3y = 30$"

Answers to Supplementary Problems

41. (1) $R = \{(2,3), (2,5), (3,2), (3,4), (3,5), (4,3), (4,5), (5,2), (5,3), (5,4)\}$

(2)

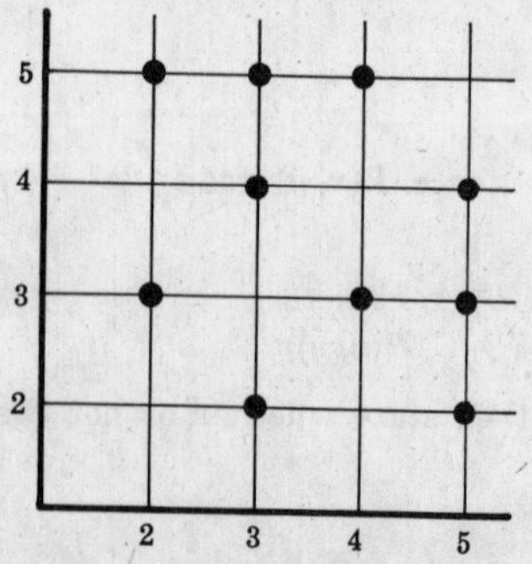

42. $R = \{(2,2), (2,5), (3,3), (3,6), (4,4), (5,5), (5,2), (6,6), (6,3)\}$

43. (1) (*a*) True, (*b*) False, (*c*) False, (*d*) True

(2) (*a*) $\{1,4,5\}$, (*b*) $\emptyset$, (*c*) $\{2,3,4\}$, (*d*) $\{3\}$

(3) $\{1,3,5\}$ (4) $\{1,2,4,5\}$

(5) $R^{-1} = \{(1,3), (1,5), (2,1), (2,5), (4,1), (4,3), (4,5), (5,3)\}$

44. (1)

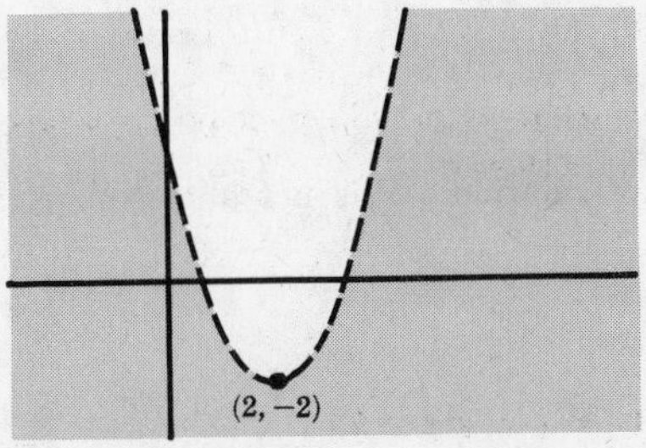

(2)

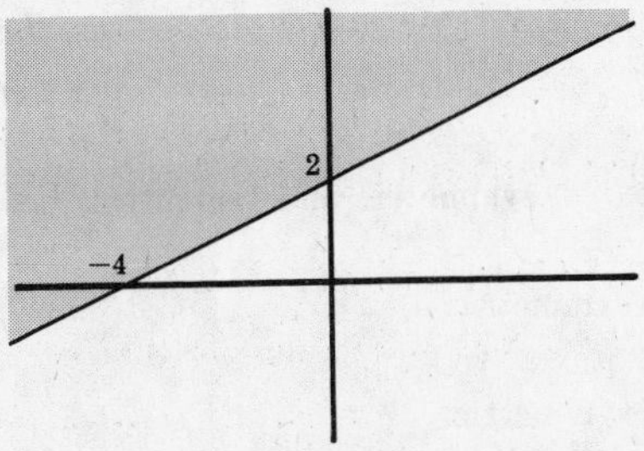

(3)

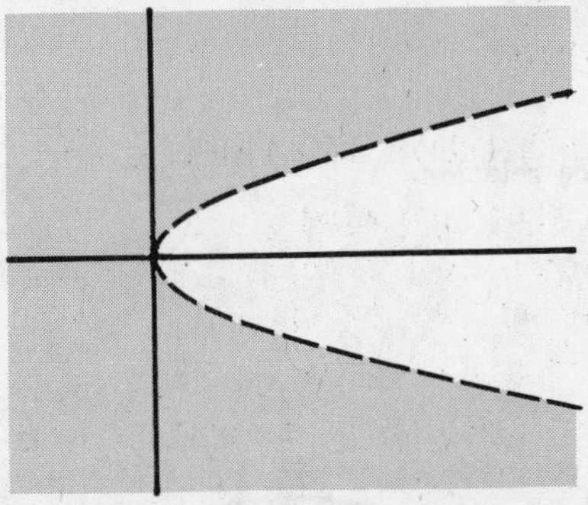

(4)

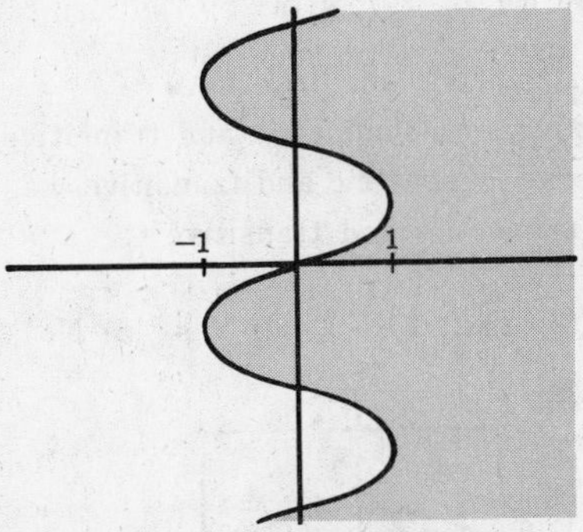

45. (1)

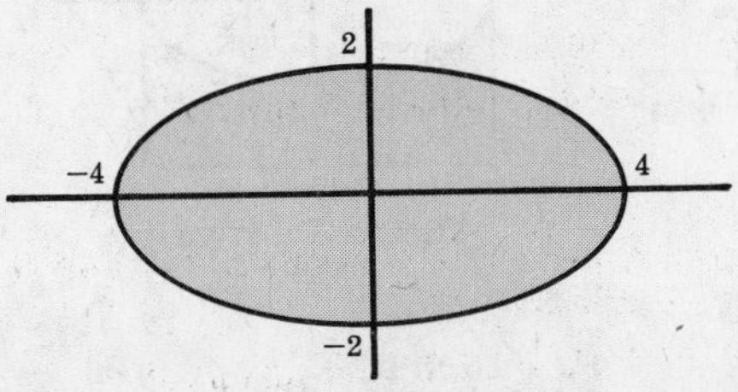

(2) The domain of R is $[-4,4]$. (3) The range of R is $[-2,2]$.

46. (1)

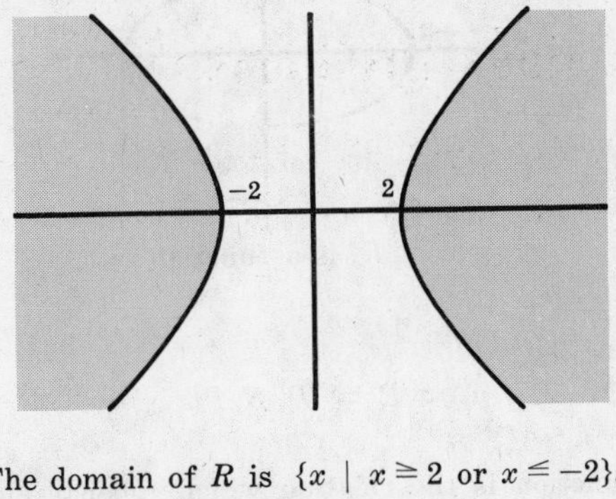

(2) The domain of R is $\{x \mid x \geq 2 \text{ or } x \leq -2\}$.

(3) The range of R is $R^{\#}$.

(4) $R^{-1} = \{(x,y) \mid x \in R^{\#},\ y \in R^{\#},\ x^2 - y^2 \leq -4\}$

47. (1) $R = \{(9,1), (6,2), (3,3)\}$
(2) $\{9, 6, 3\}$
(3) $\{1, 2, 3\}$
(4) $R^{-1} = \{(1,9), (2,6), (3,3)\}$

48. (1) $\emptyset$, (2) $\emptyset$, (3) $\emptyset$, (4) $\emptyset$.

49. (1) R_4 and R_5 are symmetric.
(2) Only R_5 is not anti-symmetric.
(3) All the relations are transitive.
(4) Only R_5 is reflexive.

50. (1) True (2) True (3) True (4) False (5) False (6) True (7) False (8) True (9) True (10) True

51. R is reflexive, symmetric and transitive, i.e. an equivalence relation. R is not anti-symmetric.

52. R is only symmetric.

53. R is only symmetric.

54. (1) anti-symmetric, (2) symmetric

55. (1) anti-symmetric and transitive
(2) reflexive, anti-symmetric and transitive
(3) reflexive, symmetric and transitive, i.e. an equivalence relation
(4) anti-symmetric and transitive

56. (1) Yes, (2) Yes, (3) No, (4) Yes, (5) No

57. (1)

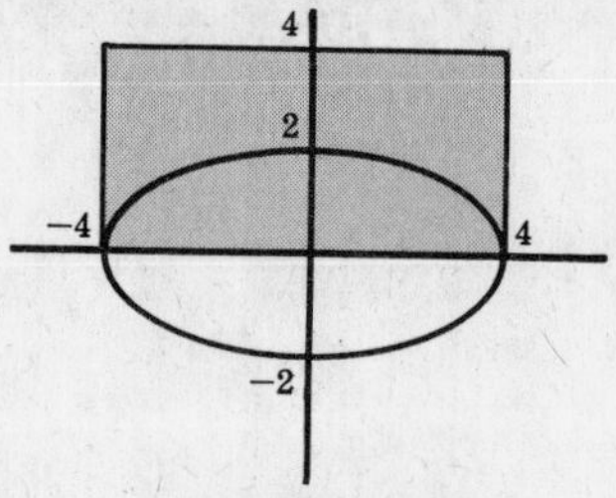

R_1 is a function

(2)

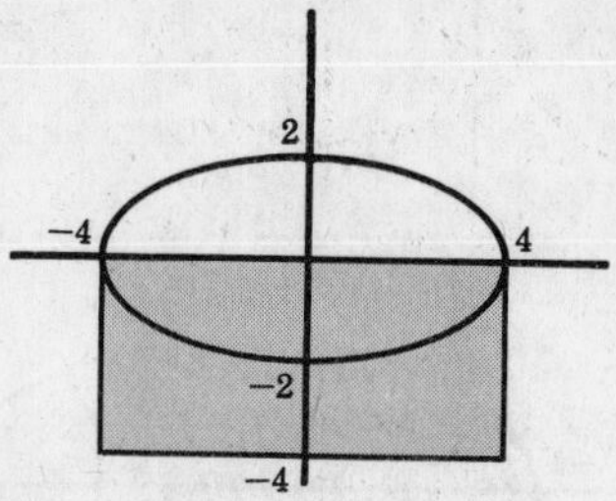

R_2 is a function

(3)

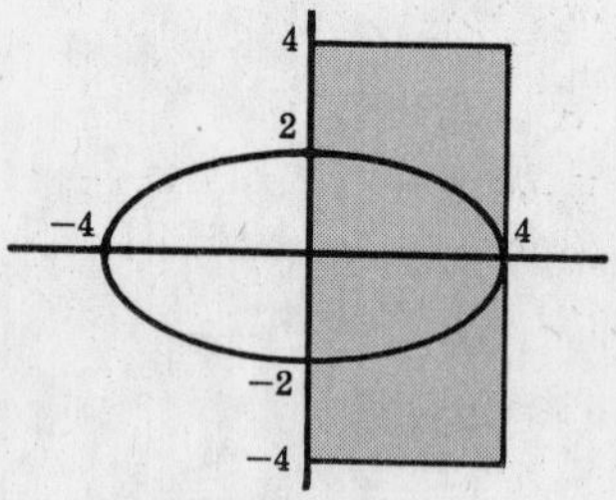

R_3 is not a function

(4)

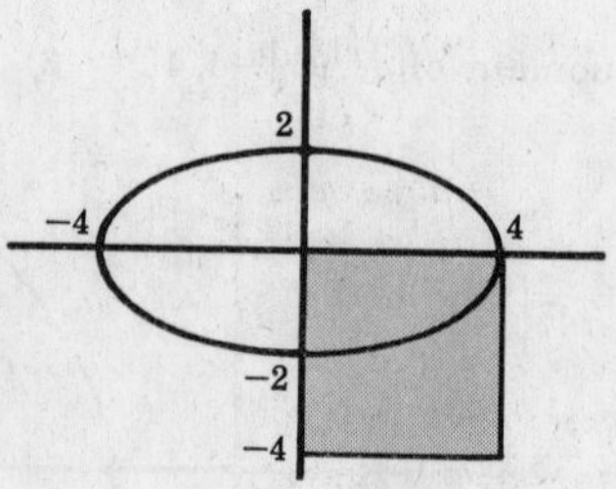

R_4 is a function

58. The only relations which are functions are:
$R = (C, A, P(x,y))$, $R = (C, B, P(x,y))$, $R = (D, A, P(x,y))$, $R = (D, B, P(x,y))$.

59. The only reflexive relation in a set A which is a function is the relation which consists only of the ordered pairs on the "diagonal line" of $A \times A$; it defines the identity function on A. Hence: (1) no, (2) yes.

60. As $A \neq \emptyset$, there is an element $a \varepsilon A$. If R is a function, then there is an ordered pair $(a, b) \varepsilon R$ such that, by hypothesis, $a \neq b$. Furthermore, since $b \varepsilon A$, there is an ordered pair $(b, c) \varepsilon R$ such that $b \neq c$. Since R is transitive

$$(a, b) \varepsilon R \text{ and } (b, c) \varepsilon R \quad \text{implies} \quad (a, c) \varepsilon R$$

Thus

$$(a, b) \varepsilon R, \ (a, c) \varepsilon R, \ b \neq c$$

R cannot be a function since it contains two different ordered pairs with the same first element.

61. (1)

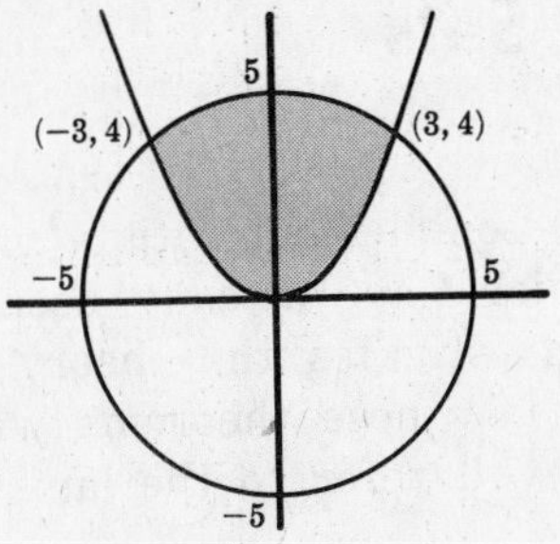

$R \cap R'$ is shaded

(2) $[-3, 3]$

(3) $[0, 5]$

62. (1)

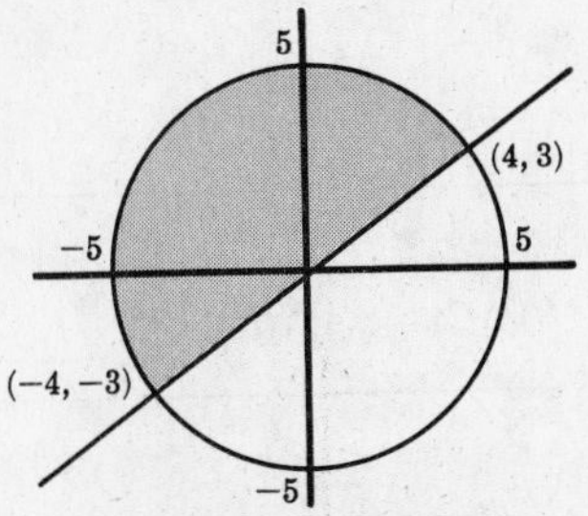

Domain is $[-5, 4]$
Range is $[-3, 5]$

(2)

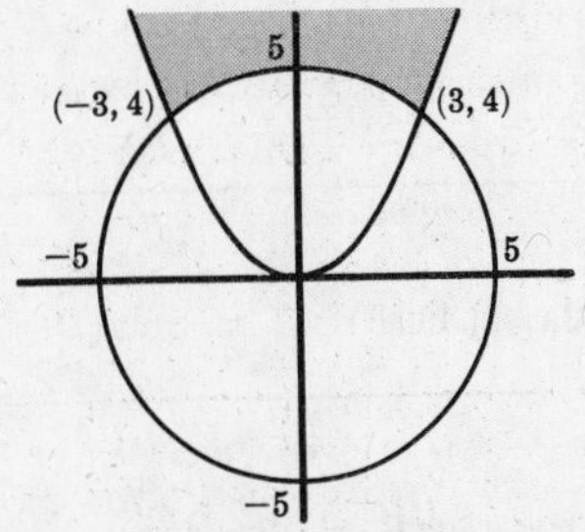

Domain is $R^\#$
Range is $[4, \infty)$

(3)

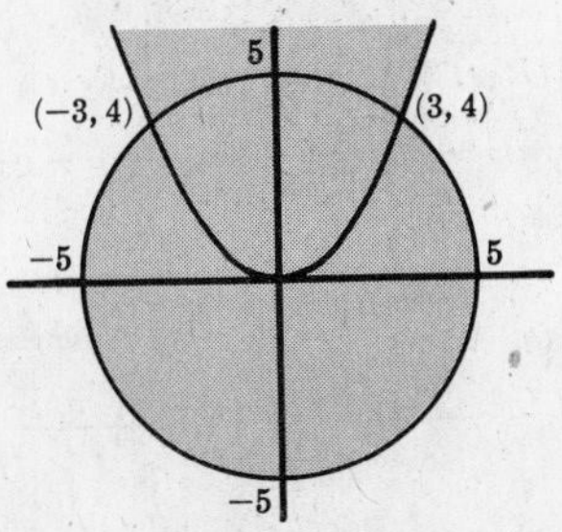

Domain is $R^\#$
Range is $[-5, \infty)$

(4)

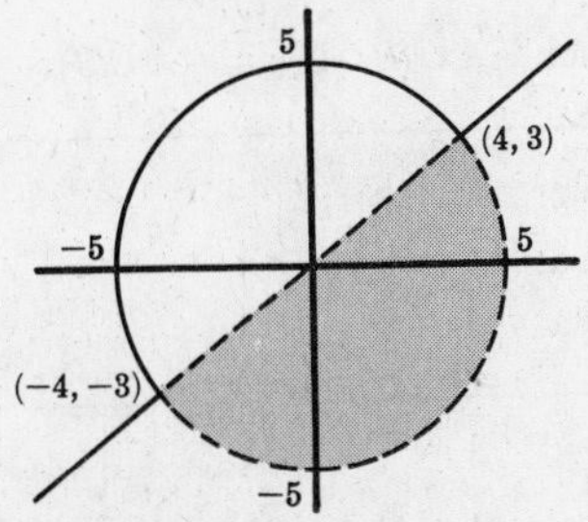

Domain is $\{x \mid -4 < x < 5\}$
Range is $\{x \mid -5 < x < 3\}$

63. (1) "x is the wife of y"
(2) "x is younger than y"
(3) "x is shorter than y"
(4) "x is poorer than y"
(5) "x is less intelligent than y"

64. (1) "x is less than y"
(2) "x is less than or equal to y"
(3) "x divides y" or "x is a factor of y"
(4) "$3x + 2y = 30$"

Chapter 7

Further Theory of Sets

ALGEBRA OF SETS

Sets under the operations of union, intersection and complement satisfy various laws, i.e. identities. Table 1 lists laws of sets, most of which have already been noted and proven in Chapter 2. One branch of mathematics investigates the theory of sets by studying those theorems that follow from these laws, i.e. those theorems whose proofs require the use of only these laws and no others. We will refer to the laws in Table 1 and their consequences as the algebra of sets.

LAWS OF THE ALGEBRA OF SETS	
Idempotent Laws	
1a. $A \cup A = A$	1b. $A \cap A = A$
Associative Laws	
2a. $(A \cup B) \cup C = A \cup (B \cup C)$	2b. $(A \cap B) \cap C = A \cap (B \cap C)$
Commutative Laws	
3a. $A \cup B = B \cup A$	3b. $A \cap B = B \cap A$
Distributive Laws	
4a. $A \cup (B \cap C) = (A \cup B) \cap (A \cup C)$	4b. $A \cap (B \cup C) = (A \cap B) \cup (A \cap C)$
Identity Laws	
5a. $A \cup \emptyset = A$	5b. $A \cap U = A$
6a. $A \cup U = U$	6b. $A \cap \emptyset = \emptyset$
Complement Laws	
7a. $A \cup A' = U$	7b. $A \cap A' = \emptyset$
8a. $(A')' = A$	8b. $U' = \emptyset, \ \emptyset' = U$
De Morgan's Laws	
9a. $(A \cup B)' = A' \cap B'$	9b. $(A \cap B)' = A' \cup B'$

Table 1

Notice that the concept of "element" and the relation "a belongs to A" do not appear anywhere in Table 1. Although these concepts were essential to our original development of the theory of sets, they do not appear in investigating the algebra of sets. The relation "A is a subset of B" is defined in our algebra of sets by

$$A \subset B \quad \text{means} \quad A \cap B = A$$

As examples, we now prove two theorems in our algebra of sets, that is, we prove two theorems which follow directly from the laws in Table 1. Other theorems and proofs are given in the problem section.

Example 1.1: Prove: $(A \cup B) \cap (A \cup B') = A$

Statement	Reason
1. $(A \cup B) \cap (A \cup B') = A \cup (B \cap B')$	1. Distributive Law
2. $B \cap B' = \emptyset$	2. Complement Law
3. $\therefore (A \cup B) \cap (A \cup B') = A \cup \emptyset$	3. Substitution
4. $A \cup \emptyset = A$	4. Identity Law
5. $\therefore (A \cup B) \cap (A \cup B') = A$	5. Substitution

Example 1.2: Prove: $A \subset B$ and $B \subset C$ implies $A \subset C$.

Statement	Reason
1. $A = A \cap B$ and $B = B \cap C$	1. Definition of subset
2. $\therefore A = A \cap (B \cap C)$	2. Substitution
3. $A = (A \cap B) \cap C$	3. Associative Law
4. $\therefore A = A \cap C$	4. Substitution
5. $\therefore A \subset C$	5. Definition of subset

PRINCIPLE OF DUALITY

If we interchange $\cap$ and $\cup$ and also U and $\emptyset$ in any statement about sets, then the new statement is called the *dual* of the original one.

Example 2.1: The dual of

$$(U \cup B) \cap (A \cup \emptyset) = A$$

is

$$(\emptyset \cap B) \cup (A \cap U) = A$$

Notice that the dual of every law in Table 1 is also a law in Table 1. This fact is extremely important in view of the following principle:

Principle of Duality: If certain axioms imply their own duals, then the dual of any theorem that is a consequence of the axioms is also a consequence of the axioms. For, given any theorem and its proof, the dual of the theorem can be proven in the same way by using the dual of each step in the original proof.

Thus the principle of duality applies to the algebra of sets.

Example 2.2: Prove: $(A \cap B) \cup (A \cap B') = A$

The dual of this theorem is proven in Example 1.1; hence this theorem is true by the Principle of Duality.

INDEXED SETS

Consider the sets

$$A_1 = \{1, 10\}, \quad A_2 = \{2, 4, 6, 10\}, \quad A_3 = \{3, 6, 9\}, \quad A_4 = \{4, 8\}, \quad A_5 = \{5, 6, 10\}$$

and the set

$$I = \{1, 2, 3, 4, 5\}$$

Notice that to each element $i \in I$ there corresponds a set A_i. In such a situation I is called the *index set*, the sets $\{A_1, \ldots, A_5\}$ are called the *indexed sets*, and the subscript i of A_i, i.e. each $i \in I$, is called an *index*. Furthermore, such an indexed family of sets is denoted by

$$\{A_i\}_{i \in I}$$

We can look at an indexed family of sets from another point of view. Since to each element $i \in I$ there is assigned a set A_i, we state

Definition 7.1: An indexed family of sets $\{A_i\}_{i \in I}$ is a function

$$f: I \rightarrow \mathcal{A}$$

where the domain of f is the index set I and the range of f is a family of sets.

Example 3.1: Define $B_n = \{x \mid 0 \leq x \leq (1/n)\}$, where $n \in N$, the natural numbers. Then

$$B_1 = [0, 1], \quad B_2 = [0, \tfrac{1}{2}], \quad \ldots$$

Example 3.2: Let I be the set of words in the English language, and let $i \in I$. Define

$$W_i = \{x \mid x \text{ is a letter in the word } i \in I\}$$

If i is the word "follow", then $W_i = \{f, l, o, w\}$.

Example 3.3: Define $D_n = \{x \mid x \text{ is a multiple of } n\}$, where $n \in N$, the natural numbers. Then

$$D_1 = \{1, 2, 3, 4, \ldots\}, \quad D_2 = \{2, 4, 6, 8, \ldots\}, \quad D_3 = \{3, 6, 9, 12, \ldots\}$$

Notice that the index set N is also D_1 and also the universal set for the indexed sets.

Remark 7.1: Any family $\mathcal{B}$ of sets can be indexed by itself. Specifically, the identity function

$$i: \mathcal{B} \rightarrow \mathcal{B}$$

is an indexed family of sets

$$\{A_i\}_{i \in \mathcal{B}}$$

where $A_i \in \mathcal{B}$ and where $i = A_i$. In other words, the index of any set in $\mathcal{B}$ is the set itself.

GENERALIZED OPERATIONS

The operations of union and intersection were defined for two sets. These definitions can easily be extended, by induction, to a finite number of sets. Specifically, for sets $A_1, \ldots, A_n$,

$$\cup_{i=1}^{n} A_i \equiv A_1 \cup A_2 \cup \ldots \cup A_n$$

$$\cap_{i=1}^{n} A_i \equiv A_1 \cap A_2 \cap \ldots \cap A_n$$

In view of the associative law, the union (intersection) of the sets may be taken in any order; thus parentheses need not be used in the above.

These concepts are generalized in the following way. Consider the indexed family of sets

$$\{A_i\}_{i \in I}$$

and let $J \subset I$. Then

$$\cup_{i \in J} A_i$$

consists of those elements which belong to at least one A_i where $i \in J$. Specifically,

$$\cup_{i \in J} A_i = \{x \mid \text{there exists an } i \in J \text{ such that } x \in A_i\}$$

In an analogous way

$$\cap_{i \in J} A_i$$

consists of those elements which belong to every A_i for $i \in J$. In other words,

$$\cap_{i \in J} A_i = \{x \mid x \in A_i \text{ for every } i \in J\}$$

Example 4.1: Let $A_1 = \{1, 10\}$, $A_2 = \{2, 4, 6, 10\}$, $A_3 = \{3, 6, 9\}$, $A_4 = \{4, 8\}$, $A_5 = \{5, 6, 10\}$; and let $J = \{2, 3, 5\}$. Then

$$\cap_{i \in J} A_i = \{6\} \quad \text{and} \quad \cup_{i \in J} A_i = \{2, 4, 6, 10, 3, 9, 5\}$$

Example 4.2: Let $B_n = [0, 1/n]$, where $n \in N$, the natural numbers. Then

$$\cap_{i \in N} B_i = \{0\} \quad \text{and} \quad \cup_{i \in N} B_i = [0, 1]$$

Example 4.3: Let $D_n = \{x \mid x \text{ is a multiple of } n\}$, where $n \in N$, the natural numbers. Then

$$\cap_{i \in N} D_i = \emptyset$$

There are also generalized distributive laws for a set B and an indexed family of sets $\{A_i\}_{i \in I}$. Specifically,

Theorem 7.1: Let $\{A_i\}_{i \in I}$ be an indexed family of sets. Then for any set B,

$$B \cap (\cup_{i \in I} A_i) = \cup_{i \in I} (B \cap A_i)$$
$$B \cup (\cap_{i \in I} A_i) = \cap_{i \in I} (B \cup A_i)$$

In those books which write $A + B$ for the union of two sets and AB for the intersection of two sets, Theorem 7.1 is then written:

$$B \sum_{i \in I} A_i = \sum_{i \in I} BA_i$$
$$B + \prod_{i \in I} A_i = \prod_{i \in I} (B + A_i)$$

PARTITIONS

Consider the set $A = \{1, 2, \ldots, 9, 10\}$ and its subsets

$$B_1 = \{1, 3\}, \quad B_2 = \{7, 8, 10\}, \quad B_3 = \{2, 5, 6\}, \quad B_4 = \{4, 9\}$$

The family of sets $\mathcal{B} = \{B_1, B_2, B_3, B_4\}$ has two important properties:

(1) A is the union of the sets in $\mathcal{B}$, i.e.,

$$A = B_1 \cup B_2 \cup B_3 \cup B_4$$

(2) For any sets B_i and B_j,

$$\text{either } B_i = B_j \text{ or } B_i \cap B_j = \emptyset$$

Such a family of sets is called a *partition* of A. Specifically, we give the

Definition 7.2: Let $\{B_i\}_{i \in I}$ be a family of non-empty subsets of A. Then $\{B_i\}_{i \in I}$ is called a *partition* of A if

P_1: $\cup_{i \in I} B_i = A$

P_2: For any B_i, B_j, either $B_i = B_j$ or $B_i \cap B_j = \emptyset$.

Furthermore, each B_i is then called an *equivalence class* of A.

Example 5.1: Let $N = \{1, 2, 3, \ldots\}$, $E = \{2, 4, 6, \ldots\}$ and $F = \{1, 3, 5, \ldots\}$. Then $\{E, F\}$ is a partition of N.

Example 5.2: Let $T = \{1, 2, \ldots, 9, 10\}$, and let $A = \{1, 3, 5\}$, $B = \{2, 6, 10\}$ and $C = \{4, 8, 9\}$. Then $\{A, B, C\}$ is not a partition of T since

$$T \neq A \cup B \cup C$$

i.e. since $7 \in T$ but $7 \notin (A \cup B \cup C)$.

Example 5.3: Let $T = \{1, 2, \ldots, 9, 10\}$, and let $F = \{1, 3, 5, 7, 9\}$, $G = \{2, 4, 10\}$ and $H = \{3, 5, 6, 8\}$. Then $\{F, G, H\}$ is not a partition of T since

$$F \cap H \neq \emptyset, \qquad F \neq H$$

Example 5.4: Let y_1, y_2, y_3 and y_4 be respectively the words "follow", "thumb", "flow" and "again", and let

$$A = \{w, g, u, o, i, m, l, a, t, f, n, h, b\}$$

Furthermore, define

$$W_i = \{x \mid x \text{ is a letter in the word } y_i\}$$

Then $\{W_1, W_2, W_3, W_4\}$ is a partition of A. Notice that W_1 and W_3 are not disjoint, but there is no contradiction since the sets are equal. (This example is highly instructive. The reader should actually find the sets W_i and then verify that they do define a partition of A.)

EQUIVALENCE RELATIONS AND PARTITIONS

Recall the following

Definition: A relation R in a set A is an *equivalence relation* if:

(1) R is reflexive, i.e. for every $a \in A$, a is related to itself;

(2) R is symmetric, i.e. if a is related to b then b is related to a;

(3) R is transitive, i.e. if a is related to b and b is related to c then a is related to c.

The reason that partitions and equivalence relations appear together is because of the

Theorem 7.2, Fundamental Theorem on Equivalence Relations: Let R be an equivalence relation in a set A and, for every $\alpha \in A$, let

$$B_\alpha = \{x \mid (x, \alpha) \in R\}$$

i.e. the set of elements related to α. Then the family of sets

$$\{B_\alpha\}_{\alpha \in A}$$

is a partition of A.

In other words, an equivalence relation R in a set A partitions the set A by putting those elements which are related to each other in the same equivalence class.

Moreover, the set B_α is called the *equivalence class* determined by α, and the set of equivalence classes $\{B_\alpha\}_{\alpha \in A}$ is denoted by

$$A/R$$

and called the *quotient set.*

The converse of the previous theorem is also true. Specifically,

Theorem 7.3: Let $\{B_i\}_{i \in I}$ be a partition of A and let R be the relation in A defined by the open sentence "x is in the same set (of the family $\{B_i\}_{i \in I}$) as y". Then R is an equivalence relation in A.

Thus there is a one to one correspondence between all partitions of a set A and all equivalence relations in A.

Example 6.1: In the Euclidean plane, similarity of triangles is an equivalence relation. Thus all triangles in the plane are partitioned into disjoint sets in which similar triangles are elements of the same set.

Example 6.2: Let R_5 be the relation in the integers defined by

$$x \equiv y \pmod 5$$

which reads "x is congruent to y modulo 5" and which means "$x - y$ is divisible by 5". Then R_5 is an equivalence relation. There are exactly five equivalence classes in Z/R_5: E_0, E_1, E_2, E_3 and E_4. Since each integer x is uniquely expressible in the form $x = 5q + r$ where $0 \leq r < 5$, then x is a member of the equivalence class E_r, where r is the remainder. Thus

$$\begin{aligned} E_0 &= \{\ldots, -10, -5, 0, 5, 10, \ldots\} \\ E_1 &= \{\ldots, -9, -4, 1, 6, 11, \ldots\} \\ E_2 &= \{\ldots, -8, -3, 2, 7, 12, \ldots\} \\ E_3 &= \{\ldots, -7, -2, 3, 8, 13, \ldots\} \\ E_4 &= \{\ldots, -6, -1, 4, 9, 14, \ldots\} \end{aligned}$$

and the quotient set $Z/R_5 = \{E_0, E_1, E_2, E_3, E_4\}$.

Solved Problems

ALGEBRA OF SETS AND DUALITY

1. Write the dual of each of the following:

(1) $(B \cup C) \cap A = (B \cap A) \cup (C \cap A)$ (2) $A \cup (A' \cap B) = A \cup B$ (3) $(A \cap U) \cap (\emptyset \cup A') = \emptyset$

Solution:

In each statement interchange $\cup$ and $\cap$, and $\emptyset$ and U; hence

(1) $(B \cap C) \cup A = (B \cup A) \cap (C \cup A)$ (2) $A \cap (A' \cup B) = A \cap B$ (3) $(A \cup \emptyset) \cup (U \cap A') = U$

2. Prove the Right Distributive Law: $(B \cup C) \cap A = (B \cap A) \cup (C \cap A)$.

Solution:

Statement	Reason
1. $(B \cup C) \cap A = A \cap (B \cup C)$	1. Commutative Law
2. $= (A \cap B) \cup (A \cap C)$	2. Distributive Law
3. $= (B \cap A) \cup (C \cap A)$	3. Commutative Law

3. Prove: $(B \cap C) \cup A = (B \cup A) \cap (C \cup A)$.

Solution:

Method 1. The dual of this theorem was proven in Problem 2. Hence the theorem is true by the Principle of Duality.

Method 2.

Statement	Reason
1. $(B \cap C) \cup A = A \cup (B \cap C)$	1. Commutative Law
2. $= (A \cup B) \cap (A \cup C)$	2. Distributive Law
3. $= (B \cup A) \cap (C \cup A)$	3. Commutative Law

4. Prove: $(A \cap B) \cup (A \cap B') = A$.

Solution:

Method 1. By the Principle of Duality, the theorem is true since its dual was proven in Example 1.1.

Method 2.

Statement	Reason
1. $(A \cap B) \cup (A \cap B') = A \cap (B \cup B')$	1. Distributive Law
2. $B \cup B' = U$	2. Complement Law
3. $(A \cap B) \cup (A \cap B') = A \cap U$	3. Substitution
4. $A \cap U = A$	4. Identity Law
5. $(A \cap B) \cup (A \cap B') = A$	5. Substitution

5. Prove: If $A \cup B = U$, then $A' \subset B$.

Solution:

Statement	Reason
1. $U \cap A' = A'$	1. Identity Law
2. $A \cup B = U$	2. Hypothesis
3. $(A \cup B) \cap A' = A'$	3. Substitution
4. $(A \cap A') \cup (B \cap A') = A'$	4. Right Distributive Law
5. $\emptyset \cup (B \cap A') = A'$	5. Complement Law
6. $A' \cap B = A'$	6. Identity Law, Commutative Law
7. $A' \subset B$	7. Definition of subset

INDEXED SETS AND GENERALIZED OPERATIONS

6. Let $A_n = \{x \mid x \text{ is a multiple of } n\}$, where $n \in N$, the natural numbers. Find: (1) $A_3 \cap A_5$; (2) $A_4 \cap A_6$; (3) $\cup_{i \in P} A_i$, where P is the set of prime numbers, 2, 3, 5, 7, 11, ...

Solution:

(1) Those numbers which are divisible by 3 and also divisible by 5 are the multiples of 15; hence

$$A_3 \cap A_5 = A_{15}$$

(2) The multiples of 12 and no other numbers are contained in both A_4 and A_6; so

$$A_4 \cap A_6 = A_{12}$$

(3) Every natural number except 1 is a multiple of at least one prime number; hence

$$\cup_{i \in P} A_i = \{2, 3, 4, \ldots\} = N - \{1\}$$

7. Let $B_i = [i, i+1]$, where $i \in Z$, the integers. Find (1) $B_1 \cup B_2$, (2) $B_3 \cap B_4$, (3) $\cup_{i=7}^{18} B_i$, (4) $\cup_{i \in Z} B_i$.

Solution:

(1) $B_1 \cup B_2$ consists of all points in the intervals [1, 2] and [2, 3]; hence

$$B_1 \cup B_2 = [1, 3]$$

(2) $B_3 \cap B_4$ consists of the points which lie in both [3, 4] and [4, 5]; thus

$$B_3 \cap B_4 = \{4\}$$

(3) $\cup_{i=7}^{18} B_i$ means the union of the sets [7, 8], [8, 9], . . ., [18, 19]; thus

$$\cup_{i=7}^{18} B_i = [7, 19]$$

(4) Since every real number belongs to at least one interval $[i, i+1]$, then $\cup_{i \in Z} B_i = R^\#$.

8. Let $D_n = (0, 1/n)$, where $n \in N$, the natural numbers. Find:

(1) $D_3 \cup D_7$ (3) $D_s \cup D_t$ (5) $\cup_{i \in A} D_i$, where A is a subset of N

(2) $D_3 \cap D_{20}$ (4) $D_s \cap D_t$ (6) $\cap_{i \in N} D_i$

Solution:

(1) Since (0, 1/3) is a superset of (0, 1/7), $D_3 \cup D_7 = D_3$.

(2) Since (0, 1/20) is a subset of (0, 1/3), $D_3 \cap D_{20} = D_{20}$.

(3) Let $m = \min(s, t)$, that is, the smaller of the two numbers s and t; then D_m is equal to D_s or D_t and contains the other as a subset. Hence $D_s \cup D_t = D_m$.

(4) Let $M = \max(s, t)$, that is, the larger of the two numbers. Then $D_s \cap D_t = D_M$.

(5) Let $a \in A$ be the smallest natural number in A. Then $\cup_{i \in A} D_i = D_a$.

(6) If x is a real number, then there is at least one number i such that $x \notin (0, 1/i)$. Hence $\cap_{i \in N} D_i = \emptyset$.

9. Prove: $B \cap (\cup_{i \in I} A_i) = \cup_{i \in I} (B \cap A_i)$.

Solution:

Let x belong to $B \cap (\cup_{i \in I} A_i)$. Then $x \in B$ and $x \in (\cup_{i \in I} A_i)$; thus there exists an i_0 such that $x \in A_{i_0}$. Hence x belongs to $B \cap A_{i_0}$, which implies x belongs to $\cup_{i \in I} (B \cap A_i)$. Since $x \in B \cap (\cup_{i \in I} A_i)$ implies $x \in \cup_{i \in I} (B \cap A_i)$,

$$B \cap (\cup_{i \in I} A_i) \subset \cup_{i \in I} (B \cap A_i)$$

Let y belong to $\cup_{i \in I} (B \cap A_i)$. Then there exists an i_0 such that $y \in B \cap A_{i_0}$; thus $y \in B$ and $y \in A_{i_0}$. Hence y is a member of $\cup_{i \in I} A_i$. Since $y \in B$ and $y \in \cup_{i \in I} A_i$, y belongs to $B \cap (\cup_{i \in I} A_i)$. Consequently,

$$\cup_{i \in I}(B \cap A_i) \subset B \cap (\cup_{i \in I} A_i)$$

By Definition 1.1,

$$B \cap (\cup_{i \in I} A_i) = \cup_{i \in I} (B \cap A_i)$$

10. Prove: Let $\{A_i\}_{i \varepsilon I}$ be any indexed family of sets and let $i_0 \varepsilon I$. Then

$$\cap_{i \varepsilon I} A_i \subset A_{i_0} \subset \cup_{i \varepsilon I} A_i$$

Solution:

Let $x \varepsilon \cap_{i \varepsilon I} A_i$; then $x \varepsilon A_i$ for every $i \varepsilon I$. In particular, $x \varepsilon A_{i_0}$. Hence

$$\cap_{i \varepsilon I} A_i \subset A_{i_0}$$

Let $y \varepsilon A_{i_0}$. Since $i_0 \varepsilon I$, $y \varepsilon \cup_{i \varepsilon I} A_i$. Consequently,

$$A_{i_0} \subset \cup_{i \varepsilon I} A_i$$

11. Let the indexed family $\{A_i\}_{i \varepsilon Z \times Z}$ of subsets of $R^\# \times R^\#$ be defined by

$$A_{(r,s)} = \{(x,y) \mid x \varepsilon R^\#,\ r \leqq x \leqq r+1,\ y \varepsilon R^\#,\ s \leqq y \leqq s+1\}$$

(1) Sketch $A_{(2,3)}$ on a coordinate diagram of $R^\# \times R^\#$.

(2) Sketch $B = \cup_{k=0}^{2} (\cup_{j=-2}^{1} A_{(j,k)}) = \cup_{k=0}^{2} \cup_{j=-2}^{1} A_{(j,k)}$ on a coordinate diagram of $R^\# \times R^\#$ and write B in set-builder notation.

(3) Sketch $C = \cup_{k \varepsilon N} \cup_{j \varepsilon N} A_{(j,k)}$ where N is the natural numbers, on a coordinate diagram of $R^\# \times R^\#$ and write C in set-builder notation.

Solution:

(1) $A_{(2,3)} = \{(x,y) \mid x \varepsilon [2,3],\ y \varepsilon [3,4]\}$, that is, the set of points whose first coordinate lies between 2 and 3 and whose second coordinate lies between 3 and 4. Thus $A_{(2,3)}$ is shaded as shown in Fig. 7-1 below.

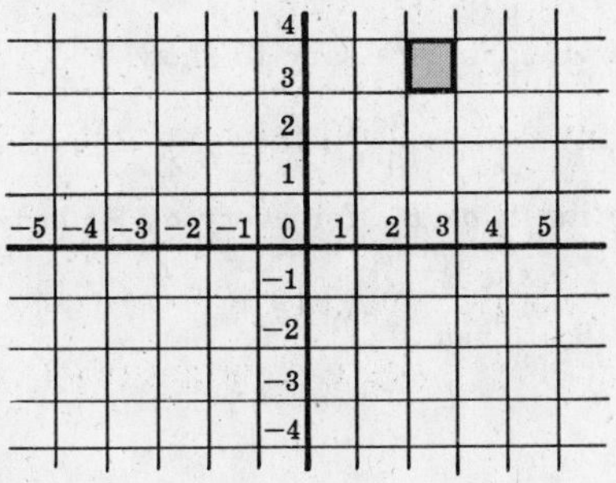

Fig. 7-1

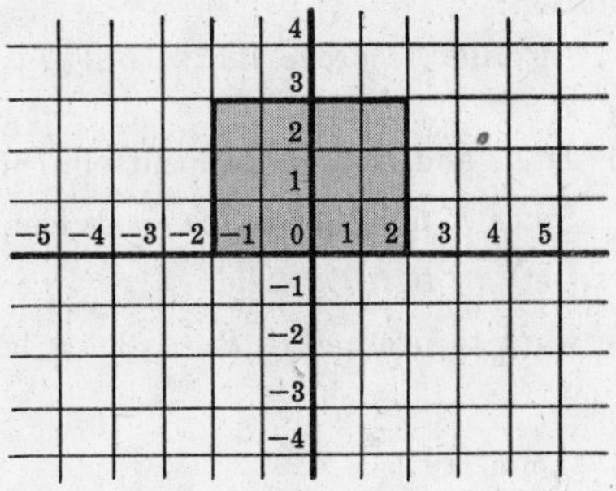

Fig. 7-2

(2) Note first that $B = \cup_{k=0}^{2} (A_{(-2,k)} \cup A_{(-1,k)} \cup A_{(0,k)} \cup A_{(1,k)})$

Hence B will consist of twelve "squares" which are shaded as shown in Fig. 7-2 above.

Thus $B = \{(x,y) \mid -2 \leqq x \leqq 2,\ 0 \leqq y \leqq 3\}$.

(3) Note first that $C = \cup_{k \varepsilon N} (A_{(1,k)} \cup A_{(2,k)} \cup A_{(3,k)} \cup \cdots)$

Thus C is shaded as shown in Fig. 7-3 below.

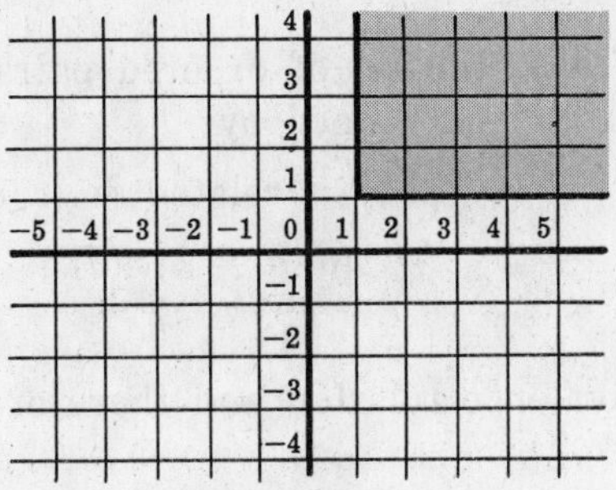

Fig. 7-3

Hence $C = \{(x,y) \mid x \geqq 1,\ y \geqq 1\}$.

PARTITIONS AND EQUIVALENCE RELATIONS

12. Let $A = \{a, b, c, d, e, f, g\}$. State whether or not each of the following families of sets is a partition of A.

(1) $\{B_1 = \{a, c, e\},\ B_2 = \{b\},\ B_3 = \{d, g\}\}$
(2) $\{C_1 = \{a, e, g\},\ C_2 = \{c, d\},\ C_3 = \{b, e, f\}\}$
(3) $\{D_1 = \{a, b, e, g\},\ D_2 = \{c\},\ D_3 = \{d, f\}\}$
(4) $\{E_1 = \{a, b, c, d, e, f, g\}\}$

Solution:

(1) Note that $A \neq B_1 \cup B_2 \cup B_3$, since $f \in A$ but $f \notin (B_1 \cup B_2 \cup B_3)$. Thus $\{B_1, B_2, B_3\}$ is not a partition of A.

(2) Note that $C_1 \neq C_3$ and, furthermore, C_1 and C_3 are not disjoint since $e \in C_1$ and $e \in C_3$. Hence $\{C_1, C_2, C_3\}$ is not a partition of A.

(3) Since $A = D_1 \cup D_2 \cup D_3$, and the sets are pairwise disjoint, $\{D_1, D_2, D_3\}$ is a partition of A.

(4) Although $\{E_1\}$ consists of only one set, it is still a partition of A. In other words, for any non-empty set A, the family $\{A\}$ is a partition of A.

13. Prove Theorem 7.2, the Fundamental Theorem on Equivalence Relations: Let R be an equivalence relation in A and, for every $\alpha \in A$, let

$$B_\alpha = \{x \mid (x, \alpha) \in R\}$$

Then the family of sets $\{B_\alpha\}_{\alpha \in A}$ is a partition of A.

Solution:

In order to prove that $\{B_\alpha\}_{\alpha \in A}$ is a partition of A, it is necessary to show:

(1) $A = \cup_{\alpha \in A} B_\alpha$

(2) If B_r and B_s have elements in common, i.e. if $B_r \cap B_s \neq \emptyset$, then $B_r = B_s$.

Since R is reflexive, i.e. each element is related to itself, $a \in B_a$, for every $a \in A$; hence (1) is true.

Let $z \in B_r \cap B_s$. Then $(z, r) \in R$ and $(z, s) \in R$

We want to prove $B_r = B_s$; so let x be any element in B_r. Then

$$(x, r) \in R$$

By symmetry, $(r, z) \in R$

and, by transitivity, $(x, r) \in R$ and $(r, z) \in R$ implies $(x, z) \in R$

and $(x, z) \in R$ and $(z, s) \in R$ implies $(x, s) \in R$

Thus x belongs to B_s. Since x was an arbitrary element in B_r, B_r is a subset of B_s. Similarly, it can be shown that B_s is a subset of B_r; hence

$$B_r = B_s$$

Consequently, $\{B_\alpha\}_{\alpha \in A}$ is a partition of A.

14. Consider the set $N \times N$, i.e. the set of ordered pairs of natural numbers. Let R be the relation in $N \times N$ which is defined by

$$(a, b) \text{ is related to } (c, d)$$

which we shall write as $(a, b) \simeq (c, d)$

if and only if $ad = bc$

Prove that R is an equivalence relation and therefore induces a partition of $N \times N$.

Solution:

Note that $(a, b) \simeq (a, b)$, since $ab = ba$; hence R is reflexive.

Suppose $(a, b) \simeq (c, d)$. Then $ad = bc$, which implies that $cb = da$. Thus $(c, d) \simeq (a, b)$ and R is symmetric.

Now suppose $(a, b) \simeq (c, d)$ and $(c, d) \simeq (e, f)$. Then $ad = bc$ and $cf = de$. Thus

$$(ad)(cf) = (bc)(de)$$

and, by cancelling from both sides,

$$af = be$$

Thus $(a, b) \simeq (e, f)$ and hence R is transitive.

Accordingly, R is an equivalence relation.

If the ordered pair (a, b) is written as a fraction $\frac{a}{b}$, then the above relation R is, in fact, the usual definition of the equality of two fractions, i.e. $\frac{a}{b} = \frac{c}{d}$, if and only if $ad = bc$.

15. Find all of the partitions of $A = \{a, b, c, d\}$.

Solution:

Note first that each partition of A contains either 1, 2, 3 or 4 different sets. The partitions are:

(1) $\{\{a, b, c, d\}\}$

(2) $\{\{a\}, \{b, c, d\}\}$, $\{\{b\}, \{a, c, d\}\}$, $\{\{c\}, \{a, b, d\}\}$, $\{\{d\}, \{a, b, c\}\}$,
$\{\{a, b\}, \{c, d\}\}$, $\{\{a, c\}, \{b, d\}\}$, $\{\{a, d\}, \{b, c\}\}$

(3) $\{\{a\}, \{b\}, \{c, d\}\}$, $\{\{a\}, \{c\}, \{b, d\}\}$, $\{\{a\}, \{d\}, \{b, c\}\}$,
$\{\{b\}, \{c\}, \{a, d\}\}$, $\{\{b\}, \{d\}, \{a, c\}\}$, $\{\{c\}, \{d\}, \{a, b\}\}$

(4) $\{\{a\}, \{b\}, \{c\}, \{d\}\}$

There are fifteen different partitions.

Supplementary Problems

ALGEBRA OF SETS AND DUALITY

16. Write the dual of each of the following:
(1) $A \cup (A \cap B) = A$, (2) $(A \cup U) \cap (A \cap \emptyset) = \emptyset$, (3) $(A \cup B) \cap (B \cup C) = (A \cap C) \cup B$.

17. Prove: $A \cup (A' \cap B) = A \cup B$.

18. Prove: $A \cap (A' \cup B) = A \cap B$.

19. Prove: $A \cup (A \cap B) = A$.

20. Prove: $A \cap (A \cup B) = A$.

INDEXED SETS AND GENERALIZED OPERATIONS

21. Let $A_n = \{x \mid x \text{ is a multiple of } n\} = \{n, 2n, 3n, \ldots\}$, where $n \in N$, the natural numbers.
Find: (1) $A_2 \cap A_7$; (2) $A_6 \cap A_8$; (3) $A_3 \cup A_{12}$; (4) $A_3 \cap A_{12}$; (5) $A_s \cup A_{st}$, where $s, t \in N$; (6) $A_s \cap A_{st}$, where $s, t \in N$.

22. Let $B_i = (i, i+1]$, a half-open interval, where $i \in Z$, the integers. Find, i.e. write in interval notation:

(1) $B_4 \cup B_5$ (3) $\cup_{i=4}^{20} B_i$ (5) $\cup_{i=0}^{15} B_{s+i}$
(2) $B_6 \cap B_7$ (4) $B_s \cup B_{s+1} \cup B_{s+2}$, $s \in Z$ (6) $\cup_{i \in Z} B_{s+i}$

23. Let $D_n = [0, 1/n]$, $S_n = (0, 1/n]$ and $T_n = [0, 1/n)$ where $n \in N$, the natural numbers. Find:
(1) $\cap_{n \in N} D_n$, (2) $\cap_{n \in N} S_n$, (3) $\cap_{n \in N} T_n$.

24. Let the indexed family $\{A_i\}_{i \in Z \times Z}$ of subsets of $R^\# \times R^\#$ be defined by

$$A_{(r,s)} = \{(x, y) \mid r \leq x \leq r+1,\ s \leq y \leq s+1\}$$

(See Problem 11.) Sketch each of the following sets on a coordinate diagram of $R^\# \times R^\#$:
(1) $A_{(1,2)} \cup A_{(1,3)} \cup A_{(1,4)}$, (2) $\cup_{i=0}^{3} A_{(2,i)}$, (3) $\cup_{i=-2}^{1} \cup_{j=0}^{3} A_{(i,j)}$, (4) $\cup_{i \in Z} (A_{(1,i)} \cup A_{(2,i)})$.

25. Prove: Let $A_n = \{n, 2n, 3n, \ldots\}$, where $n \in N$, the natural numbers, and let J be an infinite subset of N. Then $\cap_{i \in J} A_i = \emptyset$.

PARTITIONS AND EQUIVALENCE RELATIONS

26. Let $W = \{1, 2, 3, 4, 5, 6\}$. State whether or not each of the following families of sets is a partition of W.

(1) $\{\{1, 3, 5\}, \{2, 4\}, \{3, 6\}\}$
(2) $\{\{1, 5\}, \{2\}, \{3, 6\}\}$
(3) $\{\{1, 5\}, \{2\}, \{4\}, \{1, 5\}, \{3, 6\}\}$
(4) $\{\{1, 2, 3, 4, 5, 6\}\}$

27. Find all of the partitions of $V = \{1, 2, 3\}$.

28. Consider the set $N \times N$, the set of ordered pairs of natural numbers. Let R be the relation in $N \times N$ which is defined by

$$(a, b) \text{ is related to } (c, d)$$

which we shall write as

$$(a, b) \simeq (c, d)$$

if and only if

$$a + d = b + c$$

Prove that R is an equivalence relation and therefore induces a partition of $N \times N$.

Answers to Supplementary Problems

16. (1) $A \cap (A \cup B) = A$ (2) $(A \cap \emptyset) \cup (A \cup U) = U$ (3) $(A \cap B) \cup (B \cap C) = (A \cup C) \cap B$

17.

Statement	Reason
1. $A \cup (A' \cap B) = (A \cup A') \cap (A \cup B)$	1. Distributive Law
2. $A \cup A' = U$	2. Complement Law
3. $A \cup (A' \cap B) = U \cap (A \cup B)$	3. Substitution
4. $= A \cup B$	4. Identity Law

18. *Method 1.* The dual of this theorem is proven in the preceding problem; hence this theorem is true by the Principle of Duality.

Method 2.

Statement	Reason
1. $A \cap (A' \cup B) = (A \cap A') \cup (A \cap B)$	1. Distributive Law
2. $A \cap A' = \emptyset$	2. Complement Law
3. $A \cap (A' \cup B) = \emptyset \cup (A \cap B)$	3. Substitution
4. $= A \cap B$	4. Identity Law

19.

Statement	Reason
1. $A \cap U = A$	1. Identity Law
2. $A \cup (A \cap B) = (A \cap U) \cup (A \cap B)$	2. Substitution
3. $= A \cap (U \cup B)$	3. Distributive Law
4. $= A \cap U$	4. Identity Law, Substitution
5. $= A$	5. Identity Law

20. *Method 1.* The dual of this theorem is proven in the preceding problem; hence this theorem is true by the Principle of Duality.

Method 2.

	Statement		Reason
1.	$A \cup \emptyset = A$	1.	Identity Law
2.	$A \cap (A \cup B) = (A \cup \emptyset) \cap (A \cup B)$	2.	Substitution
3.	$= A \cup (\emptyset \cap B)$	3.	Distributive Law
4.	$= A \cup \emptyset$	4.	Identity Law, Substitution
5.	$= A$	5.	Identity Law

21. (1) A_{14}, (2) A_{24}, (3) A_3, (4) A_{12}, (5) A_s, (6) A_{st}

22. (1) $(4, 6]$, (2) $\emptyset$, (3) $(4, 21]$, (4) $(s, s+3]$, (5) $(s, s+16]$, (6) $(-\infty, \infty)$

23. (1) $\{0\}$, (2) $\emptyset$, (3) $\{0\}$

24.

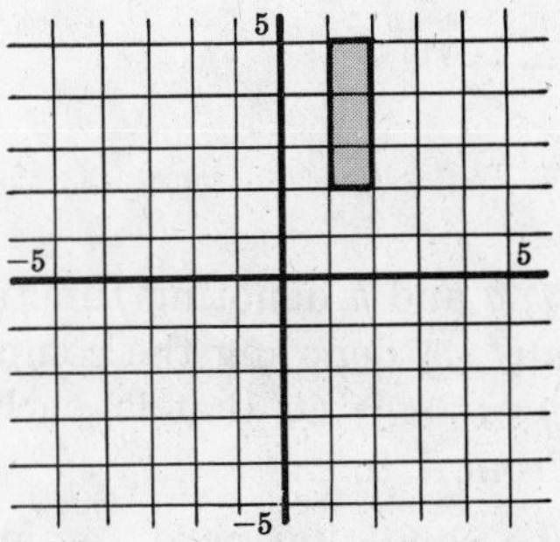

(1)

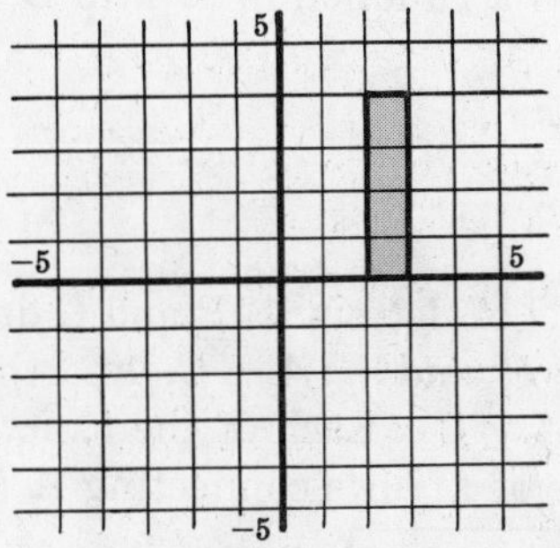

(2)

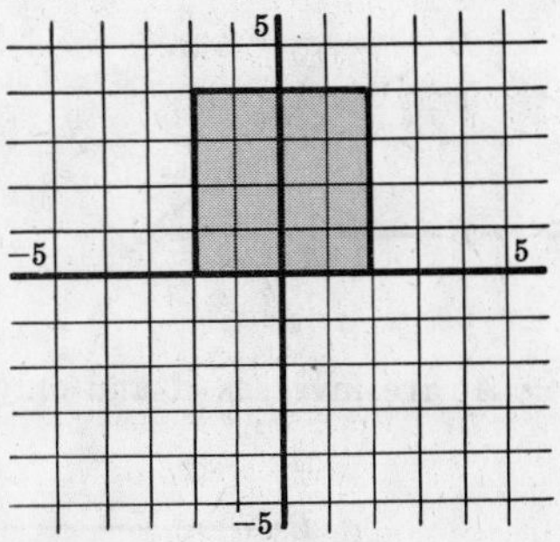

(3)

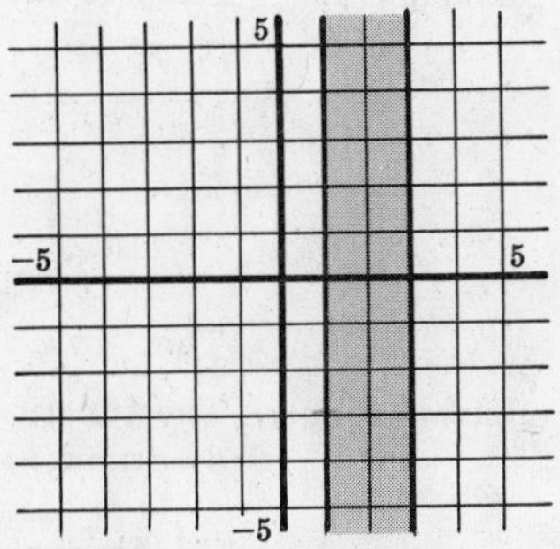

(4)

25. Let m be any natural number. Since J is infinite, there exists an $i_0 \in J$ such that $i_0 > m$. Then $m \notin A_{i_0}$ and therefore $m \notin \cap_{i \in J} A_i$.

Since m was arbitrary, $\cap_{i \in J} A_i = \emptyset$.

26. (1) No, (2) No, (3) Yes, (4) Yes

27. There are five different partitions of V:

$\{\{1, 2, 3\}\}$, $\{\{1\}, \{2, 3\}\}$, $\{\{2\}, \{1, 3\}\}$, $\{\{3\}, \{1, 2\}\}$, $\{\{1\}, \{2\}, \{3\}\}$.

28. Note that $(a, b) \simeq (a, b)$, since $a + b = b + a$; hence R is reflexive.

Suppose $(a, b) \simeq (c, d)$. Then $a + d = b + c$, which implies that $c + b = d + a$. Thus $(c, d) \simeq (a, b)$ and R is symmetric.

Now suppose $(a, b) \simeq (c, d)$ and $(c, d) \simeq (e, f)$. Then $a + d = b + c$ and $c + f = d + e$. Thus

$$(a + d) + (c + f) = (b + c) + (d + e)$$

and, subtracting $c + d$ from both sides, $a + f = b + e$. Thus $(a, b) \simeq (e, f)$ and hence R is transitive.

Accordingly, R is an equivalence relation.

Chapter 8

Further Theory of Functions, Operations

FUNCTIONS AND DIAGRAMS

As mentioned previously, the symbol

$$A \xrightarrow{f} B$$

denotes a function of A into B. In a similar manner, the diagram

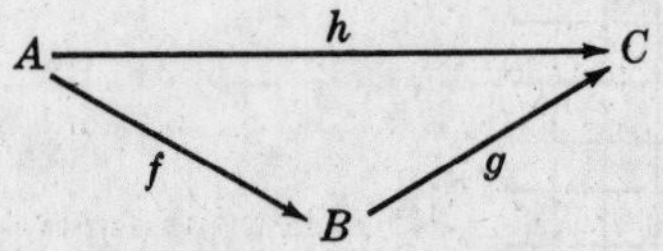

consists of letters A, B and C denoting sets, arrows f, g and h denoting functions $f: A \to B$, $g: B \to C$ and $h: A \to C$, and the sequence of arrows $\{f, g\}$ denoting the composite function $g \circ f: A \to C$. Each of the functions $h: A \to C$ and $g \circ f: A \to C$, that is, each arrow or sequence of arrows connecting A to C is called a *path* from A to C.

Definition 8.1: A diagram of functions is said to be *commutative* if for any sets X and Y in the diagram, any two paths from X to Y are equal.

Example 1.1: Suppose the following diagram of functions is commutative.

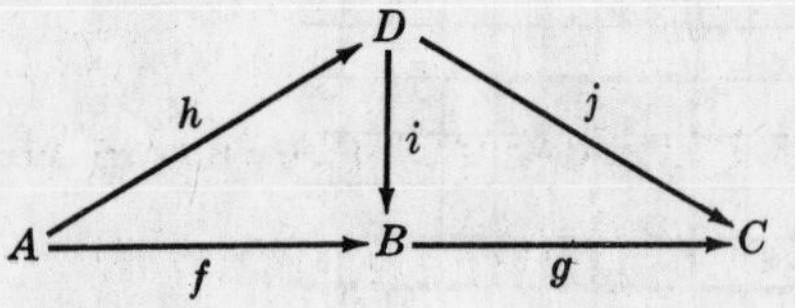

Then $i \circ h = f$, $g \circ i = j$ and $g \circ f = j \circ h = g \circ i \circ h$.

Example 1.2: The functions $f: A \to B$ and $g: B \to A$ are inverses if and only if the following diagrams are commutative:

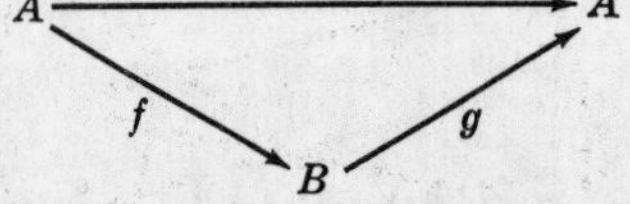

Here 1_A and 1_B are the identity functions.

RESTRICTIONS AND EXTENSIONS OF FUNCTIONS

Let f be a function of A into C, i.e. let $f: A \to C$, and let B be a subset of A. Then f induces a function $f': B \to C$ which is defined by

$$f'(b) = f(b)$$

for any $b \in B$. The function f' is called the *restriction* of f to B and is denoted by

$$f \mid B$$

Example 2.1: Let $f: R^{\#} \to R^{\#}$ be defined by $f(x) = x^2$. Then

$$f \mid N = \{(1, 1), (2, 4), (3, 9), (4, 16), \ldots\}$$

is the restriction of f to N, the natural numbers.

Example 2.2: The set $g = \{(2,5), (5,1), (3,7), (8,3), (9,5)\}$ is a function from $\{2,5,3,8,9\}$ into N. Then

$$\{(2,5), (3,7), (9,5)\}$$

a subset of g, is the restriction of g to $\{2,3,9\}$, the set of first elements of the ordered pairs in g.

We can look at this situation from another point of view. Let $f: A \to C$ and let B be a superset of A. Then a function $F: B \to C$ is called an *extension* of f if, for every $a \in A$,

$$F(a) = f(a)$$

Example 2.3: Let f be the function on the positive real numbers defined by $f(x) = x$, that is, the identity function. Then the absolute value function

$$|x| = \begin{cases} x & \text{if } x \geqq 0 \\ -x & \text{if } x < 0 \end{cases}$$

is an extension of f to all the real numbers.

Example 2.4: Consider the function

$$f = \{(1,2), (3,4), (7,2)\}$$

whose domain is $\{1,3,7\}$. Then the function

$$F = \{(1,2), (3,4), (5,6), (7,2)\}$$

which is a superset of the function f, is an extension of f.

SET FUNCTIONS

Let f be a function of A into B and let T be a subset of A, that is, $A \xrightarrow{f} B$ and $T \subset A$. Then

$$f(T)$$

which is read "f of T", is defined to be the set of image points of elements in T. In other words,

$$f(T) = \{x \mid f(a) = x,\ a \in T,\ x \in B\}$$

Notice that $f(T)$ is a subset of B.

Example 3.1: Let $A = \{a, b, c, d\}$, $T = \{b, c\}$ and $B = \{1, 2, 3\}$. Define $f: A \to B$ by

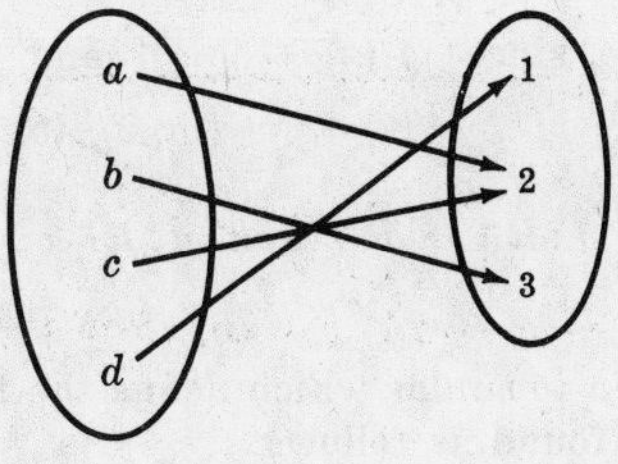

Then $f(T) = \{2, 3\}$.

Example 3.2: Let $g: R^\# \to R^\#$ be defined by $g(x) = x^2$, and let $T = [3, 4]$. Then

$$g(T) = [9, 16] = \{x \mid 9 \leqq x \leqq 16\}$$

Now let $\mathcal{A}$ be the family of subsets of A, and let $\mathcal{B}$ be the family of subsets of B. If $f: A \to B$, then f assigns to each set $T \in \mathcal{A}$ a unique set $f(T) \in \mathcal{B}$. In other words, the function $f: A \to B$ induces a function $f: \mathcal{A} \to \mathcal{B}$. Although each function is denoted by the same letter f, they are essentially two different functions. Notice that the domain of $f: \mathcal{A} \to \mathcal{B}$ consists of sets.

Generally speaking, a function is called a *set function* if its domain consists of sets.

REAL-VALUED FUNCTIONS

A function $f: A \to R^{\#}$, which maps a set into the real numbers, i.e. which assigns to each $a \in A$ a real number $f(a) \in R^{\#}$, is called a *real-valued function.* Those functions which are usually studied in elementary mathematics, e.g.,

$$\begin{aligned} p(x) &= a_0x^n + a_1x^{n-1} + \cdots + a_{n-1}x + a_n \\ t(x) &= \sin x, \ \cos x \ \text{ or } \ \tan x \\ f(x) &= \log x \ \text{ or } \ e^x \end{aligned}$$

that is, polynomials, trigonometric functions and logarithmic and exponential functions, are special examples of real-valued functions.

ALGEBRA OF REAL-VALUED FUNCTIONS

Let $\mathcal{J}_D$ be the family of all real-valued functions with the same domain D. Then many (algebraic) operations are defined in $\mathcal{J}_D$. Specifically, let $f: D \to R^{\#}$ and $g: D \to R^{\#}$, and let $k \in R^{\#}$. Then each of the following functions is defined as follows:

$$\begin{aligned} (f+k): D \to R^{\#} \quad &\text{by} \quad (f+k)(x) \equiv f(x) + k \\ (|f|): D \to R^{\#} \quad &\text{by} \quad (|f|)(x) \equiv |f(x)| \\ (f^n): D \to R^{\#} \quad &\text{by} \quad (f^n)(x) \equiv (f(x))^n \\ (f \pm g): D \to R^{\#} \quad &\text{by} \quad (f \pm g)(x) \equiv f(x) \pm g(x) \\ (kf): D \to R^{\#} \quad &\text{by} \quad (kf)(x) \equiv k(f(x)) \\ (fg): D \to R^{\#} \quad &\text{by} \quad (fg)(x) \equiv f(x)g(x) \\ (f/g): D \to R^{\#} \quad &\text{by} \quad (f/g)(x) \equiv f(x)/g(x) \qquad (\text{where } g(x) \neq 0) \end{aligned}$$

Note that $(fg): D \to R^{\#}$ is not the same as the composition function which was discussed previously.

Example 4.1: Let $D = \{a, b\}$, and let $f: D \to R^{\#}$ and $g: D \to R^{\#}$ be defined by:

$$f(a) = 1, \ f(b) = 3 \quad \text{and} \quad g(a) = 2, \ g(b) = -1$$

In other words,

$$f = \{(a, 1), \ (b, 3)\} \quad \text{and} \quad g = \{(a, 2), \ (b, -1)\}$$

Then

$$\begin{aligned} (3f - 2g)(a) &\equiv 3f(a) - 2g(a) = 3(1) - 2(2) = -1 \\ (3f - 2g)(b) &\equiv 3f(b) - 2g(b) = 3(3) - 2(-1) = 11 \end{aligned}$$

that is, $3f - 2g = \{(a, -1), (b, 11)\}$

Also, since $|g|(x) \equiv |g(x)|$ and $(g+3)(x) \equiv g(x) + 3$,

$$|g| = \{(a, 2), \ (b, 1)\} \quad \text{and} \quad g + 3 = \{(a, 5), \ (b, 2)\}$$

Example 4.2: Let $f: R^{\#} \to R^{\#}$ and $g: R^{\#} \to R^{\#}$ be defined by the formulas

$$f(x) = 2x - 1 \quad \text{and} \quad g(x) = x^2$$

Then formulas which define the functions $(3f - 2g): R^{\#} \to R^{\#}$ and $(fg): R^{\#} \to R^{\#}$ are found as follows:

$$\begin{aligned} (3f - 2g)(x) &= 3(2x - 1) - 2(x^2) = -2x^2 + 6x - 3 \\ (fg)(x) &= (2x - 1)(x^2) = 2x^3 - x^2 \end{aligned}$$

RULE OF THE MAXIMUM DOMAIN

A formula of the form

$$f(x) = 1/x, \quad g(x) = \sin x, \quad h(x) = \sqrt{x}$$

does not, in itself, define a function unless there is given, explicitly or implicitly, a domain, i.e. a set of numbers, on which the formula then defines a function. Hence the following expressions appear:

Let $f(x) = x^2$ be defined on $[-2, 4]$.
Let $g(x) = \sin x$ be defined for $0 \leq x \leq 2\pi$.

However, if the domain of a function defined by a formula is the maximum set of real numbers for which the formula yields a real number, e.g.,

$$\text{let} \quad f(x) = 1/x \text{ for } x \neq 0$$

then the domain is usually not stated explicitly. This convention is sometimes called the *rule of the maximum domain*.

Example 5.1: Consider the following functions:

$$\begin{aligned} f_1(x) &= x^2 && \text{for } x \geq 0 \\ f_2(x) &= 1/(x-2) && \text{for } x \neq 2 \\ f_3(x) &= \cos x && \text{for } 0 \leq x \leq 2\pi \\ f_4(x) &= \tan x && \text{for } x \neq \pi/2 + n\pi,\ n \in N \end{aligned}$$

The domains of f_2 and f_4 need not have been explicitly stated since each consists of all those numbers for which the formula has meaning, that is, the functions could have been defined by writing

$$f_2(x) = 1/(x-2) \quad \text{and} \quad f_4(x) = \tan x$$

Example 5.2: Consider the function $f(x) = \sqrt{1-x^2}$; its domain, unless otherwise stated, is $[-1, 1]$. Here we implicitly assume that the co-domain is $R^\#$.

CHARACTERISTIC FUNCTIONS

Let A be any subset of a universal set U. Then the real-valued function

$$\chi_A : U \to \{1, 0\}$$

defined by

$$\chi_A(x) = \begin{cases} 1 & \text{if } x \in A \\ 0 & \text{if } x \notin A \end{cases}$$

is called the *characteristic function* of A.

Example 6.1: Let $U = \{a, b, c, d, e\}$ and $A = \{a, d, e\}$. Then the function of U into $\{1, 0\}$ defined by the following diagram

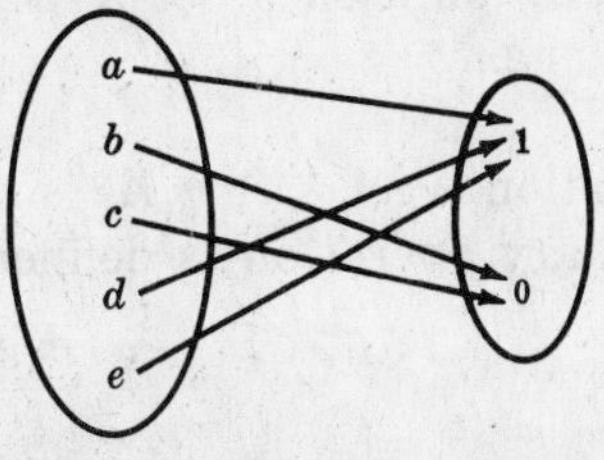

is the characteristic function χ_A of A.

Note further that any function $f : U \to \{1, 0\}$ defines a subset

$$A_f = \{x \mid x \in U,\ f(x) = 1\}$$

of U and that the characteristic function χ_{A_f} of A_f is the original function f. Thus there is a one-to-one correspondence between all subsets of U, i.e. the power set of U, and the set of all functions of U into $\{1, 0\}$.

CHOICE FUNCTIONS

Let $\{A_i\}_{i \in I}$ be a family of non-empty subsets of B. Then a function

$$f : \{A_i\}_{i \in I} \to B$$

is called a *choice function* if, for every $i \in I$,

$$f(A_i) \in A_i$$

that is, if the image of each set is an element in the set.

Example 7.1: Consider the following subsets,

$$A_1 = \{1, 2, 3\}, \quad A_2 = \{1, 3, 4\}, \quad A_3 = \{2, 5\}$$

of $B = \{1, 2, 3, 4, 5\}$ and consider the following functions of $\{A_1, A_2, A_3\}$ into B:

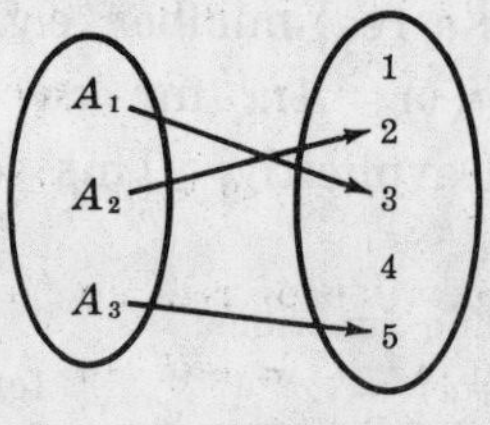

f

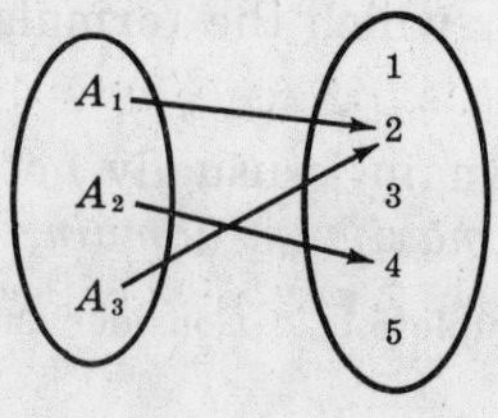

g

Note that f is not a choice function since $f(A_2) = 2$ does not belong to A_2, i.e. $f(A_2) \notin A_2$. Note further that g is a choice function since $g(A_1) \in A_1$, $g(A_2) \in A_2$ and $g(A_3) \in A_3$.

Remark 8.1: Essentially a choice function, for a given family of sets, "chooses" an element from each set in the family. Whether or not a choice function exists for an arbitrary family of sets, is a question which lies at the foundation of the theory of sets. Chapter 11 will be devoted to this question.

OPERATIONS

The reader is familiar with the operations of addition and multiplication of numbers, union and intersection of sets, and composition of functions. These operations are denoted as follows:

$$a+b = c, \quad a \cdot b = c, \quad A \cup B = C, \quad A \cap B = C, \quad g \circ f = h$$

In each situation, an element (c, C or h) is assigned to an original pair of elements. In other words, there is a function that assigns an element to each ordered pair of elements. We now introduce the precise

Definition 8.2: An operation α on a set A is a function of the Cartesian product $A \times A$ into A, i.e.,

$$\alpha : A \times A \to A$$

Remark 8.2: The operation $\alpha : A \times A \to A$ is sometimes referred to as a *binary operation*, and an n-ary operation is defined to be a function

$$\alpha : \underbrace{A \times A \times \cdots \times A}_{(n \text{ times})} \to A$$

We shall continue to use the word operation instead of binary operation.

COMMUTATIVE OPERATIONS

The operation $\alpha : A \times A \to A$ is called *commutative* if, for every $a, b \in A$,

$$\alpha(a, b) = \alpha(b, a)$$

Example 8.1: Addition and multiplication of real numbers are commutative operations since

$$a + b = b + a \quad \text{and} \quad ab = ba$$

Example 8.2: Let $\alpha : R^\# \times R^\# \to R^\#$ be the operation of subtraction defined by $\alpha : (x, y) \to x - y$. Then

$$\alpha(5, 1) = 4 \quad \text{and} \quad \alpha(1, 5) = -4$$

Hence subtraction is not a commutative operation.

Example 8.3: Union and intersection of sets are commutative operations, since

$$A \cup B = B \cup A \quad \text{and} \quad A \cap B = B \cap A$$

ASSOCIATIVE OPERATIONS

The operation $\alpha: A \times A \to A$ is called *associative* if, for every $a, b, c \in A$,

$$\alpha(\alpha(a, b), c) = \alpha(a, \alpha(b, c))$$

In other words, if $\alpha(a, b)$ is written $a * b$, then α is associative if

$$(a * b) * c = a * (b * c)$$

Example 9.1: Addition and multiplication of real numbers are associative operations, since

$$(a + b) + c = a + (b + c) \quad \text{and} \quad (ab)c = a(bc)$$

Example 9.2: Let $\alpha: R^\# \times R^\# \to R^\#$ be the operation of division defined by $\alpha: (x, y) \to x/y$. Then

$$\alpha(\alpha(12, 6), 2) = \alpha(2, 2) = 1$$
$$\alpha(12, \alpha(6, 2)) = \alpha(12, 3) = 4$$

Hence division is not an associative operation.

Example 9.3: Union and intersection of sets are associative operations, since

$$(A \cup B) \cup C = A \cup (B \cup C) \quad \text{and} \quad (A \cap B) \cap C = A \cap (B \cap C)$$

DISTRIBUTIVE OPERATIONS

Consider the following two operations:

$$\alpha: A \times A \to A$$
$$\beta: A \times A \to A$$

The operation α is said to *distribute over* the operation β if, for every $a, b, c \in A$,

$$\alpha(a, \beta(b, c)) = \beta(\alpha(a, b), \alpha(a, c))$$

In other words, if $\alpha(a, b)$ is written $a * b$, and $\beta(a, b)$ is written $a \Delta b$, then α distributes over β if

$$a * (b \Delta c) = (a * b) \Delta (a * c)$$

Example 10.1: The operation of multiplication of real numbers distributes over the operation of addition, since

$$a(b + c) = ab + ac$$

But the operation of addition of real numbers does not distribute over the operation of multiplication, since

$$a + (bc) \neq (a + b)(a + c)$$

Example 10.2: The operations of union and intersection of sets distribute over each other since

$$A \cup (B \cap C) = (A \cup B) \cap (A \cup C)$$
$$A \cap (B \cup C) = (A \cap B) \cup (A \cap C)$$

IDENTITY ELEMENTS

Let $\alpha: A \times A \to A$ be an operation written $\alpha(a, b) = a * b$. An element $e \in A$ is called an *identity element* for the operation α if, for every element $a \in A$,

$$e * a = a * e = a$$

Example 11.1: Let $\alpha: R^\# \times R^\# \to R^\#$ be the operation of addition. Then 0 is an identity element for addition since, for every real number $a \in R^\#$,

$$0 * a = a * 0 = a, \quad \text{that is,} \quad 0 + a = a + 0 = a$$

Example 11.2: Consider the operation of intersection of sets. Then U, the universal set, is an identity element, since for every set A (which is a subset of U),

$$U * A = A * U = A, \quad \text{that is,} \quad U \cap A = A \cap U = A$$

Example 11.3: Consider the operation of multiplication of real numbers. Then the number 1 is an identity element since, for every real number a,

$$1 * a = a * 1 = a, \quad \text{that is,} \quad 1 \cdot a = a \cdot 1 = a$$

Theorem 8.1: If an operation $\alpha : A \times A \rightarrow A$ has an identity element $e \in A$, then it is the only identity element.

Thus we can speak of *the* identity element for an operation instead of *an* identity element.

INVERSE ELEMENTS

Let $\alpha : A \times A \rightarrow A$ be an operation written $\alpha(a, b) = a * b$, and let $e \in A$ be the identity element for α. Then the *inverse* of an element $a \in A$, denoted by

$$a^{-1}$$

is an element in A with the following property:

$$a^{-1} * a = a * a^{-1} = e$$

Example 12.1: Consider the operation of addition of real numbers for which 0 is the identity element. Then, for any real number a, its negative $(-a)$ is its *additive inverse* since

$$-a * a = a * -a = 0, \quad \text{that is,} \quad (-a) + a = a + (-a) = 0$$

Example 12.2: Consider the operation of multiplication of rational numbers, for which 1 is the identity element. Then for any non-zero rational number p/q, where p and q are integers, its reciprocal q/p is its *multiplicative inverse*, since

$$(q/p)(p/q) = (p/q)(q/p) = 1$$

Example 12.3: Let $\alpha : N \times N \rightarrow N$ be the operation of multiplication for which 1 is the identity element; here N is the set of natural numbers. Then 2 has no multiplicative inverse, since there is no element $x \in N$ with the property

$$x \cdot 2 = 2 \cdot x = 1$$

In fact, no element in N has a multiplicative inverse except 1 which has itself as an inverse.

OPERATIONS AND SUBSETS

Consider an operation $\alpha : A \times A \rightarrow A$ and a subset B of A. Then B is said to be *closed under the operation of* α if, for every $b, b' \in B$,

$$\alpha(b, b') \in B$$

that is, if

$$\alpha(B \times B) \subset B$$

Example 13.1: Consider the operation of addition of natural numbers. Then the set of even numbers is closed under the operation of addition since the sum of any two even numbers is always even. Moreover, the set of odd numbers is not closed under the operation of addition since the sum of two odd numbers is not odd.

Example 13.2: The four complex numbers $1, -1, i, -i$ are closed under the operation of multiplication.

Solved Problems

DIAGRAMS AND FUNCTIONS

1. In the adjoining diagram of functions, how many paths are there from A to E and what are they?

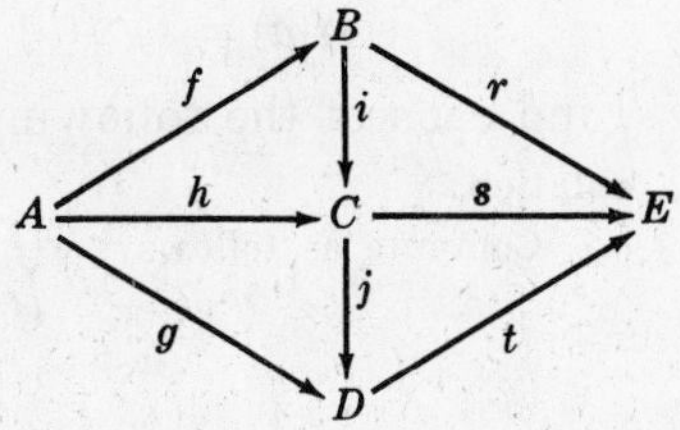

Solution:

There are six paths from A to E:

$A \xrightarrow{f} B \xrightarrow{r} E$ $\quad$ $A \xrightarrow{h} C \xrightarrow{s} E$

$A \xrightarrow{f} B \xrightarrow{i} C \xrightarrow{s} E$ $\quad$ $A \xrightarrow{h} C \xrightarrow{j} D \xrightarrow{t} E$

$A \xrightarrow{f} B \xrightarrow{i} C \xrightarrow{j} D \xrightarrow{t} E$ $\quad$ $A \xrightarrow{g} D \xrightarrow{t} E$

that is, $\quad r \circ f, \quad s \circ i \circ f, \quad t \circ j \circ i \circ f, \quad s \circ h, \quad t \circ j \circ h, \quad t \circ g$

As noted previously, the functions are written from right to left.

2. Suppose the adjoining diagram is commutative. Here 1_A is the identity function on A. State all the information that is inferred from the diagram.

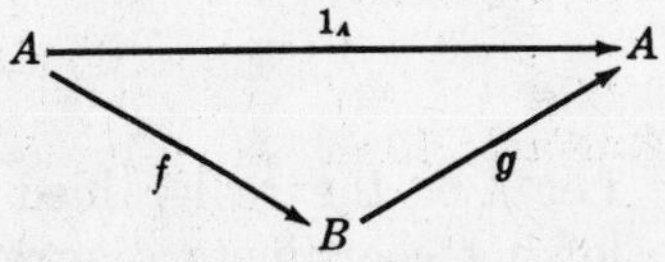

Solution:

First, since the diagram is commutative, $g \circ f = 1_A$.

Furthermore, since $g \circ f$ is one-one, f must also be one-one; and since $g \circ f$ is onto, g must also be onto. It need not be true that $g = f^{-1}$, since we do not know whether $f \circ g = 1_B$.

SET FUNCTIONS

3. Let $W = \{a, b, c, d\}$, $V = \{1, 2, 3\}$, and let $f: W \to V$ be defined by the adjoining diagram. Find: (1) $f(\{a, b, d\})$, (2) $f(\{a, c\})$.

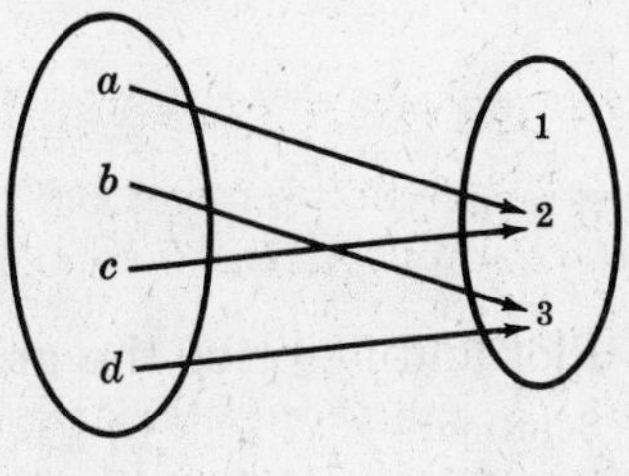

Solution:

(1) Compute as follows:

$$f(\{a, b, d\}) = \{f(a), f(b), f(d)\} = \{2, 3, 3\} = \{2, 3\}$$

(2) Similarly,

$$f(\{a, c\}) = \{f(a), f(c)\} = \{2, 2\} = \{2\}$$

Note that $f(\{a, c\}) = 2$ is an *incorrect* statement since the image of the set $\{a, c\}$, in this situation, is a subset of V, i.e. $\{2\}$, and not an element in V.

4. Prove: Let $f: A \to B$ be one-one. Then the induced set function $f: 2^A \to 2^B$ is also one-one. Here 2^A and 2^B are the power sets of A and B respectively.

Solution:

Let X and Y be two different subsets of A, i.e.,

$$X \in 2^A, \quad Y \in 2^A, \quad X \neq Y$$

Then there exists an element $z \in A$ with the property that

$$z \in X, \quad z \notin Y \quad (\text{or } z \in Y, \ z \notin X)$$

Thus $f(z) \in f(X)$ and, since f is one-one, $f(z) \notin f(Y)$ (or $f(z) \in f(Y)$ and $f(z) \notin f(X)$). Hence $f(X) \neq f(Y)$ and, by definition, the induced set function is also one-one.

5. Prove: Let $f: A \to B$ be onto. Then the induced set function $f: 2^A \to 2^B$ is also an onto function.

Solution:

It is necessary to show that each set in 2^B is the image of at least one set in 2^A. Let $Y \in 2^B$. Since f is onto,

$$f^{-1}(Y) = \{x \mid x \in A, \ f(x) \in Y\}$$

is not empty. But Y is the image of $f^{-1}(Y)$, i.e. $f(f^{-1}(Y)) = Y$. Hence $f: 2^A \to 2^B$ is onto.

REAL-VALUED FUNCTIONS

6. Let $W = \{a, b, c\}$, and let f and g be the following real-valued functions on W:

$$f(a) = 1,\ f(b) = -2,\ f(c) = 3 \qquad g(a) = -2,\ g(b) = 0,\ g(c) = 1$$

Find each of the following functions: (1) $f + 2g$, (2) $fg - 2f$.

Solution:

(1) Compute as follows:

$$(f + 2g)(a) \equiv f(a) + 2g(a) = 1 - 4 = -3$$
$$(f + 2g)(b) \equiv f(b) + 2g(b) = -2 + 0 = -2$$
$$(f + 2g)(c) \equiv f(c) + 2g(c) = 3 + 2 = 5$$

Thus $f + 2g = \{(a, -3),\ (b, -2),\ (c, 5)\}$.

(2) Similarly,

$$(fg - 2f)(a) \equiv f(a)\,g(a) - 2f(a) = (1)(-2) - 2(1) = -4$$
$$(fg - 2f)(b) \equiv f(b)\,g(b) - 2f(b) = (-2)(0) - 2(-2) = 4$$
$$(fg - 2f)(c) \equiv f(c)\,g(c) - 2f(c) = (3)(1) - 2(3) = -3$$

Hence $fg - 2f = \{(a, -4),\ (b, 4),\ (c, -3)\}$.

7. Let f be the real-valued function with domain $[-3, 3]$ which is displayed on the following coordinate diagram:

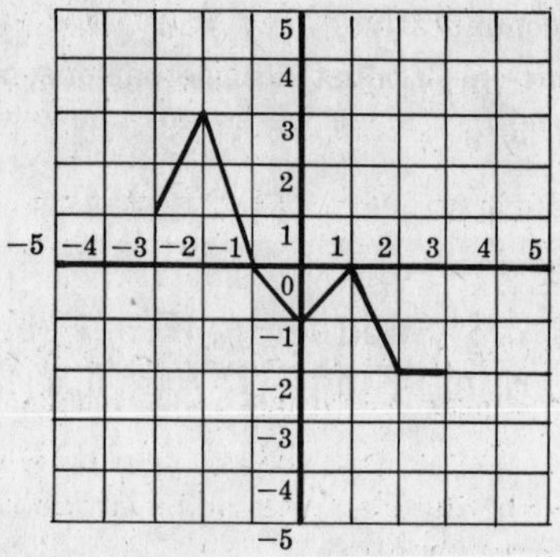

Plot and describe the graph of each of the following functions: (1) $f + 2$, (2) $|f|$.

Solution:

(1) Since, by definition, $(f + 2)(x) \equiv f(x) + 2$, each image value of the original function is increased by 2. Hence raise the entire graph of f two units in order to obtain the graph of $f + 2$, as shown in Fig. 8-1 below.

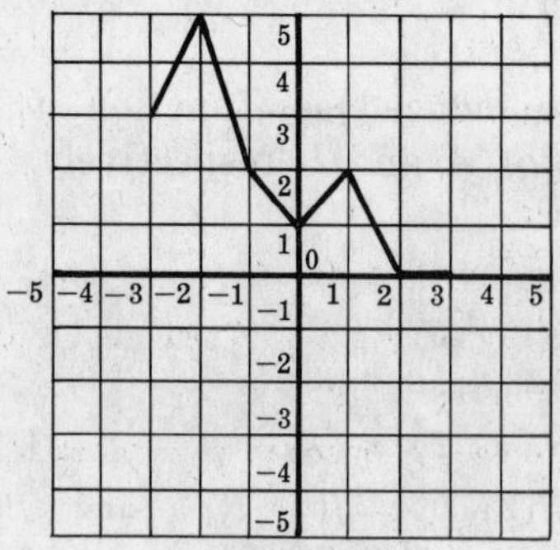

$f + 2$ is displayed

Fig. 8-1

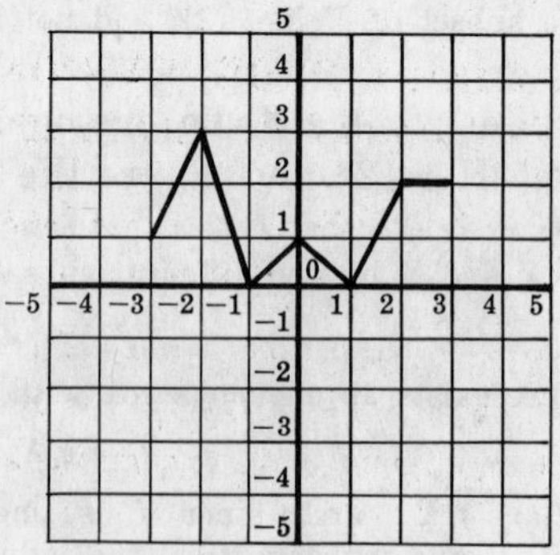

$|f|$ is displayed

Fig. 8-2

(2) Note that

$$(|f|)(x) \equiv |f(x)| = \begin{cases} f(x) & \text{if } f(x) \geqq 0 \\ -f(x) & \text{if } f(x) < 0 \end{cases}$$

Hence part of the graph of $|f|$ is identical with that part of the graph of f which appears above the x-axis; and the remainder of the graph of $|f|$ is the reflection, over the x-axis, of that portion of the graph of f which appears below the x-axis. See Fig. 8-2 above.

8. Find the domain of each of the following real-valued functions:

(1) $f_1(x) = 1/x$ where $x > 0$ (3) $f_3(x) = \log(x-1)$

(2) $f_2(x) = \sqrt{3-x}$ (4) $f_4(x) = x^2$ where $0 \leqq x \leqq 4$

Solution:

(1) The domain is explicitly given as $\{x \mid x > 0\}$.

(2) No domain is explicitly given, so we apply the rule of the maximum domain. Since f_2 assumes real values only when $3 - x \geqq 0$, that is, when $x \leqq 3$, the domain of f_2 is $\{x \mid x \leqq 3\}$.

(3) No domain is explicitly given, so we again apply the rule of the maximum domain. Note that the logarithmic function is defined only on positive numbers. Hence f_3 has meaning only if $x - 1 > 0$, i.e. if $x > 1$. Thus the domain of f_3 is $\{x \mid x > 1\}$.

(4) The domain is explicitly given as $\{x \mid 0 \leqq x \leqq 4\}$.

9. The real-valued function $0_A : A \to R^\#$ which is defined by

$$0_A(x) = 0 \quad \text{for every } x \in A$$

is called the *zero function* (on A). Prove that for any function $f : A \to R^\#$,

$$(1)\ f + 0_A = f \qquad \text{and} \qquad (2)\ f \cdot 0_A = 0_A$$

Solution:

(1) Since $(f + 0_A)(x) \equiv f(x) + 0_A(x) = f(x) + 0 = f(x)$ for every $x \in A$, $f + 0_A = f$.

(2) Also, $(f \cdot 0_A)(x) \equiv f(x) \cdot 0_A(x) = f(x) \cdot 0 = 0 \equiv 0_A(x)$ for every $x \in A$. Hence $f \cdot 0_A = 0_A$.

Note that the zero function has properties which are very similar to properties of the number 0.

10. Consider the real-valued function $f = \{(1, 2), (2, -3), (3, -1)\}$ (with domain $\{1, 2, 3\}$). Find (1) $f + 4$, (2) $|f|$, (3) f^2.

Solution:

(1) Since, by definition, $(f + 4)(x) \equiv f(x) + 4$, add 4 to each of the image values, i.e. add 4 to the second element of each ordered pair belonging to f. Hence

$$f + 4 = \{(1, 6), (2, 1), (3, 3)\}$$

(2) Since $|f|(x) = |f(x)|$, substitute the absolute value of the second element for the second element in each ordered pair in f. Accordingly,

$$|f| = \{(1, 2), (2, 3), (3, 1)\}$$

(3) Since $f^2(x) = (f(x))^2$, substitute the square of the second element for the second element in each ordered pair in f. Consequently,

$$f^2 = \{(1, 4), (2, 9), (3, 1)\}$$

MISCELLANEOUS PROBLEMS ON FUNCTIONS

11. Prove: Let A and B be subsets of a universal set U. Then $\chi_{A \cap B} = \chi_A \chi_B$.

Solution:

Let $x \in A \cap B$; therefore $x \in A$ and $x \in B$. Then

$$\chi_{A\cap B}(x) = 1, \qquad (\chi_A \chi_B)(x) \equiv \chi_A(x)\,\chi_B(x) = (1)(1) = 1$$

Let $y \in (A \cap B)'$; thus $y \in A' \cup B'$; and hence $y \in A'$ or $y \in B'$. Then $\chi_{A\cap B}(y) = 0$. Also, $(\chi_A \chi_B)(y) = \chi_A(y)\,\chi_B(y) = 0$ since $\chi_A(y) = 0$ or $\chi_B(y) = 0$.

Thus $\chi_{A\cap B}$ and $\chi_A \chi_B$ assign the same number to each element in U. Hence by definition,

$$\chi_{A\cap B} = \chi_A \chi_B$$

12. Consider the function: $f(x) = x$ where $x \geqq 0$. State whether or not each of the following functions is an extension of f.

(1) $g_1(x) = x$ where $x \geqq -2$ (3) $1 : R^\# \to R^\#$ (5) $g_4(x) = x$ where $x \in [-1, 1]$

(2) $g_2(x) = |x|$ for all $x \in R^\#$ (4) $g_3(x) = (x + |x|)/2$

Solution:

Note that a function f' is an extension of f if, first, the domain of f' is a superset of $[0, \infty)$, the domain of f, and, secondly, if $f'(x) = x$ for every $x \in [0, \infty)$.

(1) Since g_1 satisfies both of the above conditions, g_1 is an extension of f.

(2) Since $g_2(x) = |x| = \begin{cases} x & \text{if } x \in [0, \infty) \\ -x & \text{if } x < 0 \end{cases}$,

the absolute value function is an extension of f.

(3) By definition of an identity function, $1(x) = x$ for every $x \in R^\#$. Thus the identity function is an extension of f.

(4) Since $g_3(x) = (x + |x|)/2 = \begin{cases} (x+x)/2 = x & \text{if } x \in [0, \infty) \\ (x-x)/2 = 0 & \text{if } x < 0 \end{cases}$, g_3 is an extension of f.

(5) The domain of g_4 is not a superset of the domain of f; hence g_4 is not an extension of f.

13. As sets, what is the relationship between a function $f: A \to B$ and the restriction of f to a subset A' of A?

Solution:

The restriction of f to A', i.e. $f|A'$, is a subset of f. For $x \in A'$ implies $x \in A$ and, hence,

$$(x, f(x)) \in f|A' \quad \text{implies} \quad (x, f(x)) \in f$$

14. Consider the subsets $A_1 = \{1,2,3\}$, $A_2 = \{1,5\}$, $A_3 = \{2,4,5\}$ and $A_4 = \{3,4\}$ of $B = \{1,2,3,4,5\}$. State whether or not each of the following functions of $\{A_1, A_2, A_3, A_4\}$ into B is a choice function.

$$\begin{aligned} (1) \quad f_1 &= \{(A_1, 1), (A_2, 2), (A_3, 3), (A_4, 4)\} \\ (2) \quad f_2 &= \{(A_1, 1), (A_2, 1), (A_3, 4), (A_4, 4)\} \\ (3) \quad f_3 &= \{(A_1, 2), (A_2, 1), (A_3, 4), (A_4, 3)\} \\ (4) \quad f_4 &= \{(A_1, 3), (A_2, 5), (A_3, 1), (A_4, 3)\} \end{aligned}$$

Solution:

(1) Since $f_1(A_2) = 2$ is not a member of A_2, f_1 is not a choice function.

(2) Note that $f_2(A_i)$ belongs to A_i, for each i; hence f_2 is a choice function.

(3) Also, $f_3(A_i) \in A_i$ for each i; hence f_3 is a choice function.

(4) Note that $f_4(A_3) = 1$ does not belong to A_3; thus f_4 is not a choice function.

OPERATIONS

15. Let $\alpha: N \times N \to N$ be the operation of least common multiple (l.c.m.), i.e.,

$$\alpha(a, b) \equiv a * b = \text{l.c.m. of } a \text{ and } b$$

(1) Is α commutative? (2) Is α associative? (3) Find the identity element of α. (4) Which elements in N, if any, have inverses and what are they?

Solution:

(1) Since the l.c.m. of a and b is the l.c.m. of b and a, α is commutative.

(2) It is proven in number theory that $(a * b) * c = a * (b * c)$, i.e. that the operation of l.c.m. is associative.

(3) The number 1 is an identity element since the l.c.m. of 1 and any number a is a, that is, $1 * a = a$ for every $a \in N$.

(4) Since the l.c.m. of two numbers a and b is 1 if and only if $a = 1$ and $b = 1$, the only number which has an inverse is 1 and it is its own inverse.

16. Consider the operation $\alpha: Q \times Q \to Q$ which is denoted and defined by

$$\alpha(a, b) \equiv a * b \equiv a + b - ab$$

Here Q is the set of rational numbers. (1) Is α commutative? (2) Is α associative? (3) Find the identity element for α. (4) Do any of the elements in Q have an inverse and what is it?

Solution:

(1)
$$\begin{aligned} a * b &= a + b - ab \\ b * a &= b + a - ba \end{aligned}$$

Thus α is commutative since addition is associative and multiplication is commutative, i.e. $ab = ba$.

(2)
$$\begin{aligned}(a * b) * c &= (a + b - ab) * c = (a + b - ab) + c - (a + b - ab)c \\ &= (a + b - ab) + c - (a + b - ab)c \\ &= a + b - ab + c - ac - bc + abc \\ &= a + b + c - ab - ac - bc + abc\end{aligned}$$

$$\begin{aligned}a * (b * c) &= a * (b + c - bc) \\ &= a + (b + c - bc) - a(b + c - bc) \\ &= a + b + c - bc - ab - ac + abc\end{aligned}$$

Hence α is associative.

(3) An element e is an identity element for α if $a * e = a$ for every element $a \in Q$. Compute as follows:

$$a * e = a, \quad a + e - ae = a, \quad e - ea = 0, \quad e(1 - a) = 0, \quad e = 0$$

Accordingly, 0 is the identity element.

(4) In order for a to have an inverse x, we must have $a * x = 0$, since, by (3), 0 is the identity element. Compute as follows:

$$a * x = 0, \quad a + x - ax = 0, \quad a = ax - x, \quad a = x(a - 1), \quad x = a/(a - 1)$$

Thus if $a \neq 1$, then a has an inverse and it is $a/(a - 1)$.

17. Prove Theorem 8.1: If e and e' are identity elements (for the same operation), then $e = e'$.

Solution:

By hypothesis, $e * e' = e'$ and $e * e' = e$. Thus $e = e * e' = e'$.

18. Consider the operation of union of sets. (1) Find the identity element. (2) Which elements, if any, have inverses and what are they?

Solution:

(1) Note that $A \cup \emptyset = \emptyset \cup A = A$ for any set A. Hence the null set $\emptyset$ is the identity element for the operation of union of sets.

(2) In order for a set A to have an inverse X, $A \cup X = \emptyset$. Since $A \cup X = \emptyset$ implies $A = \emptyset$ and $X = \emptyset$, the only set which has an inverse is the empty set and it is its own inverse.

19. Let the operation $\alpha : A \times A \to A$, denoted by

$$\alpha(a, b) = a * b$$

be associative and have an identity element e. If b and b' are inverses of the same element a, then $b = b'$. (In other words, inverses are unique.)

Proof:

Note that
$$\begin{aligned}b * (a * b') &= b * e = b \\ (b * a) * b' &= e * b' = b'\end{aligned}$$

Since α is associative,
$$b * (a * b') = (b * a) * b'$$

and therefore $b = b'$.

20. Let $\mathcal{F}_A$ be the set of all functions of A into A. Let α be the operation of composition of functions. (1) Is α commutative? (2) Is α associative? (3) Find the identity element for α. (4) Which elements, if any, have an inverse and what is it?

Solution:

(1) If A has more than one element, then α is not commutative. For let $a \in A$, $b \in B$ and $a \neq b$, and consider the constant functions f and g defined by $f(x) = a$, $g(x) = b$. Then

$$\begin{aligned}(f \circ g)(x) &\equiv f(g(x)) = f(b) = a \\ (g \circ f)(x) &\equiv g(f(x)) = g(a) = b\end{aligned}$$

(2) By Theorem 4.1, α is an associative operation.

(3) The identity function $1_A : A \to A$ is an identity element since, as previously noted, $(1_A \circ f) = (f \circ 1_A) = f$ for any function $f \varepsilon \mathcal{J}_A$.

(4) The function $f : A \to A$ has an inverse if and only if f is one-one and onto, and its inverse is the inverse function f^{-1} which was defined in Chapter 4.

21. Let the operation $\alpha : N \times N \to N$ be defined by

$$\alpha(a, b) \equiv a * b = a$$

(1) Is α commutative? (2) Is α associative? (3) Is there an identity element? (4) Does any of the elements have an inverse and what is it?

Solution:

(1) Since $a * b = a$ and $b * a = b$, α is not commutative.

(2) Since $(a * b) * c = a * c = a$ and $a * (b * c) = a * b = a$, α is associative.

(3) If α has an identity element e then, by definition of identity element, $e * a = a$ for every $a \varepsilon N$. But by the definition of α, $e * a = e$. Hence there is no identity element.

(4) It is meaningless to talk about an inverse when there does not exist an identity element.

OPERATIONS AND SUBSETS

22. State whether or not each of the following subsets of N, the natural numbers, is closed under the operation of multiplication.

(1) $\{0, 1\}$

(2) $\{1, 2\}$

(3) $\{2, 4, 6, 8, \ldots\} = \{x \mid x \text{ is even}\}$

(4) $\{1, 3, 5, 7, \ldots\} = \{x \mid x \text{ is odd}\}$

(5) $\{x \mid x \text{ is prime}\}$

(6) $\{2, 4, 8, 16, \ldots\} = \{x \mid x = 2^n, n \varepsilon N\}$

Solution:

(1)
$$(0)(0) = 0, \quad (1)(0) = 0, \quad (0)(1) = 0, \quad (1)(1) = 1$$

Hence $\{0, 1\}$ is closed under multiplication.

(2) Since $(2)(2) = 4 \notin \{1, 2\}$, $\{1, 2\}$ is not closed under multiplication.

(3) The product of even numbers is even; hence the set is closed under multiplication.

(4) The product of odd numbers is odd; hence this set is closed under multiplication.

(5) Note that 2 and 3 are primes, but $(2)(3) = 6$ is not a prime. Hence the set is not closed under multiplication.

(6) Since $(2^r)(2^s) = 2^{r+s}$, the set is closed under the operation of multiplication.

23. State whether or not each of the sets in the preceding problem is closed under the operation of addition.

Solution:

The set of even numbers, $\{2, 4, 6, 8, \ldots\}$, is closed under addition since the sum of two even numbers is even. But each of the other sets are not closed under addition since, for example,

$$1 + 1 = 2 \notin \{0, 1\}$$
$$1 + 2 = 3 \notin \{1, 2\}$$
$$3 + 5 = 8 \notin \{1, 3, 5, 7, \ldots\}$$
$$3 + 5 = 8 \notin \{x \mid x \text{ is prime}\}$$
$$2 + 4 = 6 \notin \{2, 4, 8, 16, \ldots\}$$

24. Let (1) $\mathcal{A}$ be the family of finite sets of real numbers
(2) $\mathcal{B}$ be the family of intervals
(3) $\mathcal{C}$ be the family of supersets of the unit interval $[0,1]$.

State whether or not each of the above families of real numbers is closed under the operation of (*a*) union, (*b*) intersection.

Solution:

(1) Since the union and intersection of finite sets are finite, $\mathcal{A}$ is closed under both operations.

(2) Since $[1,2] \cup [3,4]$ is not an interval, $\mathcal{B}$ is not closed under the operation of union. As shown in Chapter 3, the intersection of two intervals is an interval; hence $\mathcal{B}$ is closed under the operation of intersection.

(3) If $[0,1] \subset A$ and $[0,1] \subset B$, then $[0,1] \subset (A \cup B)$ and $[0,1] \subset (A \cap B)$. Accordingly, $\mathcal{C}$ is closed under both operations.

25. Prove: Let $\alpha: A \times A \to A$ be an associative operation with an identity element e, and let B be the set of *units* in A, that is, the set of elements of A each of which has an inverse. Then B is closed under the operation of α.

Solution:

Let $a \in B$ and $b \in B$. Then a has an inverse a^{-1} and b has an inverse b^{-1}. We must show that

$$\alpha(a, b) \equiv a * b \in B$$

that is, $a * b$ has an inverse.

Note that

$$\begin{aligned}(a * b)(b^{-1} * a^{-1}) &= a * (b * (b^{-1} * a^{-1})) \\ &= a * ((b * b^{-1}) * a^{-1}) \\ &= a * (e * a^{-1}) \\ &= a * a^{-1} \\ &= e\end{aligned}$$

Thus $a * b$ has an inverse $b^{-1} * a^{-1}$, and therefore $a * b \in B$.

Supplementary Problems

DIAGRAMS AND FUNCTIONS

26. In the diagram of functions, shown in Fig. 8-3 below, how many paths are there from A to D and what are they?

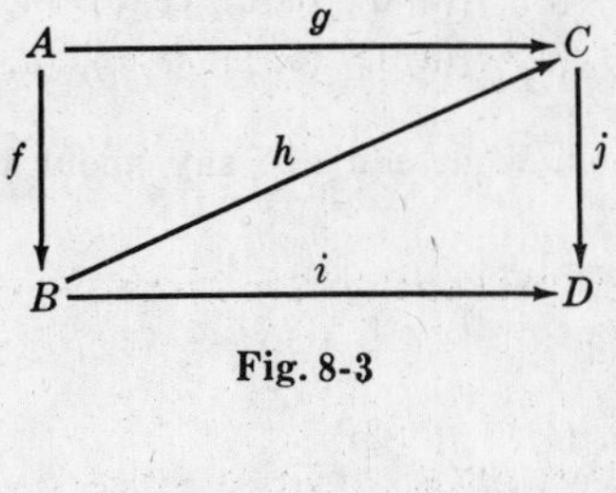

Fig. 8-3

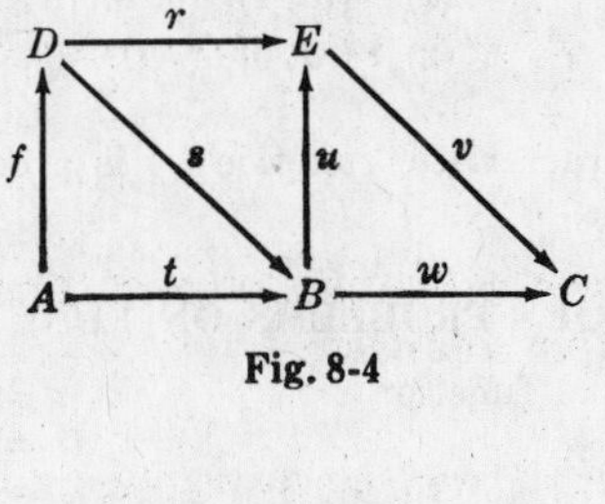

Fig. 8-4

27. If the diagram shown in Fig. 8-4 above is commutative, which functions are equal?

SET FUNCTIONS

28. Let $W = \{1,2,3,4,5\}$ and let $f: W \to W$ be the following set of ordered pairs:

$$f = \{(1,3), (2,2), (3,5), (4,3), (5,1)\}$$

Find: (1) $f(\{1,2,3\})$, (2) $f(\{1,4\})$, (3) $f(\{2,5\})$.

29. Let $S = \{a, b, c\}$, $T = \{1, 2, 3\}$ and

$$f = \{(a, 1), (b, 3), (c, 1)\}$$

Find the induced function $f: 2^S \to 2^T$.

REAL-VALUED FUNCTIONS

30. Let $V = \{a, b, c\}$ and let f and g be the following real valued functions on V:

$$f = \{(a, 2), (b, -3), (c, -1)\}, \qquad g = \{(a, -2), (b, 0), (c, 1)\}$$

Find: (1) $3f$, (2) $g + 2$, (3) $f + g$, (4) $2f - 5g$, (5) fg, (6) $|f|$, (7) f^3, (8) $|3f - fg|$.

31. Find the domain of each of the following real-valued functions:

(1) $f(x) = x/(x+3)$ (3) $f(x) = \sqrt{4 - x^2}$
(2) $f(x) = x/(x+3)$ where $x > 0$ (4) $f(x) = \log(x^2)$

32. Let f be the real-valued function with domain $[-4, 4]$ which is displayed on the following coordinate diagram:

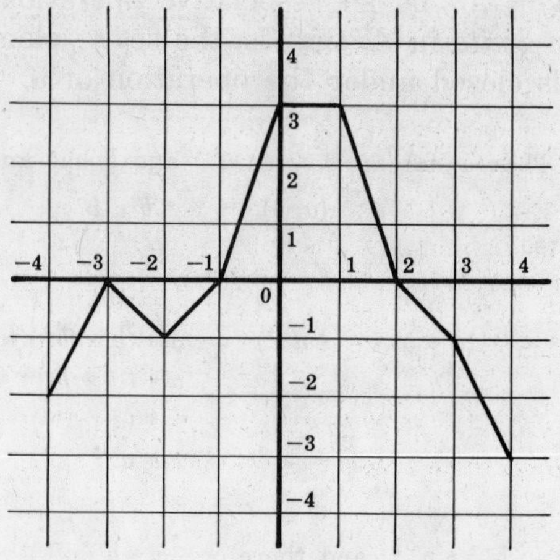

Sketch each of the following functions on a coordinate diagram: (1) $f - 3$, (2) $|f|$.

CHARACTERISTIC FUNCTIONS

33. Let $U = \{a, b, c, d, e\}$, and let $A = \{a, b, e\}$, $B = \{c, d\}$ and $C = \{a, d, e\}$.
Find: (1) χ_A, (2) χ_B, (3) χ_C.

34. Let $U = \{a, b, c, d\}$. Each of the following functions of U into $\{1, 0\}$ is a characteristic function of a subset of U. Find each subset.

(1) $\{(a, 1), (b, 0), (c, 0), (d, 1)\}$ (3) $\{(a, 0), (b, 0), (c, 0), (d, 0)\}$
(2) $\{(a, 0), (b, 1), (c, 0), (d, 0)\}$ (4) $\{(a, 1), (b, 1), (c, 0), (d, 1)\}$

35. If the characteristic function χ_A is a constant function, what can one say about the set A?

MISCELLANEOUS PROBLEMS ON FUNCTIONS

36. Consider the function

$$f = \{(1, 2), (3, 5), (4, 6), (8, 3)\}$$

whose domain is $\{1, 3, 4, 8\}$. In each of the following functions x and y appear as unknowns. For which values of x and y, if any, will each function be an extension of f?

(1) $f_1 = \{(1, 2), (2, 4), (3, x), (4, 6), (5, 8), (8, y)\}$
(2) $f_2 = \{(1, 2), (x, 3), (2, 3), (y, 6), (3, 5)\}$
(3) $f_3 = \{(4, 6), (x, 3), (3, 5), (5, y), (1, 2)\}$

37. Under what conditions can the restriction of a characteristic function χ_A to a set B, i.e. $\chi_A | B$, be a constant function?

38. Consider the subsets $A_1 = \{a, b, d\}$, $A_2 = \{c, d, e\}$, $A_3 = \{b\}$ of $B = \{a, b, c, d, e\}$. State whether or not each of the following functions of $\{A_1, A_2, A_3\}$ into B is a choice function.

(1) $f_1 = \{(A_1, a), (A_2, b), (A_3, b)\}$
(2) $f_2 = \{(A_1, d), (A_2, d), (A_3, b)\}$
(3) $f_3 = \{(A_1, b), (A_2, e), (A_3, b)\}$

OPERATIONS

39. Let α be the operation of intersection of sets. (1) Is α commutative? (2) Is α associative? (3) Find the identity element for α. (4) Which elements, if any, have inverses and what are they?

40. Let α be the operation of difference of sets. Note that

$$A - B = A \cap B'$$

(1) Is α commutative? (2) Is α associative?

(3) Show, by Venn diagrams, that the operation of union does not distribute over α, i.e. in general,

$$A \cup (B - C) \neq (A \cup B) - (A \cup C)$$

(4) Prove that the operation of intersection does distribute over α, i.e.,

$$A \cap (B - C) = (A \cap B) - (A \cap C)$$

41. Let α be the operation on sets defined (and denoted) by

$$A \Delta B \equiv (A \cup B) - (A \cap B)$$

This operation is called the *symmetric difference*. (1) Is α commutative? (2) Is α associative? (3) Find the identity element. (4) Find the inverse of an arbitrary set A. (5) Show by Venn diagrams that the operation of union does not distribute over α, i.e. in general,

$$A \cup (B \Delta C) \neq (A \cup B) \Delta (A \cup C)$$

(6) Prove that the operation of intersection does distribute over α, i.e.,

$$A \cap (B \Delta C) = (A \cap B) \Delta (A \cap C)$$

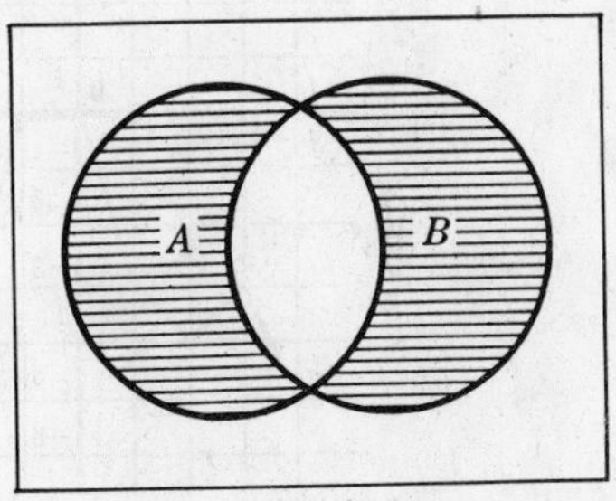

$A \Delta B$ is shaded

42. Let α be the operation on $Q \times Q$, the set of ordered pairs of rational numbers, defined (and denoted) by

$$(a, b) * (x, y) = (ax, ay + b)$$

(1) Is α commutative? (2) Is α associative? (3) Find the identity element for α. (4) Which elements, if any, have inverses and what are they?

43. Let α be the operation on the real numbers defined (and denoted) by

$$a * b \equiv a + b + 2ab$$

(1) Is α commutative? (2) Is α associative? (3) Find the identity element for α. (4) Which elements, if any, have inverses and what are they?

OPERATIONS AND SUBSETS

44. Let $E = \{\ldots, -4, -2, 0, 2, 4, \ldots\}$, i.e. the even integers. State whether or not E is closed under the operation of (1) addition, (2) subtraction, (3) multiplication, (4) division (except by zero).

45. Let $F = \{\ldots, -5, -3, -1, 1, 3, 5, \ldots\}$, i.e. the odd integers. State whether or not F is closed under the operation of (1) addition, (2) subtraction, (3) multiplication, (4) division (except by zero).

46. Let $\mathcal{B}$ be the family of all bounded sets of real numbers. State whether or not $\mathcal{B}$ is closed under the operation of (1) union, (2) intersection, (3) difference.

47. Let $\mathcal{A}$ be the family of all open-closed intervals $(a, b]$, together with the null set. State whether or not $\mathcal{A}$ is closed under the operation of (1) union, (2) intersection, (3) difference.

48. Consider the family $\mathcal{A}$ of sets of real numbers which are supersets of $\{0\}$, i.e. $A \in \mathcal{A}$ if and only if $0 \in A$. State whether or not $\mathcal{A}$ is closed under the operation of (1) union, (2) intersection, (3) difference.

Answers to Supplementary Problems

26. Three: $i \circ f$, $j \circ g$ and $j \circ h \circ f$.

27. $t = s \circ f$, $r = u \circ s$, $r \circ f = u \circ s \circ f = u \circ t$, $w = v \circ u$, $w \circ s = v \circ r = v \circ u \circ s$, and $w \circ t = w \circ s \circ f = v \circ u \circ t = v \circ u \circ s \circ f = v \circ r \circ f$

28. (1) $\{3, 2, 5\}$, (2) $\{3\}$, (3) $\{2, 1\}$

29. $f(\{a, b, c\}) = \{1, 3\}$, $f(\{a, b\}) = \{1, 3\}$, $f(\{a, c\}) = \{1\}$, $f(\emptyset) = \emptyset$, $f(\{b, c\}) = \{1, 3\}$, $f(\{a\}) = \{1\}$, $f(\{b\}) = \{3\}$, $f(\{c\}) = \{1\}$

30. (1) $3f = \{(a, 6), (b, -9), (c, -3)\}$
(2) $g + 2 = \{(a, 0), (b, 2), (c, 3)\}$
(3) $f + g = \{(a, 0), (b, -3), (c, 0)\}$
(4) $2f - 5g = \{(a, 14), (b, -6), (c, -7)\}$
(5) $fg = \{(a, -4), (b, 0), (c, -1)\}$
(6) $|f| = \{(a, 2), (b, 3), (c, 1)\}$
(7) $f^3 = \{(a, 8), (b, -27), (c, -1)\}$
(8) $|3f - fg| = \{(a, 10), (b, 9), (c, 2)\}$

31. (1) $\{x \mid x \in R^\#, x \neq -3\}$ (2) $\{x \mid x > 0\}$ (3) $\{x \mid -2 \leq x \leq 2\}$ (4) $\{x \mid x \in R^\#, x \neq 0\}$

32. (1)

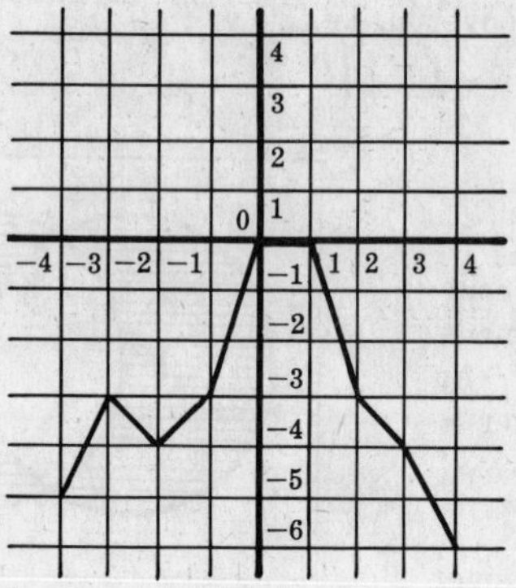

$f - 3$

(2)

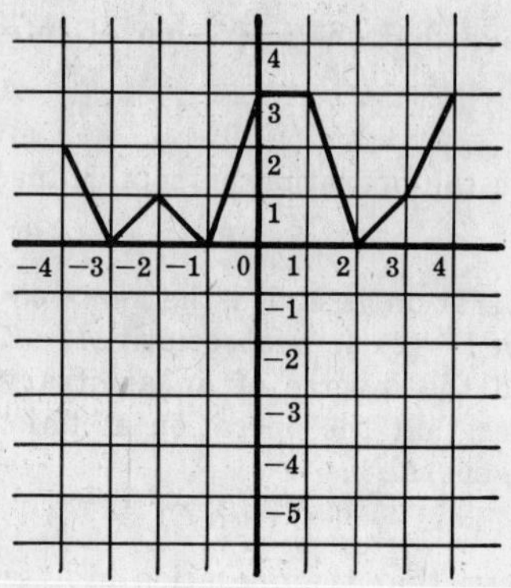

$|f|$

33. (1) $\chi_A = \{(a, 1), (b, 1), (c, 0), (d, 0), (e, 1)\}$
(2) $\chi_B = \{(a, 0), (b, 0), (c, 1), (d, 1), (e, 0)\}$
(3) $\chi_C = \{(a, 1), (b, 0), (c, 0), (d, 1), (e, 1)\}$

34. (1) $\{a, d\}$, (2) $\{b\}$, (3) $\emptyset$, (4) $\{a, b, d\}$

35. Either $A = \emptyset$ or $A = U$.

36. (1) $x = 5$, $y = 3$ (2) $x = 8$, $y = 4$ (3) $x = 8$, y can be any element

37. Either B is a subset of A, or B is a subset of the complement of A.

38. (1) No, (2) Yes, (3) Yes

39. (1) Yes (2) Yes (3) U, the universal set, is the identity element. (4) Only U has an inverse which is itself.

40. (1) No, (2) No
(3)

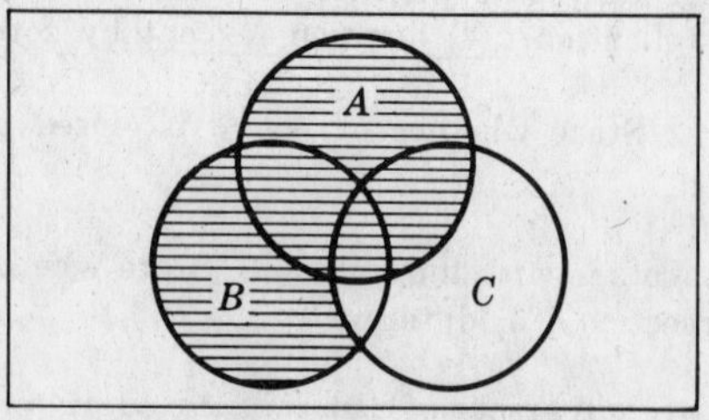

$A \cup (B - C)$ is shaded

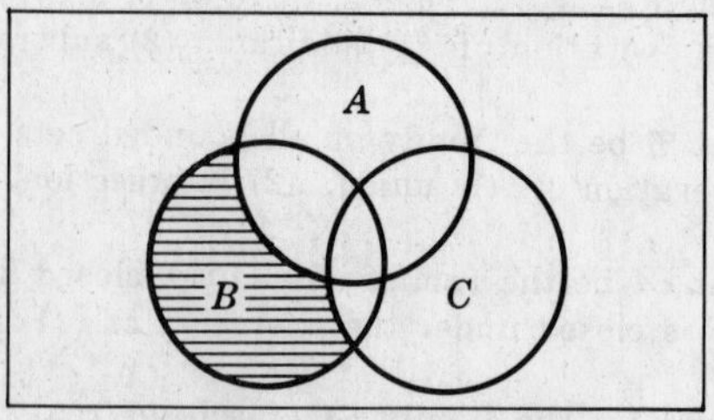

$(A \cup B) - (A \cup C)$ is shaded

(4)

	Statement	Reason
1.	$(A \cap B) - (A \cap C) = (A \cap B) \cap (A \cap C)'$	1. Definition of difference
2.	$= (A \cap B) \cap (A' \cup C')$	2. De Morgan's Law
3.	$= [(A \cap B) \cap A'] \cup [(A \cap B) \cap C']$	3. Distributive Law
4.	But $(A \cap B) \cap A' = (A \cap A') \cap B$	4. Associative Law, Commutative Law
5.	$= \emptyset \cap B$	5. Complement Law
6.	$= \emptyset$	6. Identity Law
7.	$\therefore (A \cap B) - (A \cap C) = \emptyset \cup [(A \cap B) \cap C']$	7. Substitution
8.	$= (A \cap B) \cap C'$	8. Identity Law
9.	$= A \cap (B \cap C')$	9. Associative Law
10.	$= A \cap (B - C)$	10. Definition of difference

41. (1) Yes. (2) Yes. (3) $\emptyset$ is the identity element. (4) The inverse of any set is itself.

(5)

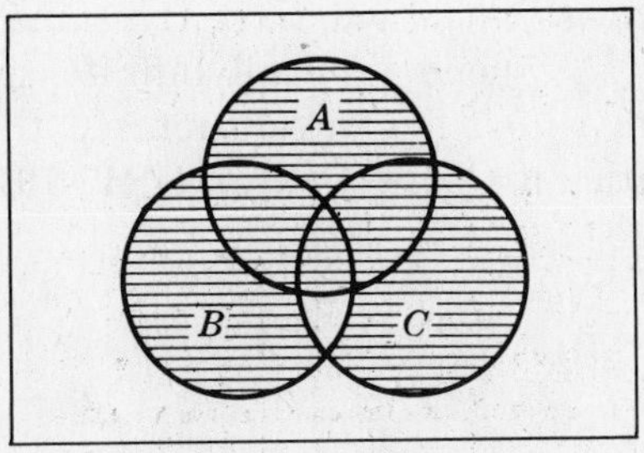

$A \cup (B \Delta C)$ is shaded

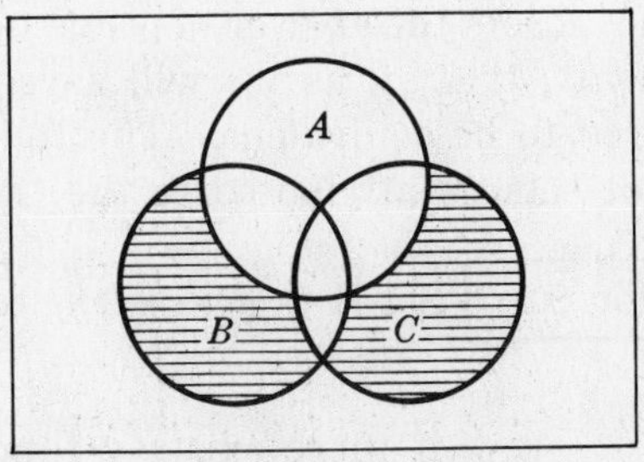

$(A \cup B) \Delta (A \cup C)$ is shaded

(6)

	Statement	Reason
1.	$(A \cap B) \Delta (A \cap C) = [(A \cap B) \cup (A \cap C)] - [(A \cap B) \cap (A \cap C)]$	1. Definition
2.	But $(A \cap B) \cup (A \cap C) = A \cap (B \cup C)$	2. Distributive Law
3.	$(A \cap B) \cap (A \cap C) = (A \cap A) \cap (B \cap C)$	3. Associative Law, Commutative Law
4.	$= A \cap (B \cap C)$	4. Idempotent Law
5.	$\therefore (A \cap B) \Delta (A \cap C) = [A \cap (B \cup C)] - [A \cap (B \cap C)]$	5. Substitution
6.	$= A \cap [(B \cup C) - (B \cap C)]$	6. Intersection distributes over difference
7.	$= A \cap (B \Delta C)$	7. Definition

42. (1) No. (2) Yes. (3) The ordered pair $(1, 0)$ is the identity element. (4) The ordered pair (a, b) has an inverse if $a \neq 0$, and its inverse is $(1/a, -b/a)$.

43. (1) Yes. (2) Yes. (3) Zero is the identity element. (4) If $a \neq 1/2$ then a has an inverse, and its inverse is $-a/(1+2a)$.

44. (1) Yes (2) Yes (3) Yes (4) No

45. (1) No (2) No (3) Yes (4) No

46. (1) Yes (2) Yes (3) Yes

47. (1) No (2) Yes (3) Yes

48. (1) Yes (2) Yes (3) No

Part II — Cardinals, Ordinals, Transfinite Induction

Chapter 9

Cardinal Numbers

EQUIVALENT SETS

It is natural to ask whether or not any two sets have the same number of elements. For finite sets the answer can be found by counting the number of elements in each set. For infinite sets the answer depends upon how one defines two sets to have the same number of elements, that is, as we will say, to be *equivalent*. At one time all infinite sets were considered to be equivalent. The following definition, which has revolutionized the entire theory of sets, is attributed to the German mathematician Georg Cantor (1845-1918).

Definition 9.1: Set A is *equivalent* to set B, denoted by

$$A \sim B$$

if there exists a function

$$f: A \to B$$

which is both one-one and onto.

The function f is then said to define a *one-to-one correspondence* between the sets A and B.

Example 1.1: Let $R = \{1, 2, 5, 8\}$ and $T = \{\text{Marc, Erik, Paul, Betty}\}$. The following diagram defines a function of R into T which is both one-one and onto. Hence R is equivalent to T.

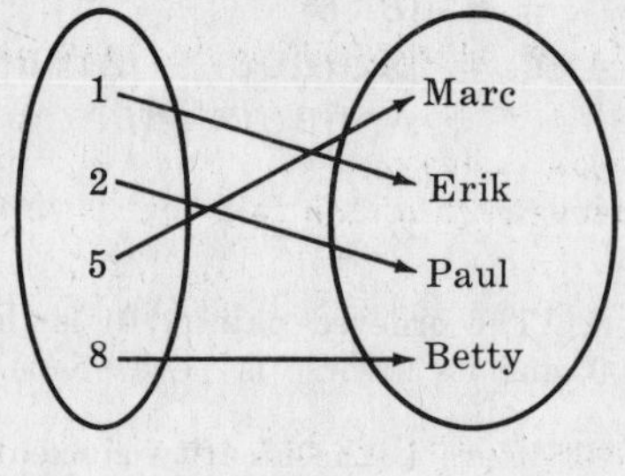

Example 1.2: Let $M = \{1, 2, 3\}$ and $N = \{1, 2\}$. If we list all the functions of M into N, none of them will be both one-one and onto. Hence M is not equivalent to N.

If we examine the above two examples, it is not difficult to see that, in general, two finite sets are equivalent if and only if they contain the same number of elements. Hence, for finite sets, Definition 9.1 corresponds to the usual meaning of two sets containing the same number of elements.

Example 1.3: Let $G = [0, 1]$ and $H = [2, 5]$, and let $f: G \to H$ be the function defined by

$$f(x) = 3x + 2$$

Note that f is both one-one and onto. Hence $G \sim H$, i.e. G is equivalent to H.

Example 1.4: Let $N = \{1, 2, 3, \ldots\}$ and $E = \{2, 4, 6, \ldots\}$. The function $f: N \to E$, defined by $f(x) = 2x$, is both one-one and onto. Therefore $N \sim E$.

In Example 1.4 we see that the infinite set N, the natural numbers, is equivalent to a proper subset of itself. This property is characteristic of infinite sets. In fact, we now state formally

Definition 9.2: A set is *infinite* if it is equivalent to a proper subset of itself. Otherwise a set is *finite*.

Example 1.5: Let A and B be any two sets. Then

$$A \sim A \times \{1\}$$
$$B \sim B \times \{2\}$$

since the functions

$$f : a \to (a, 1), \quad a \in A$$
$$g : b \to (b, 2), \quad b \in B$$

are both one-one and onto. Moreover, although A and B need not be disjoint, note that

$$A \times \{1\} \cap B \times \{2\} = \emptyset$$

since each ordered pair in $A \times \{1\}$ contains 1 as a second element, and each ordered pair in $B \times \{2\}$ contains 2 as a second element.

We conclude this section with a theorem which we shall use later in the chapter.

Theorem 9.1: The relation in sets defined by $A \sim B$, is an equivalence relation. Specifically,

(1) $A \sim A$ for any set A,
(2) if $A \sim B$, then $B \sim A$,
(3) if $A \sim B$ and $B \sim C$, then $A \sim C$.

DENUMERABLE SETS

The reader is familiar with the set of natural numbers, $N = \{1, 2, 3, \ldots\}$.

Definition 9.3: If a set D is equivalent to N, the set of natural numbers, then D is called *denumerable* and is said to have cardinality $\mathfrak{a}$.

Definition 9.4: A set is called *countable* if it is finite or denumerable, and a set is called *non-denumerable* if it is infinite and if it is not equivalent to N, i.e. if it is not countable.

Example 2.1: Any infinite sequence

$$a_1, a_2, a_3, \ldots$$

of distinct elements is denumerable, for a sequence is essentially a function

$$f(n) = a_n$$

whose domain is N. So if the a_n are distinct, the function is one-one and onto. Hence each of the following sets is denumerable:

$$\{1, 1/2, 1/3, \ldots, 1/n, \ldots\}$$
$$\{1, -2, 3, -4, \ldots, (-1)^{n-1}n, \ldots\}$$
$$\{(1, 1), (4, 8), (9, 27), \ldots, (n^2, n^3), \ldots\}$$

Example 2.2: Consider the product set $N \times N$ as exhibited in Fig. 9-1.

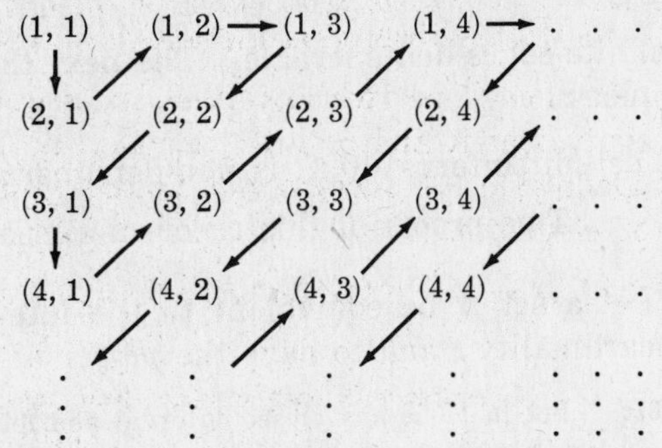

Fig. 9-1

The set $N \times N$ can be written in an infinite sequence of distinct elements as follows:

$$\{(1, 1), (2, 1), (1, 2), (1, 3), (2, 2), \ldots\}$$

(Note that the sequence is determined by "following the arrows" in Fig. 9-1.) Thus, for reasons stated in Example 2.1, $N \times N$ is denumerable.

Example 2.3: Let $M = \{0, 1, 2, \ldots\} = N \cup \{0\}$. Now each natural number $a \in N$ can be written uniquely in the form

$$a = 2^r(2s+1)$$

where $r, s \in M$. Consider the function $f: N \to M \times M$ defined by

$$f(a) = (r, s)$$

where r and s are as above. Then f is one-one and onto. Hence $M \times M$ is denumerable. Note that $N \times N$ is a subset of $M \times M$.

The following theorems concern denumerable sets.

Theorem 9.2: Every infinite set contains a subset which is denumerable.

Theorem 9.3: A subset of a denumerable set is either finite or denumerable.

Corollary 9.3: A subset of a countable set is countable.

Theorem 9.4: Let $A_1, A_2, A_3, \ldots$ be a denumerable family of pairwise disjoint sets, each of which is denumerable. Then the union of the sets

$$\cup_{i \in N} A_i$$

is denumerable.

Corollary 9.4: Let $\{A_i\}_{i \in I}$ be a countable family of countable sets. Then $\cup_{i \in I} A_i$ is countable.

Next follows a very important, and not entirely obvious, example of a denumerable set.

Example 2.4: Let Q^+ be the set of positive rational numbers and let Q^- be the set of negative rational numbers. Then

$$Q = Q^- \cup \{0\} \cup Q^+$$

is the set of rational numbers.

Let the function $f: Q^+ \to N \times N$ be defined by

$$f(p/q) = (p, q)$$

where p/q is any member of Q^+ expressed as the ratio of two relatively prime positive integers. Note f is one-one, and therefore Q^+ is equivalent to a subset of $N \times N$. By Theorem 9.3 and Example 2.2, Q^+ is denumerable. Similarly Q^- is denumerable. Hence the set of rational numbers, which is the union of Q^+, $\{0\}$ and Q^-, is denumerable.

THE CONTINUUM

Not every infinite set is denumerable. The next theorem gives a specific and extremely important example.

Theorem 9.5: The unit interval $[0, 1]$ is non-denumerable.

Two proofs of this theorem appear in the section on solved problems.

Definition 9.5: Let a set A be equivalent to the interval $[0, 1]$. Then A is said to have cardinality $\mathfrak{c}$ and to have the *power of the continuum*.

Example 3.1: Let $[a, b]$ be any closed interval and let

$$f: [0, 1] \to [a, b]$$

be the function defined by

$$f(x) = a + (b - a)x$$

Note that f is one-one and onto. Thus $[a, b]$ has cardinality $\mathfrak{c}$. Furthermore, we will prove that any open or half-open interval also has cardinality $\mathfrak{c}$.

Example 3.2: The function $f:(-\pi/2, \pi/2) \to R^{\#}$, defined by $f(x) = \tan x$, is one-one and onto; hence

$$R^{\#} \sim (-\pi/2, \pi/2)$$

Therefore the set of real numbers $R^{\#}$ has the power of the continuum, i.e. has cardinality $\mathfrak{c}$.

CARDINAL NUMBERS

Note once again, by Theorem 9.1, that the relation in sets defined by

$$A \sim B$$

is an equivalence relation. Hence by the Fundamental Theorem on Equivalence Relations, all sets are partitioned into disjoint classes of equivalent sets.

Definition 9.6: Let A be any set and let α denote the family of sets which are equivalent to A. Then α is called a *cardinal number* (or, simply, cardinal) and is denoted by

$$\alpha = \#(A)$$

Definition 9.7: The cardinal number of each of the sets

$$\emptyset, \{1\}, \{1, 2\}, \{1, 2, 3\}, \ldots$$

is denoted by $0, 1, 2, 3, \ldots$, respectively, and is called a *finite cardinal.*

Definition 9.8: The cardinal numbers of N, the set of natural numbers, and the unit interval $[0, 1]$ are denoted by

$$\#(N) = \mathfrak{a}, \quad \#([0,1]) = \mathfrak{c}$$

Remark 9.1: The symbol $\aleph_0$ (read aleph-null) is also used to denote the cardinality of denumerable sets, i.e. $\#(N)$, since this is the symbol originally used by Cantor.

CARDINAL ARITHMETIC

In view of Definition 9.7, the cardinal numbers can be considered to be a superset of the finite cardinals

$$0, 1, 2, 3, \ldots$$

that is, the natural numbers N and 0. The next definition essentially extends the ordinary operations of addition and multiplication for natural numbers to operations for all the cardinal numbers.

Definition 9.9: Let α and β be cardinal numbers and let A and B be disjoint sets such that

$$\alpha = \#(A), \quad \beta = \#(B)$$

Then

$$\alpha + \beta = \#(A \cup B)$$
$$\alpha\beta = \#(A \times B)$$

Theorem 9.6: Definition 9.9 is well-defined, that is, the definitions of $\alpha + \beta$ and $\alpha\beta$ do not depend upon the particular sets A and B. In other words, if

$$A \sim A', \ B \sim B', \ A \cap B = \emptyset, \ A' \cap B' = \emptyset$$

then

$$\#(A \cup B) = \#(A' \cup B')$$
$$\#(A \times B) = \#(A' \times B')$$

The sets A and B are assumed to be disjoint in Definition 9.9. Since the sets $A \times \{1\}$ and $B \times \{2\}$ are always disjoint, regardless of A and B, the following definition can be substituted for Definition 9.9.

Definition 9.9′: Let $\alpha = \#(A)$ and $\beta = \#(B)$, then

$$\alpha + \beta \;=\; \#(A \times \{1\} \cup B \times \{2\})$$
$$\alpha\beta \;=\; \#(A \times B)$$

Example 4.1: Note that $3 = \#(\{a, b, c\})$ and $4 = \#(\{1, 3, 5, 7\})$. Then

$$3 + 4 \;=\; \#(\{a, b, c, 1, 3, 5, 7\}) \;=\; 7$$
$$(3)(4) \;=\; \#(\{a, b, c\} \times \{1, 3, 5, 7\}) \;=\; 12$$

In other words, the operations of addition and multiplication of finite cardinal numbers correspond to the ordinary operations of addition and multiplication of natural numbers.

Example 4.2: Note that $\mathfrak{a} = \#(\{1, 3, 5, \ldots\}) = \#(\{2, 4, 6, \ldots\})$. Then

$$\mathfrak{a} + \mathfrak{a} \;=\; \#(N) \;=\; \mathfrak{a} \qquad \text{and} \qquad \mathfrak{a}\mathfrak{a} \;=\; \#(N \times N) \;=\; \mathfrak{a}$$

Theorem 9.7: The operations of addition and multiplication of cardinal numbers is associative and commutative; and addition distributes over multiplication, i.e. for any cardinal numbers α, β and γ,

(1) $(\alpha + \beta) + \gamma \;=\; \alpha + (\beta + \gamma)$

(2) $(\alpha\beta)\gamma \;=\; \alpha(\beta\gamma)$

(3) $\alpha + \beta \;=\; \beta + \alpha$

(4) $\alpha\beta \;=\; \beta\alpha$

(5) $\alpha(\beta + \gamma) \;=\; \alpha\beta + \alpha\gamma$

Not every property of addition and multiplication of natural numbers holds for cardinal numbers in general. For example, for the natural numbers, the cancellation law is true, i.e.,

$$a + b = a + c \quad \text{implies } b = c$$
$$ab = ac \quad \text{implies } b = c$$

Since, by Example 4.2,

$$\mathfrak{a} + \mathfrak{a} = \mathfrak{a} = 1 + \mathfrak{a} \quad \text{does not imply } \mathfrak{a} = 1$$
$$\mathfrak{a}\mathfrak{a} = \mathfrak{a} = 1\mathfrak{a} \quad \text{does not imply } \mathfrak{a} = 1$$

the cancellation law is not true for the operations of addition and multiplication of cardinal numbers.

Remark 9.2: Exponents can also be introduced into the arithmetic of cardinal numbers as follows: Let $\alpha = \#(A)$ and $\beta = \#(B)$, and let

$$B^A$$

denote the family of all functions from A (the exponent) into B. Then

$$\beta^\alpha \equiv \#(B^A)$$

In fact, the following properties of exponents, which are known to hold for the natural numbers, also hold for any cardinals α, β and γ:

(1) $\alpha^\beta \alpha^\gamma \;=\; \alpha^{\beta+\gamma}$

(2) $(\alpha^\beta)^\gamma \;=\; \alpha^{\beta\gamma}$

(3) $(\alpha\beta)^\gamma \;=\; \alpha^\gamma \beta^\gamma$

Example 4.3: Let $A = \{a, b, c\}$ and $B = \{1, 2\}$. Then $\#(A) = 3$, $\#(B) = 2$ and $2^3 = \#(B^A)$. But B^A consists of exactly 8 functions:

$\{(a, 1), (b, 1), (c, 1)\}$, $\{(a, 1), (b, 1), (c, 2)\}$, $\{(a, 1), (b, 2), (c, 1)\}$, $\{(a, 1), (b, 2), (c, 2)\}$
$\{(a, 2), (b, 1), (c, 1)\}$, $\{(a, 2), (b, 1), (c, 2)\}$, $\{(a, 2), (b, 2), (c, 1)\}$, $\{(a, 2), (b, 2), (c, 2)\}$

Therefore, as cardinals,

$$2^3 = 8$$

In other words, if m and n are finite cardinals then m^n denotes the same number regardless of whether we consider m and n to be cardinals or natural numbers.

INEQUALITIES AND CARDINAL NUMBERS

An inequality relation is defined for the cardinal numbers as follows:

Definition 9.10: Let $\alpha = \#(A)$ and $\beta = \#(B)$. Furthermore, let A be equivalent to a subset of B; that is, suppose there exists a function $f: A \to B$ which is one-one. Then we write

$$A \precsim B$$

which reads "*A precedes B*", and

$$\alpha \leqq \beta$$

which reads "α is *less than or equal to* β".

The following additional notation will also be used:

$$A \prec B \text{ means } A \precsim B \text{ and } A \not\sim B$$
$$\alpha < \beta \text{ means } \alpha \leqq \beta \text{ and } \alpha \neq \beta$$

Example 5.1: Let A and B be finite sets, say $n = \#(A)$ and $m = \#(B)$. Then $n \leqq m$ as cardinal numbers if and only if $n \leqq m$ as natural numbers. In other words, the inequality relation in the set of cardinal numbers is an extension of the inequality relation in the set of natural numbers.

Example 5.2: Since N, the natural numbers, is a subset of $R^\#$, the real numbers,

$$\mathfrak{a} \leqq \mathfrak{c}$$

Furthermore, since $R^\#$ is not denumerable, i.e. $\mathfrak{a} \neq \mathfrak{c}$,

$$\mathfrak{a} < \mathfrak{c}$$

Example 5.3: For any set A, the identity function $1_A: A \to A$ is one-one; hence $A \precsim A$. Also, therefore, $\alpha \leqq \alpha$ for any cardinal number α.

Example 5.4: If $f: A \to B$ is one-one and $g: B \to C$ is one-one, then the composition function $g \circ f: A \to C$ is also one-one. Therefore,

$$A \precsim B \text{ and } B \precsim C \quad \text{implies} \quad A \precsim C$$

and, for any cardinal numbers α, β and γ,

$$\alpha \leqq \beta \text{ and } \beta \leqq \gamma \quad \text{implies} \quad \alpha \leqq \gamma$$

In view of the preceding two examples, the following theorem is true.

Theorem 9.8: The relation in sets defined by $A \precsim B$ is reflexive and transitive, and the relation in the cardinal numbers defined by $\alpha \leqq \beta$ is also reflexive and transitive.

CANTOR'S THEOREM

The only infinite cardinal numbers we have seen are $\mathfrak{a}$ and $\mathfrak{c}$. It is natural to ask if there are any others. The answer is yes. In fact, Cantor's Theorem, which follows, tells us that for any cardinal number α there exists a cardinal number which is greater than α. Specifically,

Theorem 9.9 (Cantor's Theorem): For any set A,

$$A \prec 2^A$$

and, therefore, for any cardinal number α,

$$\alpha < 2^\alpha$$

Here, if $\alpha = \#(A)$ then $2^\alpha = \#(2^A)$, the cardinal number of the family of subsets of A.

SCHRÖDER-BERNSTEIN THEOREM

For any pair of sets A and B, at least one of the following must be true:

(1) A is equivalent to B, i.e.,

$$\#(A) = \#(B)$$

(2) A is not equivalent to B but A is equivalent to a subset of B (or vice versa), i.e.,

$$\#(A) < \#(B) \quad (\text{or } \#(B) < \#(A))$$

(3) A is equivalent to a subset of B and B is equivalent to a subset of A, i.e.,

$$\#(A) \leqq \#(B) \quad \text{and} \quad \#(B) \leqq \#(A)$$

(4) A is not equivalent to a subset of B and B is not equivalent to a subset of A, i.e.,

$$\#(A) \nless \#(B), \quad \#(A) \neq \#(B) \quad \text{and} \quad \#(A) \ngtr \#(B)$$

The celebrated Schröder-Bernstein Theorem states that, in case (3) above, A is equivalent to B, i.e. $\#(A) = \#(B)$. Specifically

Theorem 9.10 (Schröder-Bernstein): If $A \lesssim B$ and $B \lesssim A$, then $A \sim B$; hence for any cardinal numbers α and β,

$$\alpha \leqq \beta \text{ and } \beta \leqq \alpha \quad \text{implies} \quad \alpha = \beta$$

The Schröder-Bernstein Theorem can be stated in the following equivalent form:

Theorem 9.10′: Let $X \supset Y \supset X_1$ and let $X \sim X_1$; then $X \sim Y$.

We conclude this chapter with the statement that case (4) above is impossible. Specifically,

Theorem 9.11 (Law of Trichotomy): For any pair of cardinal numbers α and β, either

$$\alpha < \beta, \quad \alpha = \beta \quad \text{or} \quad \alpha > \beta$$

The proof of this last theorem uses tools of transfinite induction which is discussed in Chapter 12; hence the proof will be postponed until then.

CONTINUUM HYPOTHESIS

By Cantor's Theorem, $\mathfrak{a} < 2^{\mathfrak{a}}$ and, as noted previously, $\mathfrak{a} < \mathfrak{c}$. The next theorem tells us the relationship between $2^{\mathfrak{a}}$ and $\mathfrak{c}$.

Theorem 9.12: $2^{\mathfrak{a}} = \mathfrak{c}$.

It is natural to ask if there exists a cardinal number β which lies "between" $\mathfrak{a}$ and $\mathfrak{c}$. Originally Cantor supported the conjecture, which is known as the Continuum Hypothesis, that the answer to the above question is in the negative. Specifically,

Continuum Hypothesis: There exists no cardinal number β such that $\mathfrak{a} < \beta < \mathfrak{c}$.

In 1963 it was shown that the Continuum Hypothesis is independent of our axioms of set theory in somewhat the same sense that Euclid's Fifth Postulate on parallel lines is independent of the other axioms of geometry.

Solved Problems

EQUIVALENT SETS, DENUMERABLE SETS, CONTINUUM

1. Consider the concentric circles

$$C_1 = \{(x, y) \mid x^2 + y^2 = a^2\}, \qquad C_2 = \{(x, y) \mid x^2 + y^2 = b^2\}$$

where, say, $0 < a < b$. Establish, geometrically, a one-to-one correspondence between C_1 and C_2.

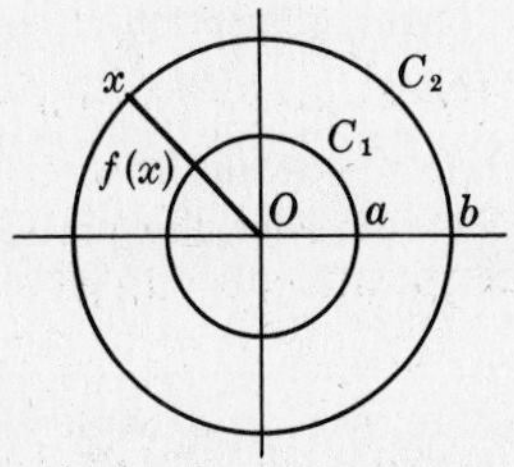

Solution:

Let $x \in C_2$. Consider the function $f: C_2 \to C_1$ where $f(x)$ is the point of intersection of the radius from the center of C_2 (and C_1) to x, and C_1, as shown in the adjacent diagram.

Note that f is both one-one and onto. Thus f defines a one-to-one correspondence between C_1 and C_2.

2. Prove: (*a*) $[0,1] \sim (0,1)$, (*b*) $[0,1] \sim [0,1)$, (*c*) $[0,1] \sim (0,1]$.

Solution:

(*a*) Note that

$$[0, 1] = \{0, 1, 1/2, 1/3, \ldots\} \cup A$$
$$(0, 1) = \{1/2, 1/3, 1/4, \ldots\} \cup A$$

where

$$A = [0, 1] - \{0, 1, 1/2, 1/3, \ldots\} = (0, 1) - \{1/2, 1/3, \ldots\}$$

Consider the function $f: [0,1] \to (0,1)$ defined by the following diagram

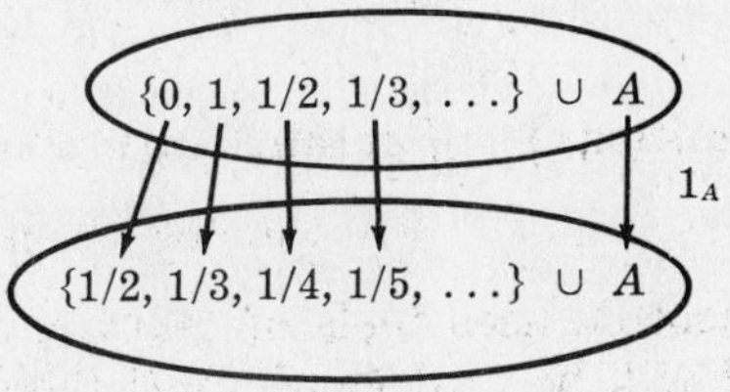

In other words,

$$f(x) = \begin{cases} 1/2 & \text{if } x = 0 \\ 1/(n+2) & \text{if } x = 1/n, \ n \in N \\ x & \text{if } x \neq 0, 1/n, \ n \in N \end{cases}$$

The function f is one-one and onto. Consequently, $[0,1] \sim (0,1)$.

(*b*) The function $f: [0,1] \to [0,1)$ defined by

$$f(x) = \begin{cases} 1/(n+1) & \text{if } x = 1/n, \ n \in N \\ x & \text{if } x \neq 1/n, \ n \in N \end{cases}$$

is one-one and onto. (It is similar to the function in Part (*a*)). Hence $[0,1] \sim [0,1)$.

(*c*) Let $f: [0,1) \to (0,1]$ be the function defined by $f(x) = 1 - x$. Then f is one-one and onto and, therefore, $[0,1) \sim (0,1]$. By Part (*b*) and Theorem 9.1, $[0,1] \sim (0,1]$.

3. Prove that each of the following intervals has the power of the continuum, i.e. has cardinality $\mathfrak{c}$: (1) $[a,b]$, (2) (a,b), (3) $[a,b)$, (4) $(a,b]$. Here $a < b$.

Solution:

The formula $f(x) = a + (b-a)x$ defines each of the following functions:

$$[0,1] \xrightarrow{f} [a,b] \qquad [0,1) \xrightarrow{f} [a,b) \qquad (0,1) \xrightarrow{f} (a,b) \qquad (0,1] \xrightarrow{f} (a,b]$$

Each function is one-one and onto. Hence, by Problem 2 and Theorem 9.1, each interval is equivalent to $[0,1]$, that is, has the power of the continuum.

4. Prove Theorem 9.1: The relation $A \sim B$ in sets is an equivalence relation. Specifically,

(1) $A \sim A$ for any set A,
(2) if $A \sim B$ then $B \sim A$,
(3) if $A \sim B$ and $B \sim C$ then $A \sim C$.

Solution:

(1) The identity function $1_A : A \to A$ is one-one and onto; hence $A \sim A$.

(2) If $A \sim B$, then there exists a function $f : A \to B$ which is one-one and onto. Hence f has an inverse function $f^{-1} : B \to A$ which is also one-one and onto. Therefore,

$$A \sim B \text{ implies } B \sim A$$

(3) If $A \sim B$ and $B \sim C$, then there exists functions $f : A \to B$ and $g : B \to C$ which are one-one and onto. Then the product function $g \circ f : A \to C$ is also one-one and onto. Therefore,

$$A \sim B \text{ and } B \sim C \text{ implies } A \sim C$$

5. Prove Theorem 9.2: Every infinite set A contains a subset D which is denumerable.

Solution:

Let $f : 2^A \to A$ be a choice function. Consider the following sequence:

$$\begin{aligned} a_1 &= f(A) \\ a_2 &= f(A - \{a_1\}) \\ a_3 &= f(A - \{a_1, a_2\}) \\ &\cdots\cdots \\ a_n &= f(A - \{a_1, \ldots, a_{n-1}\}) \\ &\cdots\cdots \end{aligned}$$

Since A is infinite, $A - \{a_1, \ldots, a_{n-1}\}$ is not empty for any $n \in N$. Furthermore, since f is a choice function,

$$a_n \neq a_i \quad \text{where } i < n$$

Thus the a_n are distinct and, therefore, $D = \{a_1, a_2, \ldots\}$ is denumerable.

Essentially, the choice function f "chooses" an element $a_1 \in A$, then chooses an element a_2 from the elements which "remain" in A, etc. Since A is infinite, the set of elements which "remain" in A is non-empty.

6. Prove: For any sets A and B, $A \times B \sim B \times A$.

Solution:

The function $f : A \times B \to B \times A$ defined by

$$f((a, b)) = (b, a), \qquad (a \in A, b \in B)$$

is one-one and onto; hence $A \times B \sim B \times A$.

7. Prove: For any sets A, B and C,

$$(A \times B) \times C \sim A \times B \times C \sim A \times (B \times C)$$

Solution:

The function $f : (A \times B) \times C \to A \times B \times C$ defined by

$$f((a, b), c) = (a, b, c), \qquad (a \in A, b \in B, c \in C)$$

is one-one and onto; hence $(A \times B) \times C \sim A \times B \times C$. Similarly, $A \times (B \times C) \sim A \times B \times C$. Thus,

$$(A \times B) \times C \sim A \times B \times C \sim A \times (B \times C)$$

8. Prove: Let X be any set and let $C(X)$ be the family of characteristic functions of X, that is, the family of functions $f : X \to \{1, 0\}$. Then the family of subsets of X is equivalent to $C(X)$, i.e. $2^X \sim C(X)$.

Solution:

Let A be any subset of X, i.e. $A \in 2^X$. Let $f : 2^X \to C(X)$ be defined by

$$f(A) = \chi_A$$

that is, f maps each subset A of X into χ_A, the characteristic function of A (relative to X). Then f is one-one and, as noted previously, onto. Hence $2^X \sim C(X)$.

9. Prove Theorem 9.3: A subset of a denumerable set is either finite or denumerable.

Solution:

Let

$$A = \{a_1, a_2, \ldots\} \tag{1}$$

be any denumerable set and let B be a subset of A. If $B = \emptyset$, then B is finite. If $B \neq \emptyset$, then let a_{n_1} be the first element in the sequence in (1) such that $a_{n_1} \in B$; let a_{n_2} be the first element which follows a_{n_1} in the sequence in (1) such that $a_{n_2} \in B$; etc. Then

$$B = \{a_{n_1}, a_{n_2}, \ldots\}$$

If the set of integers $\{n_1, n_2, \ldots\}$ is bounded, then B is finite. Otherwise B is denumerable.

10. Prove Theorem 9.5: The unit interval $A = [0,1]$ is not denumerable.

Solution:

Method 1. Assume the contrary; then

$$A = \{x_1, x_2, x_3, \ldots\}$$

i.e. the elements of A can be written in a sequence.

Each element in A can be written in the form of an infinite decimal as follows:

$$\begin{aligned} x_1 &= 0.\,a_{11}\,a_{12}\,a_{13}\,\ldots\,a_{1n}\,\ldots \\ x_2 &= 0.\,a_{21}\,a_{22}\,a_{23}\,\ldots\,a_{2n}\,\ldots \\ x_3 &= 0.\,a_{31}\,a_{32}\,a_{33}\,\ldots\,a_{3n}\,\ldots \\ &\cdots\cdots\cdots\cdots\cdots\cdots \\ x_n &= 0.\,a_{n1}\,a_{n2}\,a_{n3}\,\ldots\,a_{nn}\,\ldots \end{aligned} \tag{1}$$

where $a_{ij} \in \{0, 1, \ldots, 9\}$ and where each decimal contains an infinite number of non-zero elements. Here write 1 as .999... and, for those numbers which can be written in the form of a decimal in two ways, for example,

$$1/2 = .5000\ldots = .4999\ldots$$

(in one of them there is an infinite number of nines and in the other all except a finite set of digits are zeros), write the infinite decimal in which an infinite number of nines appear.

Now construct the real number

$$y = 0.\,b_1\,b_2\,b_3\,\ldots\,b_n\,\ldots$$

which will belong to A, in the following way: choose b_1 so $b_1 \neq a_{11}$ and $b_1 \neq 0$, choose b_2 so $b_2 \neq a_{22}$ and $b_2 \neq 0$, etc.

Note $y \neq x_1$ since $b_1 \neq a_{11}$ (and $b_1 \neq 0$), $y \neq x_2$ since $b_2 \neq a_{22}$ (and $b_2 \neq 0$), etc., that is, $y \neq x_n$, for $n \in N$; hence $y \notin A$, which contradicts the fact that $y \in A$. Thus the assumption that A is denumerable has led to a contradiction. Consequently, A is non-denumerable.

Method 2. (In this second proof of Theorem 9.5, we use the following property of the real numbers: Let $I_1 = [a_1, b_1]$, $I_2 = [a_2, b_2]$, ... be a sequence of closed intervals for which $I_1 \supset I_2 \supset \cdots$. Then there exists a real number y with the property that y belongs to every interval.)

Assume the contrary. Then, as above,

$$A = \{x_1, x_2, \ldots\}$$

Now construct a sequence of closed intervals $I_1, I_2, \ldots$ as follows. Consider the following three closed sub-intervals of $[0,1]$,

$$[0, 1/3], \quad [1/3, 2/3], \quad [2/3, 1] \tag{1}$$

each having length $\frac{1}{3}$. Now x_1 cannot belong to all three intervals. (If x_1 is one of the endpoints, then it could belong to two of the intervals.) Let $I_1 = [a_1, b_1]$ be one of the intervals in (1) such that $x_1 \notin I_1$.

Now consider the following three closed sub-intervals of $I_1 = [a_1, b_1]$,

$$[a_1, a_1 + 1/9], \quad [a_1 + 1/9, a_1 + 2/9], \quad [a_1 + 2/9, b_1] \tag{2}$$

each having length $\frac{1}{9}$. Similarly, let I_2 be one of the intervals in (2) with the property that x_2 does not belong to I_2. Continue in this manner.

Thus we have a sequence of closed intervals

$$I_1 \supset I_2 \supset \cdots \tag{3}$$

such that $x_n \notin I_n$ for every $n \in N$.

By the above property of real numbers, there exists a real number $y \in [0,1]$ such that y belongs to every interval in (3). But since

$$y \in A = \{x_1, x_2, \ldots\}$$

$y = x_m$ for some $m \in N$. Then by our construction $y = x_m \notin I_m$, which contradicts the fact that y belongs to every interval in (3). Thus our assumption that A is denumerable has led to a contradiction. Accordingly, A is non-denumerable.

CARDINAL NUMBERS AND CARDINAL ARITHMETIC

11. Let A_1, A_2, A_3 and A_4 be any sets. Define sets B_1, B_2, B_3 and B_4 such that

$$\#(A_1) + \#(A_2) + \#(A_3) + \#(A_4) = \#(B_1 \cup B_2 \cup B_3 \cup B_4)$$

Solution:

Let $B_1 = A_1 \times \{1\}$, $B_2 = A_2 \times \{2\}$, $B_3 = A_3 \times \{3\}$ and $B_4 = A_4 \times \{4\}$. Then $B_i \sim A_i$, $i = 1, 2, 3, 4$; and $B_i \cap B_j = \emptyset$ if $i \neq j$. Consequently, the above will be true.

12. Let $\{A_i\}_{i \in I}$ be any family of sets. Define a family of sets $\{B_i\}_{i \in I}$ such that

$$B_i \sim A_i,\ i \in I \text{ and } B_i \cap B_j = \emptyset \text{ if } i \neq j$$

Solution:

Let $B_i = A_i \times \{i\}$, $(i \in I)$. Then the family $\{B_i\}_{i \in I}$ has the required properties.

13. Prove Theorem 9.7: For any cardinal numbers α, β and γ,

(1) $(\alpha + \beta) + \gamma = \alpha + (\beta + \gamma)$

(2) $(\alpha\beta)\gamma = \alpha(\beta\gamma)$

(3) $\alpha + \beta = \beta + \alpha$

(4) $\alpha\beta = \beta\alpha$

(5) $\alpha(\beta + \gamma) = \alpha\beta + \alpha\gamma$

Solution:

Let A, B and C be pairwise disjoint sets such that $\alpha = \#(A)$, $\beta = \#(B)$ and $\gamma = \#(C)$.

(1)
$$(\alpha + \beta) + \gamma = \#(A \cup B) + \#(C) = \#((A \cup B) \cup C)$$
$$\alpha + (\beta + \gamma) = \#A + \#(B \cup C) = \#(A \cup (B \cup C))$$

Since union of sets is associative, i.e. $(A \cup B) \cup C = A \cup (B \cup C)$, then

$$(\alpha + \beta) + \gamma = \alpha + (\beta + \gamma)$$

(2)
$$(\alpha\beta)\gamma = \#(A \times B)\,\#(C) = \#((A \times B) \times C)$$
$$\alpha(\beta\gamma) = \#A\,\#(B \times C) = \#(A \times (B \times C))$$

By Problem 7, $(A \times B) \times C \sim A \times (B \times C)$. Therefore $(\alpha\beta)\gamma = \alpha(\beta\gamma)$.

(3) $\alpha + \beta = \#(A \cup B) = \#(B \cup A) = \beta + \alpha$, since $A \cup B = B \cup A$.

(4) Note $\alpha\beta = \#(A \times B)$ and $\beta\alpha = \#(B \times A)$. By Problem 6, $A \times B \sim B \times A$; hence $\alpha\beta = \beta\alpha$.

(5) Note first that $B \cap C = \emptyset$ implies $(A \times B) \cap (A \times C) = \emptyset$. Then

$$\alpha(\beta + \gamma) = \#(A) \cdot \#(B \cup C) = \#(A \times (B \cup C))$$
$$\alpha\beta + \alpha\gamma = \#(A \times B) + \#(A \times C) = \#((A \times B) \cup (A \times C))$$

But $A \times (B \cup C) = (A \times B) \cup (A \times C)$. Therefore $\alpha(\beta + \gamma) = \alpha\beta + \alpha\gamma$.

14. Prove: $\mathfrak{a}\mathfrak{c} = \mathfrak{c}$.

Solution:

Let $Z = \{\ldots, -1, 0, 1, \ldots\}$ and $A = [0, 1)$. Furthermore, let $f : Z \times A \to R^{\#}$ be defined by

$$f(i, a) = i + a$$

in other words, $f(\{i\} \times [0, 1))$ is mapped onto $[i, i+1)$. Then f is one-one and onto. Hence

$$(Z \times A) \sim R^{\#}$$

Since $\#(Z) = \mathfrak{a}$, $\#(A) = \mathfrak{c}$ and $\#(R^{\#}) = \mathfrak{c}$, then $\mathfrak{a}\mathfrak{c} = \mathfrak{c}$.

15. Prove: Let β be any infinite cardinal number. Then

$$\aleph_0 + \beta = \beta$$

Solution:

Let A be an infinite set, $B = \{b_1, b_2, \ldots\}$ be denumerable and $A \cap B = \emptyset$. Then the theorem is true if we show that

$$A \cup B \sim A$$

Since A is infinite, A contains a denumerable subset

$$D = \{d_1, d_2, \ldots\}$$

Let $f : A \cup B \to A$ be defined by the following diagram:

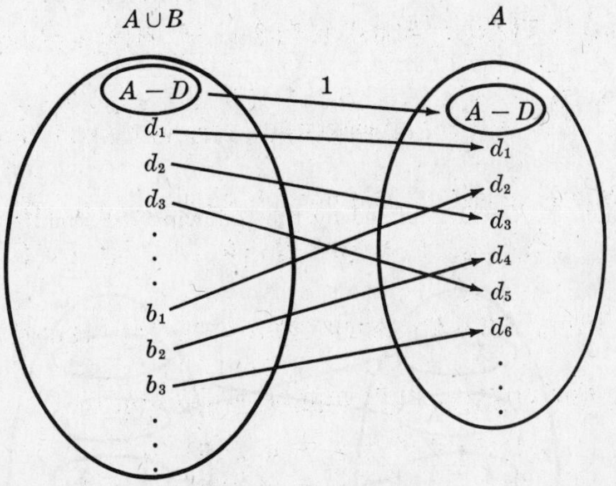

In other words, $f : A \cup B \to A$ is defined by

$$f(x) = \begin{cases} x & \text{if } x \in A - D \\ d_{2n-1} & \text{if } x = d_n \\ d_{2n} & \text{if } x = b_n \end{cases}$$

Then f is one-one and onto. Consequently, $A \cup B \sim A$ and our theorem is true.

INEQUALITIES AND CARDINAL NUMBERS

16. Prove Cantor's Theorem: For any set A, $A \prec 2^A$ and hence $\#(A) < \#(2^A)$.

Solution:

The function $g : A \to 2^A$ which sends each element $a \in A$ into the set consisting of a alone, i.e. which is defined by $g(a) = \{a\}$, is one-one; hence $A \precsim 2^A$.

If we now show that A is not equivalent to 2^A, then the theorem will follow. Suppose the contrary, i.e. let there exist a function $f : A \to 2^A$ which is one-one and onto. Let $a \in A$ be called a "bad" element if a is not a member of the set which is its image, i.e. if $a \notin f(a)$, and let B be the set of "bad" elements. Specifically,

$$B = \{x \mid x \in A,\ x \notin f(x)\}$$

Note B is a subset of A, that is, $B \in 2^A$. Hence, since $f : A \to 2^A$ is onto, there exists an element $b \in A$ with the property that $f(b) = B$. Is b "bad" or "good"? If $b \in B$ then, by definition of B, $b \notin f(b) = B$, which is impossible. Likewise, if $b \notin B$, then $b \in f(b) = B$ which is also impossible. Thus the original assumption, that $A \sim 2^A$, has led to a contradiction. Hence the assumption is false and, therefore, the theorem is true.

17. Prove the Schröder-Bernstein Theorem: Let $X \supset Y \supset X_1$ and let $X \sim X_1$; then $X \sim Y$.

Solution:

Since $X \sim X_1$, there exists a function $f : X \to X_1$ which is one-one and onto. Furthermore, since $X \supset Y$, the restriction of f to Y, which we shall also denote by f, is also one-one; hence Y is equivalent to a subset of X_1, i.e. $Y \sim Y_1$ where

$$X \supset Y \supset X_1 \supset Y_1$$

and $f: Y \to Y_1$ is one-one and onto. But now $X_1 \subset Y$; for similar reasons, $X_1 \sim X_2$ where

$$X \supset Y \supset X_1 \supset Y_1 \supset X_2$$

and $f: X_1 \to X_2$ is one-one and onto. Consequently, there exist equivalent sets $X, X_1, X_2, \ldots$ and equivalent sets $Y, Y_1, Y_2, \ldots$ such that

$$X \supset Y \supset X_1 \supset Y_1 \supset X_2 \supset Y_2 \supset \cdots$$

Let

$$B = X \cap Y \cap X_1 \cap Y_1 \cap X_2 \cap Y_2 \cap \cdots$$

Then

$$X = (X-Y) \cup (Y-X_1) \cup (X_1-Y_1) \cup \cdots \cup B$$
$$Y = (Y-X_1) \cup (X_1-Y_1) \cup (Y_1-X_2) \cup \cdots \cup B$$

Note further that

$$(X-Y) \sim (X_1-Y_1) \sim (X_2-Y_2) \sim \cdots$$

Specifically, the function

$$f: (X_n - Y_n) \to (X_{n+1} - Y_{n+1})$$

is one-one and onto.

Consider the function $g: X \to Y$ defined by the following diagram:

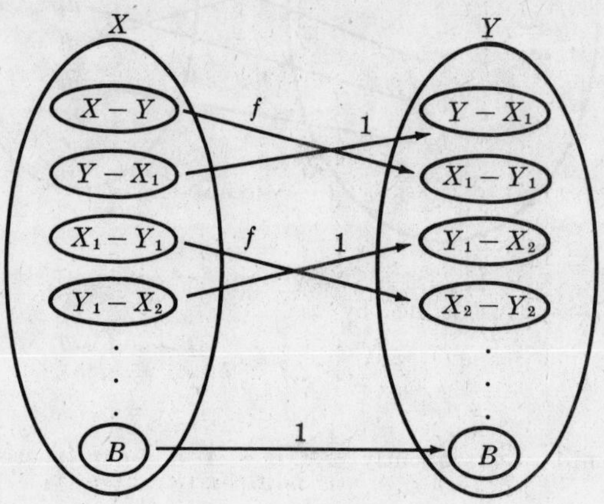

In other words,

$$g(x) = \begin{cases} f(x) & \text{if } x \in X_i - Y_i \text{ or } x \in X - Y \\ x & \text{if } x \in Y_i - X_i \text{ or } x \in B \end{cases}$$

Then g is one-one and onto. Therefore $X \sim Y$.

18. Prove Theorem 9.12: $\mathfrak{c} = 2^{\mathfrak{a}}$.

Solution:

Let $R^\#$ be the set of real numbers and 2^Q be the family of subsets of Q, the set of rational numbers. Furthermore, let the function $f: R^\# \to 2^Q$ be defined by

$$f(a) = \{x \mid x \in Q,\ x < a\}$$

i.e. f maps each real number a into the set of rational numbers less than a. We shall show that f is one-one. Let $a, b \in R^\#$, $a \neq b$ and, say, $a < b$. By a property of the real numbers, there exists a rational number r such that

$$a < r < b$$

Then $r \in f(b)$ and $r \notin f(a)$; hence $f(b) \neq f(a)$ and, therefore, f is one-one. Thus $R^\# \precsim 2^Q$ and, since $\#(R^\#) = \mathfrak{c}$ and $\#(Q) = \mathfrak{a}$,

$$\mathfrak{c} \leq 2^{\mathfrak{a}}$$

Now let $C(N)$ be the family of characteristic functions $f: N \to \{0, 1\}$ which, as proven in Problem 8, is equivalent to 2^N. Here N is the set of natural numbers. Consider the function $F: C(N) \to [0,1]$ defined by

$$F(f) = 0.\,f(1)\,f(2)\,f(3)\,\ldots$$

an infinite decimal consisting of zeros or ones. If $f, g \in C(N)$ and $f \neq g$, then $F(f) \neq F(g)$ since the decimals would be different; hence F is one-one. Therefore $2^N \sim C(N) \precsim [0,1]$ and hence

$$2^{\mathfrak{a}} \leq \mathfrak{c}$$

Consequently,

$$\mathfrak{c} = 2^{\mathfrak{a}}$$

MISCELLANEOUS PROBLEMS

19. Prove: The set P of all polynomials

$$p(x) = a_0 + a_1 x + \cdots + a_m x^m \qquad (1)$$

with integral coefficients, that is, where $a_0, a_1, \ldots, a_m$ are integers, is denumerable.

Solution:

For each pair of natural numbers (n, m), let $P_{(n,m)}$ be the set of polynomials in (1) of degree m in which

$$|a_0| + |a_1| + \cdots + |a_m| = n$$

Note that $P_{(n,m)}$ is finite. Therefore

$$P = \cup_{i \in N \times N} P_i$$

is countable since it is a countable family of countable sets. But P is not finite; hence P is denumerable.

20. A real number r is called an *algebraic number* if r is a solution to a polynomial equation

$$p(x) = a_0 + a_1 x + \cdots + a_n x^n = 0$$

with integral coefficients. Prove that the set A of algebraic numbers is denumerable.

Solution:

Note, by the preceding problem, that the set E of polynomial equations is denumerable:

$$E = \{p_1(x) = 0,\ p_2(x) = 0,\ p_3(x) = 0,\ \ldots\}$$

Define

$$A_i = \{x \mid x \text{ is a solution of } p_i(x) = 0\}$$

Since a polynomial of degree n can have at most n roots, each A_i is finite. Therefore

$$A = \cup_{i \in N} A_i$$

is a countable family of countable sets. Accordingly, A is countable and, since A is not finite, therefore denumerable.

Supplementary Problems

EQUIVALENT SETS, DENUMERABLE SETS, CONTINUUM

21. The integers, Z, can be put into a one-to-one correspondence with N, the natural numbers, as follows:

$$\begin{array}{cccccccc} 1 & 2 & 3 & 4 & 5 & 6 & 7 & \\ \downarrow & \downarrow & \downarrow & \downarrow & \downarrow & \downarrow & \downarrow & \cdots \\ 0 & 1 & -1 & 2 & -2 & 3 & -3 & \end{array}$$

Find a formula that defines a function $f: N \to Z$ which gives the above correspondence between N and Z.

22. $N \times N$ was written as a sequence by considering the diagram in Fig. 9-1. This is not the only way to write $N \times N$ as a sequence. Write $N \times N$ as a sequence in two other ways by drawing appropriate diagrams.

23. Prove Theorem 9.4: Let $A_1, A_2, A_3, \ldots$ be a denumerable family of pairwise disjoint sets, each of which is denumerable. Then the union of the sets $\cup_{i \in N} A_i$ is denumerable.

24. Prove: If A and B are denumerable then $A \times B$ is denumerable.

25. Prove that the set of points in the plane with rational coordinates is denumerable.

26. Let $\{T_i\}_{i \in I}$ be a family of pairwise disjoint intervals. Prove that the family is countable.

27. A real number x is called transcendental if x is not algebraic, i.e. if x is not a solution to a polynomial equation

$$p(x) \equiv a_0 + a_1 x + \cdots a_n x^n = 0$$

with integral coefficients. (See Problem 20.) For example, π and e are transcendental numbers. Prove that the set of transcendental numbers is non-denumerable.

28. Prove: $\mathfrak{c}^2 \equiv \mathfrak{c}\mathfrak{c} = \mathfrak{c}$. (Hence $R^\# \times R^\#$ has the power of the continuum.)

CARDINAL ARITHMETIC

29. Prove: Let $\alpha \leqq \beta$; then there exists a set B with a subset A such that $\alpha = \#(A)$ and $\beta = \#(B)$.

30. Prove: Let $\alpha \leqq \beta$; then for any cardinal γ, (1) $\alpha + \gamma \leqq \beta + \gamma$, (2) $\alpha\gamma \leqq \beta\gamma$.

31. Let T be the set of transcendental (real) numbers. Prove that $\#(T) = \mathfrak{c}$. (Note that T was only shown to be non-denumerable in Problem 27.)

32. Let $\alpha = \#(A)$. Then 2^α was defined as the cardinal number of the family of subsets of A, i.e. $2^\alpha = \#(2^A)$. Also 2^α was defined to be the cardinal number of the family of all functions from A into a set B where $\#(B) = 2$. Show that these definitions are equivalent.

33. Prove: For any cardinal numbers α, β and γ, $\alpha^\beta \alpha^\gamma = \alpha^{\beta+\gamma}$.

34. Prove: Let $\alpha \leqq \beta$; then for any cardinal number γ, (1) $\alpha^\gamma \leqq \beta^\gamma$, (2) $\gamma^\alpha \leqq \gamma^\beta$.

Answers to Supplementary Problems

21. The function $f: N \to Z$ defined by

$$f(x) = \begin{cases} -x/2 + 1/2 & \text{if } x \text{ is odd} \\ x/2 & \text{if } x \text{ is even} \end{cases}$$

has the required property.

22. Consider the following diagrams which contain $N \times N$.

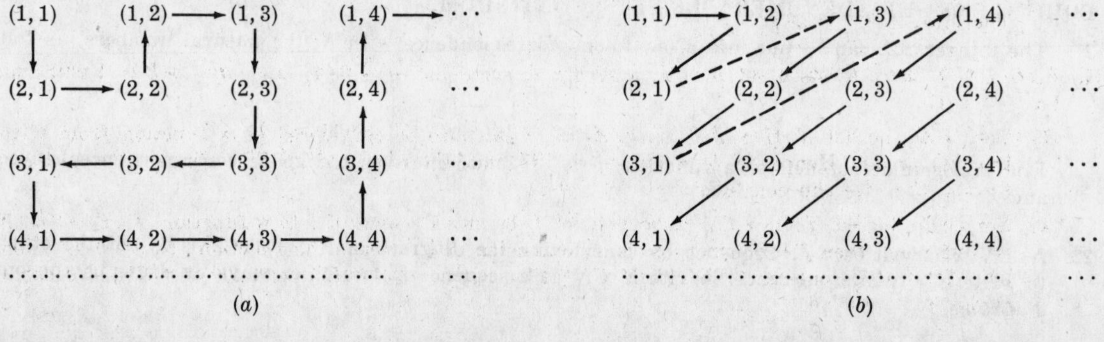

(a) (b)

In view of these diagrams, $N \times N$ can be written as an infinite sequence of distinct elements as follows:

$$N \times N = \{(1,1),\ (2,1),\ (2,2),\ (1,2),\ (1,3),\ (2,3),\ \ldots\}$$
$$N \times N = \{(1,1),\ (1,2),\ (2,1),\ (1,3),\ (2,2),\ (3,1),\ (1,4),\ \ldots\}$$

23. *Hint.* Show that $\cup_{i \in N} A_i$ is equivalent to $N \times N$.

25. Let S be the set of points in the plane with rational coordinates and let Q be the set of rational numbers. Note that $S \sim Q \times Q$, since each point $x \in S$ corresponds to a unique ordered pair $(q_1, q_2) \in Q \times Q$ and vice versa. But $Q \times Q$ is denumerable since Q is denumerable. Therefore S is denumerable.

26. Any interval T_i, $i \in I$, contains at least one rational number q_i. Furthermore, if $T_i \neq T_j$ then $q_i \neq q_j$, since T_i and T_j are disjoint. Therefore $\{T_i\}_{i \in I}$ is equivalent to a subset $\{q_i\}_{i \in I}$ of the rational numbers. Hence $\{T_i\}_{i \in I}$ is countable.

27. *Hint.* $R^\#$, which is non-denumerable, is the union of the algebraic and transcendental numbers.

28. Let $A = [0,1]$. We show that $A \times A$ has the power of the continuum. Let $x, y \in [0,1]$. Then x and y can be written uniquely in the form of an infinite decimal,

$$x = 0.\,x_1\,x_2\,x_3\ldots, \qquad y = 0.\,y_1\,y_2\,y_3\ldots$$

which contains an infinite number of non-zero digits, (e.g. for $\frac{1}{2}$ write 0.4999... instead of 0.5000...).

Let $f: A \times A \to A$ be defined by

$$f((x,y)) = 0.\,x_1\,y_1\,x_2\,y_2\,x_3\,y_3\ldots$$

Then f is one-one. Hence $A \times A$ has cardinality at most $\mathfrak{c}$. But $A \times A$ has cardinality at least $\mathfrak{c}$ since, for example,

$$\{(0,x) \mid x \in [0,1]\}$$

which is a subset of $A \times A$, is equivalent to A. Consequently, $A \times A$ has cardinality $\mathfrak{c}$, i.e. has the power of the continuum.

Note that $\#(A) = \mathfrak{c}$; hence $\mathfrak{c}^2 = \#(A \times A) = \mathfrak{c}$.

32. Let $B = \{0,1\}$. Then $\#(B) = 2$ and $B^A = C(A)$, the set of characteristic functions of A. By Problem 8, $2^A \sim B^A$. Hence $\#(2^A) = \#(B^A)$.

33. Let $\alpha = \#(A)$, $\beta = \#(B)$ and $\gamma = \#(C)$ where B and C are disjoint. Then $\beta + \gamma = \#(B \cup C)$. Note that

$$\alpha^{\beta+\gamma} = \#(A^{B \cup C}) \quad \text{and} \quad \alpha^\beta \alpha^\gamma = \#(A^B \times A^C)$$

Note also that $A^{B \cup C}$ consists of all the functions with domain $B \cup C$ and co-domain A. A^B and A^C have similar meaning. The theorem is proven if we show that

$$A^{B \cup C} \sim A^B \times A^C$$

Let $f \in A^{B \cup C}$ correspond to the ordered pair of functions

$$(f \mid B,\ f \mid C)$$

the restriction of f to B and the restriction of f to C. Note that $(f \mid B,\ f \mid C)$ belongs to $A^B \times A^C$. The function $F: A^{B \cup C} \to A^B \times A^C$ defined by

$$F(f) = (f \mid B,\ f \mid C)$$

is one-one and onto. Hence $A^{B \cup C} \sim A^B \times A^C$.

Therefore $\alpha^{\beta+\gamma} = \alpha^\beta \alpha^\gamma$.

34. Since $\alpha \leq \beta$, there exists a set B with a subset A such that $\alpha = \#(A)$ and $\beta = \#(B)$. Furthermore, let $\gamma = \#(C)$.

(1) Let $f \in A^C$, i.e. let $f: C \to A$. Since $A \subset B$, f can also be considered as a function from C to B, that is, $f \in B^C$. Hence A^C is a subset of B^C and, therefore, $A^C \lesssim B^C$. Since $\alpha^\gamma = \#(A^C)$ and $\beta^\gamma = \#(B^C)$, we can conclude that $\alpha^\gamma \leq \beta^\gamma$.

(2) Let $f \in C^A$, i.e. let $f: A \to C$. Moreover, let f' be an extension of f to a function $f': B \to C$. Note that if $f \neq g$, then $f' \neq g'$ where g' is an extension of g. Hence the function $F: C^A \to C^B$ defined by $F(f) = f'$ is one-one. Therefore $C^A \lesssim C^B$. Since $\gamma^\alpha = \#(C^A)$ and $\gamma^\beta = \#(C^B)$, we can conclude that $\gamma^\alpha \leq \gamma^\beta$.

Chapter 10

Partially and Totally Ordered Sets

PARTIALLY ORDERED SETS

A *partial order* in a set A is a relation R in A which is

(1) reflexive, i.e. $(a, a) \in R$ for every $a \in A$,

(2) anti-symmetric, i.e. $(a, b) \in R$ and $(b, a) \in R$ implies $a = b$,

(3) transitive, i.e. $(a, b) \in R$ and $(b, c) \in R$ implies $(a, c) \in R$.

Furthermore, if a relation R in A defines a partial order in A, then $(a, b) \in R$ is denoted by

$$a \precsim b$$

which reads "*a precedes b*".

Example 1.1: Let $\mathcal{A}$ be a family of sets. Then the relation in $\mathcal{A}$ defined by "x is a subset of y" is a partial order in $\mathcal{A}$.

Example 1.2: Let A be any subset of the real numbers. Then the relation in A defined by "$x \leq y$" is a partial order in A. It is called the *natural order* in A.

Example 1.3: Let R be the relation in the natural numbers N defined by "x is a multiple of y"; then R is a partial order in N. Moreover, $6 \precsim 2$, $15 \precsim 3$ and $17 \precsim 17$.

Example 1.4: Let $W = \{a, b, c, d, e\}$. Then the diagram

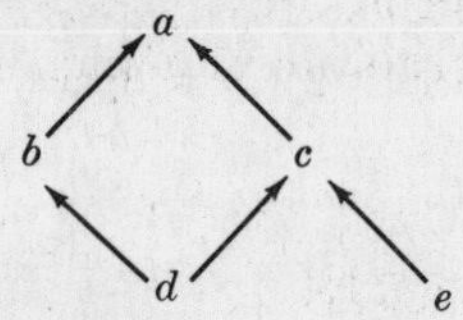

defines a partial order in W in the following way: $x \precsim y$ if $x = y$ or if one can go from x to y in the diagram, always moving in the indicated direction, i.e. upward. Note that $b \precsim a$, $d \precsim a$ and $e \precsim c$.

Example 1.5: Let R be the relation in $V = \{1, 2, 3, 4, 5, 6\}$ which is defined by "x divides y". Then R is a partial order in V. This partial order in V can also be described by the following diagram which is similar to the diagram in the preceding example, and similar to the line diagrams which were constructed for families of sets:

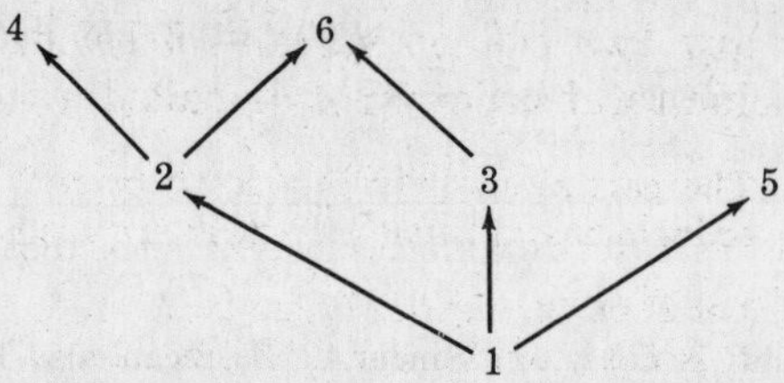

Example 1.6: Let R be the relation in a family of sets defined by "X is equivalent to a subset of Y" (i.e. $X \precsim Y$). By Theorem 9.8, R is reflexive and transitive; and by the Schröder-Bernstein Theorem 9.10, R is anti-symmetric. Hence R is a partial order in the family of sets.

Although the symbol $\precsim$ was previously used to denote a relation in sets, the relation, as seen in this example, is a partial order.

Definition 10.1: A set A together with a specific partial order relation R in A is called a *partially ordered set.*

Remark 10.1: Note that a partially ordered set consists of a set A and a specific type of a relation R in A; for this reason, a partially ordered set is sometimes denoted by the ordered pair

$$(A, R) \quad \text{or} \quad (A, \precsim)$$

Usually, though, the same symbol, say A, is used to denote both the partially ordered set and the underlying set on which the partial order is defined.

Remark 10.2: In this and the next chapter we shall assume that any set of real numbers is ordered by the natural order unless otherwise stated, explicitly or implicitly.

The following additional notation is used with respect to partially ordered sets:

$a \prec b$ means $a \precsim b$ and $a \neq b$; read "*a strictly precedes b*".
$b \succsim a$ means $a \precsim b$; read "*b dominates a*".
$b \succ a$ means $a \prec b$; read "*b strictly dominates a*".
$\not\precsim$, $\not\prec$, $\not\succsim$ and $\not\succ$ are self-explanatory.

Two elements a and b in a partially ordered set are said to be *not comparable* if

$$a \not\precsim b \quad \text{and} \quad b \not\precsim a$$

that is, if neither element precedes the other. In Example 1.3, the numbers 3 and 5 are not comparable since neither number is a multiple of the other.

Remark 10.3: If a relation R in a set A is reflexive, anti-symmetric and transitive, then the inverse relation R^{-1} is also reflexive, anti-symmetric and transitive. In other words, if R defines a partial order in A then R^{-1} also defines a partial order in A which is called the *inverse order.*

TOTALLY ORDERED SETS

The word "partial" is used in defining a partial order in a set A because some elements in A need not be comparable. If, on the other hand, every two elements in a partially ordered set A are comparable, then the partial order in A is called a *total order* in A. Specifically,

Definition 10.2: A *total order* in a set A is a partial order in A with the additional property that

$$a \prec b, \; a = b \text{ or } a \succ b$$

for any two elements a and b belonging to A. A set A together with a specific total order in A is called a *totally ordered set.*

Example 2.1: The partial order in any set A of real numbers (with the natural order) is a total order since any two numbers are comparable.

Example 2.2: Let R be the partial order in $V = \{1, 2, 3, 4, 5, 6\}$ defined by "x divides y". Then R is not a total order in V since 3 and 5 are not comparable.

Example 2.3: Let A and B be totally ordered sets. Then the Cartesian product $A \times B$ can be totally ordered as follows:

$$(a, b) \prec (a', b') \quad \text{if } a \prec a' \quad \text{or if } a = a' \text{ and } b \prec b'$$

This order is called the *lexicographical order* of $A \times B$ since it is similar to the way words are arranged in a dictionary.

Example 2.4: Let $\{A_i\}_{i \varepsilon I}$ be a totally ordered family (i.e. I is totally ordered) of pairwise disjoint totally ordered sets. Then the union $\cup_{i \varepsilon I} A_i$ is totally ordered (unless otherwise stated) as follows: Let $a, b \varepsilon \cup_{i \varepsilon I} A_i$; hence there exist $j, k \varepsilon I$ such that $a \varepsilon A_j$, $b \varepsilon A_k$. Now if $j \prec k$, $a \precsim b$; and if $j = k$, then a and b are ordered by the ordering of A_j.

Remark 10.4: The word "order" will frequently be used instead of either partial order or total order.

SUBSETS OF ORDERED SETS

Suppose a relation R defines a partial order in a set A, i.e. suppose (A, R) is an ordered set. Now let B be a subset of A. Then the partial order R in A induces a partial order R' in B in the following natural way: If $a, b \varepsilon B$, then

$$(a, b) \varepsilon R', \text{ that is, } a \precsim b$$

as elements of B, if and only if $(a, b) \varepsilon R$, i.e. $a \precsim b$ as elements of A. In such a situation the ordered set (B, R') is called a (partially ordered) *subset* of the ordered set (A, R).

Example 3.1: Let $W = \{a, b, c, d, e\}$ be ordered as follows:

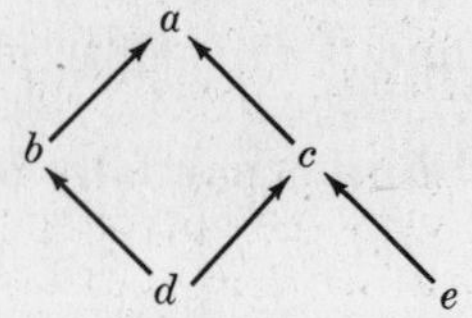

Then $V = \{a, d, e\}$, with the order

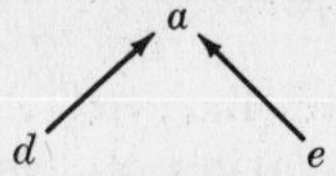

is a subset of the ordered set W. But V with the order

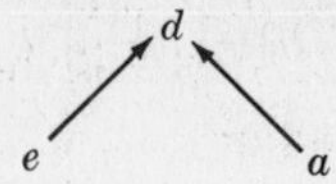

is not a subset of the *ordered set* W.

TOTALLY ORDERED SUBSETS

Let A be a partially ordered set. Then, as noted previously, the partial order in A induces a partial order in every subset of A. Some of the subsets of A will, in fact, be totally ordered.

Note that if A is a totally ordered set, then every subset of A will also be totally ordered.

Example 4.1: Let N, the natural numbers, be ordered by "x is a multiple of y". Then N is not totally ordered, since 4 and 7 are not comparable. But the set

$$M = \{2, 4, 8, \ldots, 2^n, \ldots\}$$

is a totally ordered subset of N.

Example 4.2: Consider the partial order in $W = \{a, b, c, d, e\}$ defined by the diagram

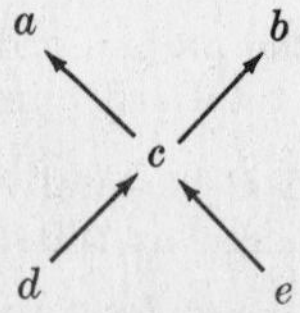

Each of the sets $\{a, c, d\}$, $\{b, d\}$, $\{b, c, e\}$, $\{a, c, e\}$ and $\{a, c\}$ is a totally ordered subset. The sets $\{a, b, c\}$ and $\{d, e\}$ are not totally ordered.

FIRST AND LAST ELEMENTS

Let A be an ordered set. The element $a \in A$ is called a *first* element of A if, for every element $x \in A$,

$$a \precsim x$$

that is, if a precedes *every* element in A. Analogously, an element $b \in A$ is called a *last* element of A if, for every $x \in A$,

$$x \precsim b$$

that is, if b dominates every element belonging to A.

Example 5.1: Let $W = \{a, b, c, d, e\}$ be ordered by the following diagram:

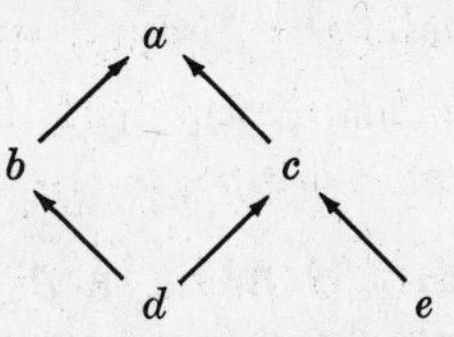

Then a is a last element in W since a dominates every element. Note that W has no first element. The element d is not a first element since d does not precede e.

Example 5.2: Consider N, the natural numbers (with the natural order). Then 1 is a first element of N. There is no last element.

Example 5.3: Let A be any set. Let $\mathcal{A}$ be the family of subsets of A, i.e. the power set of A. Let $\mathcal{A}$ be ordered by "x is a subset of y". Then the null set is a first element and A is a last element of $\mathcal{A}$.

Example 5.4: Let $A = \{x \mid 0 < x < 1\}$ be ordered by "$x \leq y$". Then A, which is totally ordered, contains no first element and no last element.

Remark 10.5: A partially ordered set can have at most one first element and one last element.

Remark 10.6: If a and b are first and last elements respectively in a partially ordered set A, then a and b will be the last and first elements respectively in the inverse order in A.

MAXIMAL AND MINIMAL ELEMENTS

Let A be an ordered set. An element $a \in A$ is called a *maximal* element if

$$a \precsim x \quad \text{implies} \quad a = x$$

In other words, a is a maximal element if there is no element in A which strictly dominates a. Similarly, an element $b \in A$ is called a minimal element if

$$x \precsim b \quad \text{implies} \quad b = x$$

that is, if there is no element in A which strictly precedes b.

Example 6.1: Let $W = \{a, b, c, d, e\}$ be ordered by the following diagram in the usual way:

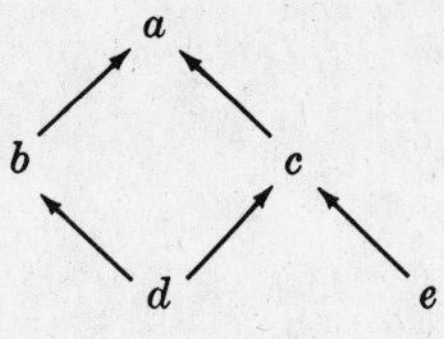

Both d and e are minimal elements since no element in W strictly precedes either of them. The element a is a maximal element.

Example 6.2: Let $W = \{a, b, c, d, e\}$ be ordered by the following diagram:

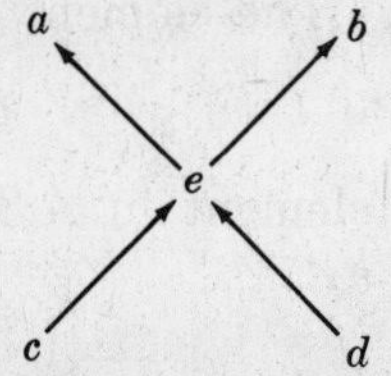

Then a and b are maximal elements, and c and d are minimal elements. Note that W has no first and no last element.

Example 6.3: Let $V = \{x \mid 0 < x < 1\}$. Then V has no maximal element and no minimal element.

The following remarks show the relationships between our previous concepts. Here A is a partially ordered set.

Remark 10.7: If a is a first element in A, then a is a minimal element in A and the only one. Similarly, a last element in A is a maximal element and the only one.

Remark 10.8: If A is totally ordered, it can contain at most one minimal element which would then be a first element. Likewise, it can contain at most one maximal element which would then be a last element.

Remark 10.9: Every finite partially ordered set has at least one maximal element and at least one minimal element. An infinite ordered set, as in Example 6.3, need not have any maximal nor minimal elements, even if it is totally ordered.

UPPER AND LOWER BOUNDS

Let B be a subset of a partially ordered set A. An element m in A is called a *lower bound* of B if, for every $x \in B$,

$$m \precsim x$$

that is, m precedes *every* element in B. If a lower bound of B dominates every other lower bound of B, then it is called the *greatest lower bound* (g.l.b.) or *infimum* of B and it is denoted by

$$\inf(B)$$

In general B may have no, one or many lower bounds, but there can be at most one $\inf(B)$.

Similarly, an element M in A is called an *upper bound* of B if M dominates every element in B, i.e. if, for every $x \in B$,

$$x \precsim M$$

If an upper bound of B precedes every other upper bound of B, then it is called the *least upper bound* (l.u.b.) or *supremum* of B and it is denoted by

$$\sup(B)$$

There can be at most one $\sup(B)$.

Example 7.1: Let $V = \{a, b, c, d, e, f, g\}$ be ordered by the following diagram:

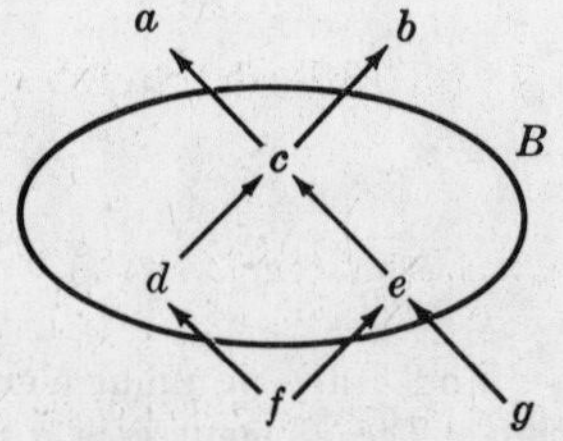

Let $B = \{c, d, e\}$. Then a, b and c are upper bounds of B, and f is the only lower bound of B. Note g is not a lower bound of B since g does not precede d; g and d are not comparable. Moreover, $c = \sup(B)$ belongs to B, while $f = \inf(B)$ does not belong to B.

Example 7.2: Let A be any bounded set of real numbers. Then a fundamental theorem about real numbers states that (in the natural ordering of $R^{\#}$) $\inf(A)$ and $\sup(A)$ exist.

Example 7.3: Let Q be the set of rational numbers. Let

$$B = \{x \mid x \in Q,\ 2 < x^2 < 3\}$$

that is, B consists of those rational numbers which lie between $\sqrt{2}$ and $\sqrt{3}$ on the real line. Then B has an infinite number of upper and lower bounds, but $\inf(B)$ and $\sup(B)$ do not exist. In other words, B has no least upper bound and no greatest lower bound. Note that the real numbers $\sqrt{2}$ and $\sqrt{3}$ do not belong to Q and cannot be considered as upper or lower bounds of B.

SIMILAR SETS

Two ordered sets are said to be *similar* if there exists a one-to-one correspondence between the elements which preserve the order relation. Specifically,

Definition 10.3: An ordered set A is similar to an ordered set B, denoted by

$$A \simeq B$$

if there exists a function $f: A \to B$ which is one-one and onto and which has the property that, for any elements $a, a' \in A$,

$$a \prec a' \quad \text{if and only if} \quad f(a) \prec f(a')$$

The function f is called a *similarity mapping* of A into B.

Example 8.1: Let $V = \{1, 2, 6, 8\}$ be ordered by "x divides y", and let $W = \{a, b, c, d\}$ be ordered by the following diagram:

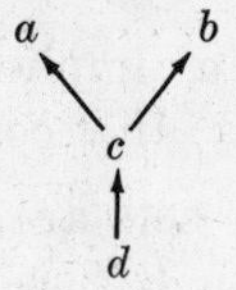

Note that a diagram of V is as follows:

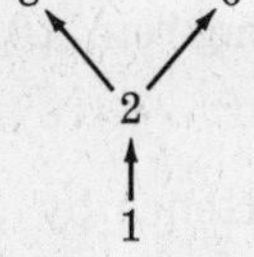

Then $V \simeq W$ since the function $f: V \to W$ defined by

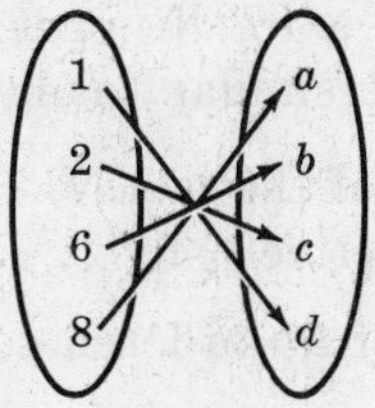

is a similarity mapping of V into W, i.e. establishes a one-to-one correspondence between the elements which preserves the order relation. Note

$$g = \{(1, d),\ (2, c),\ (6, a),\ (8, b)\}$$

is also a similarity mapping of V into W.

Example 8.2: Consider the natural numbers $N = \{1, 2, \ldots\}$ and the negative integers $M = \{-1, -2, \ldots\}$, each ordered by the natural order "$x \leq y$". Then N is not similar to M. For if $f: N \to M$ is a similarity mapping then, for every $a \in N$,

$$1 \precsim a \text{ should imply } f(1) \precsim f(a)$$

for every $f(a) \in M$. Since M has no first element, f cannot exist.

Example 8.3: The natural numbers $N = \{1, 2, 3, \ldots\}$ is similar to the even numbers $E = \{2, 4, 6, \ldots\}$ since the function $f: N \to E$ defined by $f(x) = 2x$ is a similarity of N into E.

The following theorems follow directly from the definition of similar sets.

Theorem 10.1.1: If A is totally ordered and $B \simeq A$, then B is totally ordered.

Theorem 10.1.2: Let $f: A \to B$ be a similarity mapping. Then $a \in A$ is a first (last, minimal or maximal) element if and only if $f(a)$ is a first (last, minimal or maximal) element of B.

Theorem 10.1.3: If A is similar to B then, in particular, A is equivalent to B.

The next theorem will play an important part in the subsequent theory.

Theorem 10.2: The relation in ordered sets defined by $A \simeq B$ is an equivalence relation, that is,

(1) $A \simeq A$ for any ordered set A
(2) if $A \simeq B$, then $B \simeq A$
(3) if $A \simeq B$ and $B \simeq C$, then $A \simeq C$

Remark 10.10: The condition in Definition 10.3 that

$$a \prec a' \text{ if and only if } f(a) \prec f(a')$$

is equivalent to the following two conditions:

(1) $a \prec a'$ implies $f(a) \prec f(a')$; (hence $a \succ a'$ implies $f(a) \succ f(a')$),
(2) $a \parallel a'$ (not comparable) implies $f(a) \parallel f(a')$.

Hence, if the sets are totally ordered, only (1) above is necessary.

ORDER TYPES

Note, by the previous Theorem 10.2, that the relation in ordered sets defined by

$$A \simeq B$$

is an equivalence relation. Hence by the Fundamental Theorem on Equivalence Relations, all partially ordered sets and, in particular, all totally ordered sets are partitioned into disjoint classes of similar sets.

Definition 10.4: Let A be a totally ordered set and let ξ denote the family of those sets which are similar to A. Then ξ is called the *order type* of A.

The order type of each of the sets N, Z and Q, the natural numbers, integers and rational numbers, is denoted, respectively, by ω, π and η.

If ξ is the order type of an ordered set A then ξ^* will denote the order type of A with the inverse order.

Example 9.1: The order type of $E = \{2, 4, 6, \ldots\}$ is ω, since E is similar to N.

Example 9.2: Note that $N = \{1, 2, 3, \ldots\}$ with the natural order is not similar to N with the inverse order; hence $\omega \neq \omega^*$. But $Z = \{\ldots, -2, -1, 0, 1, 2, \ldots\}$ with the natural order is similar to Z with the inverse order. Hence $\pi = \pi^*$.

Solved Problems

ORDERED SETS AND SUBSETS

1. The relation in N, the natural numbers, defined by "x divides y" is a partial order.

(1) Insert the correct symbol, $\prec$, $\succ$, or $\parallel$ (not comparable) between each pair of numbers:

$$(a)\ 2\_\_\_8, \quad (b)\ 18\_\_\_24, \quad (c)\ 9\_\_\_3, \quad (d)\ 5\_\_\_15$$

(2) State whether or not each of the following subsets of N is totally ordered:

(a) $\{24, 2, 6\}$, (b) $\{3, 15, 5\}$, (c) $\{15, 5, 30\}$, (d) $\{2, 8, 32, 4\}$, (e) $\{1, 2, 3, \ldots\}$, (f) $\{7\}$

Solution:

(1) (a) Since 2 divides 8, 2 precedes 8, i.e. $2 \prec 8$.
(b) 18 does not divide 24, and 24 does not divide 18; hence $18 \parallel 24$.
(c) Since 9 is divisible by 3, $9 \succ 3$.
(d) Since 5 divides 15, $5 \prec 15$.

(2) (a) Since 2 divides 6 which divides 24, the set is totally ordered.
(b) Since 3 and 5 are not comparable, the set is not totally ordered.
(c) The set is totally ordered since 5 divides 15 which divides 30.
(d) The set is totally ordered since $2 \prec 4 \prec 8 \prec 32$.
(e) The set is not totally ordered since 2 and 3 are not comparable.
(f) Any set consisting of one element is totally ordered.

2. Let $V = \{a, b, c, d, e\}$ be ordered by the following diagram:

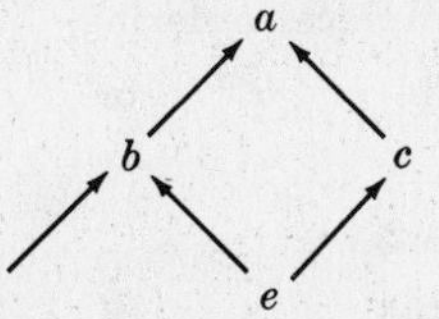

(1) Insert the correct symbol, $\prec$, $\succ$, or $\parallel$ (not comparable) between each pair of elements:

$$(a)\ a\_\_\_e, \quad (b)\ b\_\_\_c, \quad (c)\ d\_\_\_a, \quad (d)\ c\_\_\_d$$

(2) Construct a diagram of the elements in V which defines the inverse order.

Solution:

(1) (a) Since there is a "path" from e to c to a, e precedes a; hence $a \succ e$.
(b) There is no path from b to c, or vice versa; hence $b \parallel c$.
(c) There is a path from d to b to a; hence $d \prec a$.
(d) Neither $d \prec c$ nor $c \prec d$; hence $c \parallel d$.

(2) The inverse order is found by inverting the original diagram and reversing the arrows as follows:

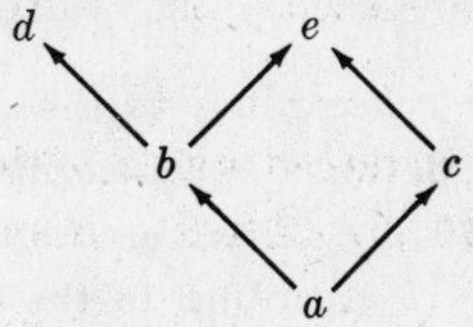

3. Let N, the natural numbers, be ordered as follows: Each pair of elements $a, a' \in N$ can be written uniquely in the form

$$a = 2^r(2s+1), \quad a' = 2^{r'}(2s'+1)$$

where $r, r', s, s' \in \{0, 1, 2, 3, \ldots\}$. Let

$$a \prec a' \text{ if } r < r' \text{ or if } r = r' \text{ but } s < s'$$

Insert the correct symbol, $\prec$ or $\succ$, between each of the following pairs of numbers:

(*a*) 5____14, (*b*) 6____9, (*c*) 3____20, (*d*) 14____21

Solution:

The elements in N can be written as follows:

s (columns), r (rows)

$r \backslash s$	0	1	2	3	4	5	6	7	
0	1	3	5	7	9	11	13	15	...
1	2	6	10	14	18	22	26	30	...
2	4	12	20	28	36	44	52	60	...
	.	.	.	.	.	.	.	.	
	.	.	.	.	.	.	.	.	
	.	.	.	.	.	.	.	.	

Then a number in a higher row precedes a number in a lower row and, if two numbers are in the same row, the number to the left precedes the number to the right. Consequently,

(*a*) $5 \prec 14$, (*b*) $6 \succ 9$, (*c*) $3 \prec 20$, (*d*) $14 \succ 21$

4. Let $N \times N$ be ordered lexicographically. Insert the correct symbol, $\prec$ or $\succ$, between each of the following pairs of elements of $N \times N$:

(*a*) (5, 78)____(7, 1), (*b*) (4, 6)____(4, 2), (*c*) (5, 5)____(4, 23), (*d*) (1, 3)____(1, 2)

Solution:

Note that according to the lexicographical ordering,

$$(a, b) \prec (a', b') \quad \text{if } a < a' \quad \text{or} \quad \text{if } a = a' \text{ but } b < b'$$

(*a*) $(5, 78) \prec (7, 1)$, since $5 < 7$.

(*b*) $(4, 6) \succ (4, 2)$, since $4 = 4$ but $6 > 2$.

(*c*) $(5, 5) \succ (4, 23)$, since $5 > 4$.

(*d*) $(1, 3) \succ (1, 2)$, since $1 = 1$ but $3 > 2$.

5. Let $A = (N, \preceq)$, the natural numbers with the natural order, and let $B = (N, \succeq)$, the natural numbers with the inverse order. Furthermore, let $A \times B$ denote the lexicographical ordering of $N \times N$ according to the order of A and then B. Insert the correct symbol, $\prec$ or $\succ$, between each of the following pairs of elements of $N \times N$.

(*a*) (3, 8)____(1, 1), (*b*) (2, 1)____(2, 8), (*c*) (3, 3)____(3, 1), (*d*) (4, 9)____(7, 15)

Solution:

The following rule applies: $(a, b) \prec (a', b')$ if $\begin{cases} a < a' \\ \text{or } a = a' \text{ but } b > b' \end{cases}$.

(*a*) $(3, 8) \succ (1, 1)$, since $3 > 1$; i.e. $3 \succ 1$ according to the order of A.

(*b*) $(2, 1) \succ (2, 8)$, since $2 = 2$ but $1 < 8$; i.e. $1 \succ 8$ according to the order of B.

(*c*) $(3, 3) \prec (3, 1)$, since $3 = 3$ but $3 > 1$; i.e. $3 \prec 1$ according to the order of B.

(*d*) $(4, 9) \prec (7, 15)$, since $4 < 7$; i.e. $4 \prec 7$ according to the order of A.

6. Let $\mathcal{A}$ be the family of all subsets A of the natural numbers N where A has the following properties: A is finite and the greatest common divisor of the elements of A is 1.

(1) State whether or not each of the following subsets of N belongs to $\mathcal{A}$:

(*a*) {2, 3, 8}
(*b*) {2, 3, 5, 8}
(*c*) {2, 5}
(*d*) {2, 3, 4, 5, . . .}
(*e*) {4, 6, 8}
(*f*) {2, 3}

(2) Now order $\mathcal{A}$ by set inclusion, i.e. $X \lesssim Y$ if $X \subset Y$, and let $\mathcal{B}$ be the subfamily of $\mathcal{A}$ which consists of the sets in (1) which belong to $\mathcal{A}$. Construct a diagram of $\mathcal{B}$.

Solution:

(1) The greatest common divisor of $\{4, 6, 8\}$ is 2, and $\{2, 3, 4, 5, \ldots\}$ is not finite; hence these sets do not belong to $\mathcal{A}$. All the other sets do belong to $\mathcal{A}$ and therefore belong to $\mathcal{B}$.

(2) A diagram of $\mathcal{B}$ is as follows:

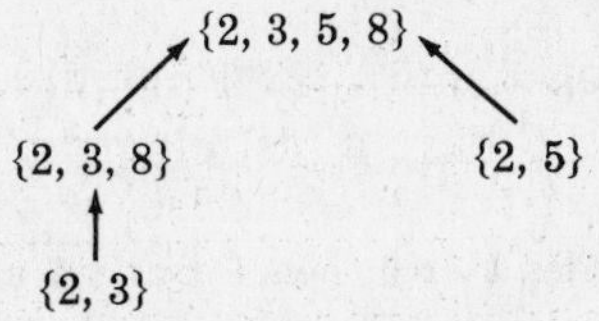

7. Let $A = \{a, b, c\}$ be ordered as follows:

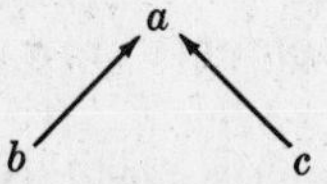

Let $\mathcal{A}$ be the family of all non-empty totally ordered subsets of A, and let $\mathcal{A}$ be partially ordered by set inclusion. Construct a diagram of $\mathcal{A}$.

Solution:

The totally ordered subsets of A are: $\{a\}$, $\{b\}$, $\{c\}$, $\{a, b\}$, $\{a, c\}$. Hence a diagram of $\mathcal{A}$ is as follows:

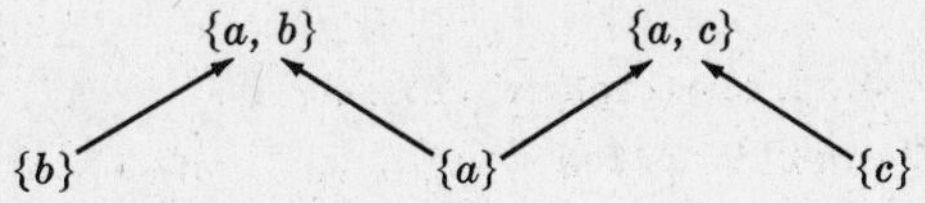

8. Let $B = \{1, 2, 3, 4, 5\}$ be ordered as follows:

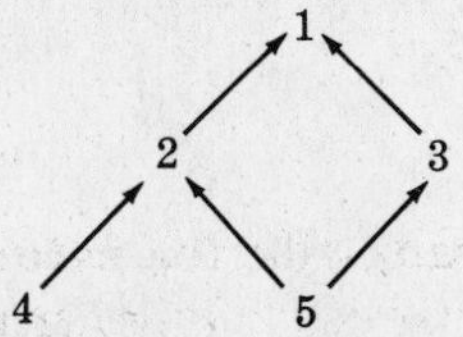

Let $\mathcal{B}$ be the family of all totally ordered subsets of B which contain 2 or more elements, and let $\mathcal{B}$ be partially ordered by set inclusion. Construct a diagram of $\mathcal{B}$.

Solution:

The elements in $\mathcal{B}$ are: $\{1,2,4\}$, $\{1,2,5\}$, $\{1,3,5\}$, $\{1,2\}$, $\{1,4\}$, $\{2,4\}$, $\{1,5\}$, $\{2,5\}$, $\{1,3\}$, $\{3,5\}$. Hence the diagram of $\mathcal{B}$ is as follows:

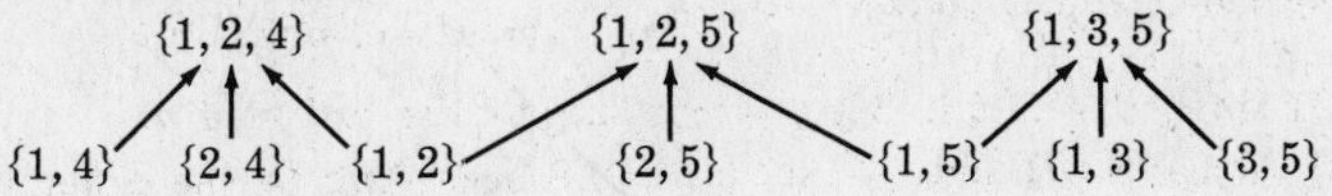

MINIMAL, MAXIMAL, FIRST AND LAST ELEMENTS

9. Let $A = \{2, 3, 4, 5, \ldots\}$ be ordered by "x divides y". (1) Find all minimal elements. (2) Find all maximal elements.

Solution:

(1) If p is a prime number, then only p divides p (since $1 \notin A$); hence all the prime numbers are minimal elements. Furthermore, if $a \in A$ is not prime, then there is a number $b \in A$ such that b divides a, i.e. $b \lesssim a$, and $b \neq a$. Therefore the only minimal elements are the prime numbers.

(2) There are no maximal elements since, for every $a \in A$, a divides, in particular, $2a$.

10. Let $B = \{1,2,3,4,5\}$ be ordered as follows:

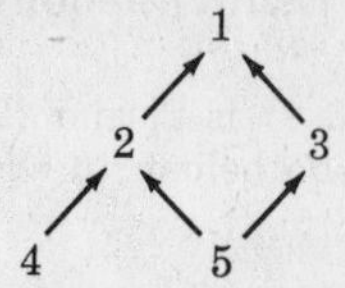

(1) Find all the minimal elements. (2) Find all the maximal elements. (3) Does B have a first element? (4) Does B have a last element?

Solution:

(1) No element strictly precedes 4 or 5; hence 4 and 5 are minimal elements.

(2) The only maximal element is 1.

(3) There is no first element. Note 5 is not a first element since 5 does not precede 4.

(4) The number 1 is a last element since it dominates every element in B.

11. Prove: Let a and b be minimal elements in a totally ordered set A; then $a = b$.

Solution:

The elements a and b are comparable since A is totally ordered; hence $a \precsim b$ or $b \precsim a$. Since b is a minimal element, $a \precsim b$ implies $a = b$; and since a is a minimal element, $b \precsim a$ implies $a = b$. In either case, $a = b$.

12. Let $B = \{2,3,4,5,6,8,9,10\}$ be ordered by "x is a multiple of y". (1) Find all maximal elements of B. (2) Find all minimal elements of B. (3) Does B have a first or a last element?

Solution:

First construct a diagram of B as follows:

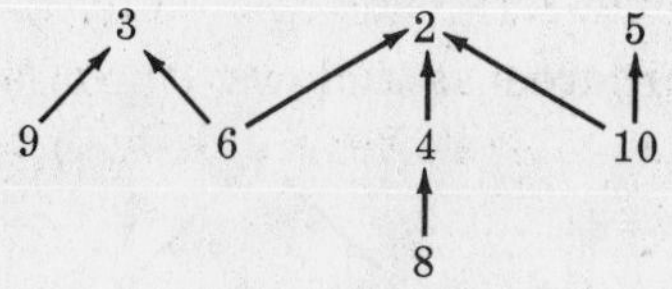

(1) The maximal elements are 3, 2 and 5. (2) The minimal elements are 9, 6, 8 and 10. (3) There is no first and no last element.

UPPER AND LOWER BOUNDS

13. Let $W = \{1, 2, \ldots, 7, 8\}$ be ordered as follows:

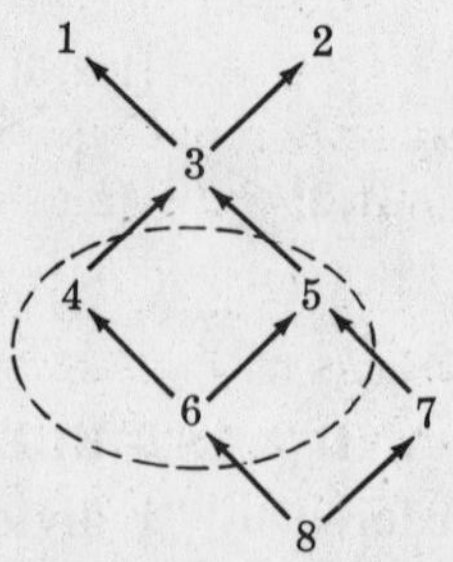

Consider the subset $V = \{4,5,6\}$ of W. (1) Find the set of upper bounds of V. (2) Find the set of lower bounds of V. (3) Does sup(V) exist? (4) Does inf(V) exist?

Solution:

(1) Each of the elements in $\{1,2,3\}$, and only these elements, dominates every element in V and therefore is an upper bound.

(2) Only 6 and 8 precede every element in V; hence $\{6, 8\}$ is the set of lower bounds. Note that 7 is not a lower bound since 7 does not precede 4 or 6.

(3) Since 3 is a first element in the set of upper bounds of V, $\sup(V) = 3$. Notice that 3 does not belong to V.

(4) Since 6 is a last element in the set of lower bounds of V, $\inf(V) = 6$. Notice here that 6 does belong to V.

14. Let $D = \{1, 2, 3, 4, 5, 6\}$ be ordered as follows:

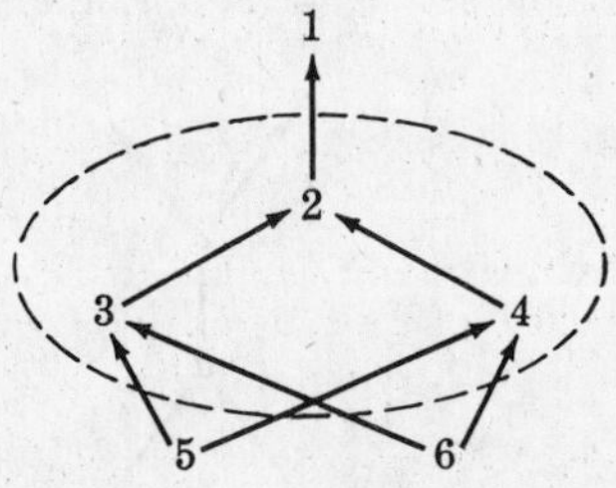

Consider the subset $E = \{2, 3, 4\}$ of D. (1) Find the set of upper bounds of E. (2) Find the set of lower bounds of E. (3) Does $\sup(E)$ exist? (4) Does $\inf(E)$ exist?

Solution:

(1) Both 1 and 2, and no other elements, dominate every number in E; hence $\{1, 2\}$ is the set of upper bounds of E.

(2) Both 5 and 6, and no other numbers, precede every number in E; hence $\{6, 5\}$ is the set of lower bounds of E.

(3) Since 2 is a first element in $\{1, 2\}$, the set of upper bounds of E, then $\sup(E) = 2$.

(4) Since $\{5, 6\}$, the set of lower bounds of E, has no last element, then $\inf(E)$ does not exist.

15. Consider Q, the set of rational numbers, and its subset $A = \{x \mid x \in Q,\ x^3 < 3\}$.

(1) Is A bounded above, i.e. does A have an upper bound?

(2) Is A bounded below, i.e. does A have a lower bound?

(3) Does $\sup(A)$ exist?

(4) Does $\inf(A)$ exist?

Solution:

(1) A is bounded above since, for example, 50 is an upper bound.

(2) There are no lower bounds of A; hence A is not bounded below.

(3) Sup (A) does not exist. Considering A as a subset of $R\#$, the real numbers, then $\sqrt[3]{3}$ would be the least upper bound of A; but as a subset of Q, $\sup(A)$ does not exist.

(4) Inf (A) does not exist since the set of lower bounds is empty.

16. Let $\mathcal{A}$ be a family of sets which is partially ordered by set inclusion, and let $\mathcal{B} = \{A_i\}_{i \in I}$ be a subset of $\mathcal{A}$.

(1) Prove that if $B \in \mathcal{A}$ is an upper bound of $\mathcal{B}$, then $(\cup_{i \in I} A_i) \subset B$.

(2) Is $\cup_{i \in I} A_i$ an upper bound of $\mathcal{B}$?

Solution:

(1) Let x belong to $\cup_{i \in I} A_i$. Then there exists an A_j, where $j \in I$, such that $x \in A_j$. Since B is an upper bound, $A_j \subset B$; hence x belongs to B. Since $x \in \cup_{i \in I} A_i$ implies $x \in B$, then $(\cup_{i \in I} A_i) \subset B$.

(2) Even though $\{A_i\}_{i \in I}$ is a subfamily of $\mathcal{A}$, it need not be true that the union $\cup_{i \in I} A_i$ is an element in $\mathcal{A}$. Therefore $\cup_{i \in I} A_i$ is an upper bound of $\mathcal{B}$ if, and only if, $\cup_{i \in I} A_i$ belongs to $\mathcal{A}$.

17. Let N, the natural numbers, be ordered by "x divides y", and let $A = \{a_1, a_2, \ldots, a_m\}$ be a finite subset of N. (1) Does $\inf(A)$ exist? (2) Does $\sup(A)$ exist?

Solution:

(1) The greatest common divisor of the elements in A is $\inf(A)$ and it always exists.

(2) The least common multiple of the elements in A is $\sup(A)$ and it always exists.

SIMILAR SETS

18. Give an example of an ordered set $X = (A, R)$ which is similar to $Y = (A, R^{-1})$, the set A with the inverse order.

Solution:

The set of rational numbers Q, with the natural order, is similar to Q with the inverse order. In fact the function $f: Q \to Q$ defined by $f(x) = -x$ is a similarity mapping since, for any real numbers,

$$x \leqq y \quad \text{if and only if} \quad -x \geqq -y$$

As a second example, consider the set $W = \{a, b, c, d, e, f\}$ ordered as follows:

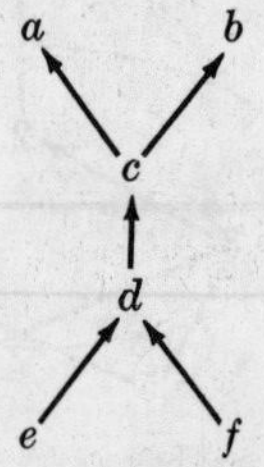

The following diagram, which is derived by inverting the original diagram and reversing the arrows, defines the inverse order:

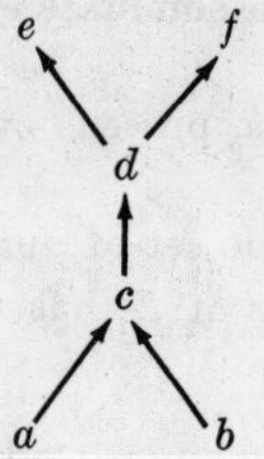

Notice that the two diagrams are similar. The function

$$f = \{(a, e), (b, f), (c, d), (d, c), (e, a), (f, b)\}$$

is a similarity mapping.

19. Let A be an ordered set and, for any element $a \in A$, let $S(a)$ be the set of elements which precede a, i.e.,

$$S(a) = \{x \mid x \in A, x \precsim a\}$$

Furthermore, let $\mathcal{A} = \{S(a)\}_{a \in A}$, the family of all sets $S(a)$, be partially ordered by set inclusion. Prove that A is similar to $\mathcal{A}$.

Solution:

We show that the function $f: A \to \mathcal{A}$ defined by $f: x \to S(x)$ is a similarity mapping. If $a \precsim b$, then $x \precsim a$ implies $x \precsim b$; hence $a \precsim b$ implies $S(a) \subset S(b)$. Also if $S(a) \subset S(b)$, then $a \in S(a)$ also belongs to $S(b)$; hence $S(a) \subset S(b)$ implies $a \precsim b$. Thus f preserves order.

By definition, f is onto. We now show that it is one-one. If $a \neq b$, then either $a \prec b$, $b \prec a$ or a and b are not comparable. In the first and last case, $b \in S(b)$ does not belong to $S(a)$. In the second case, $a \in S(a)$ does not belong to $S(b)$. Thus in either of the cases, $S(a) \neq S(b)$. Therefore f is one-one.

Consequently, f is a similarity mapping.

For example, consider the set $A = \{a, b, c, d, e\}$ ordered as follows:

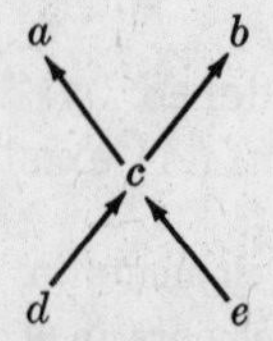

Note that $\mathcal{A} = \{S(a), S(b), S(c), S(d), S(e)\}$ is ordered as follows:

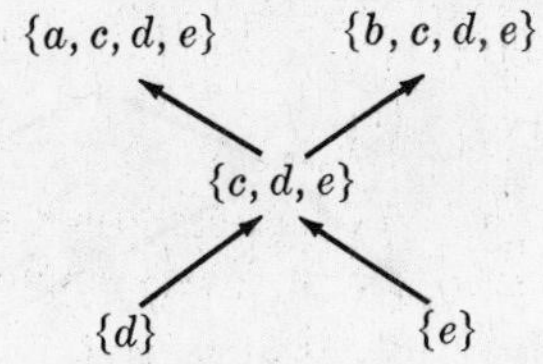

Note further that the two diagrams are similar.

Supplementary Problems

ORDERED SETS AND SUBSETS

20. The relation in N, the natural numbers, defined by "x is a multiple of y" is a partial order.

(1) Insert the correct symbol, $\prec$, $\succ$, or $\|$ (not comparable) between each pair of numbers:

(*a*) 3 ___ 7, (*b*) 2 ___ 8, (*c*) 6 ___ 1, (*d*) 3 ___ 33

(2) State whether or not each of the following subsets of N is totally ordered:

(*a*) $\{8, 2, 24\}$ (*c*) $\{5, 1, 9\}$ (*e*) $\{2, 4, 6, 8, \ldots\}$

(*b*) $\{5\}$ (*d*) $\{2, 4, 8, 24\}$ (*f*) $\{15, 3, 9\}$

21. Let $W = \{1, 2, 3, 4, 5, 6\}$ be ordered as follows:

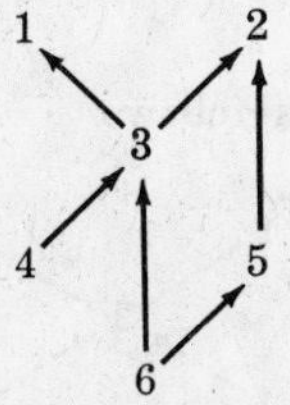

(1) Insert the correct symbol, $\prec$, $\succ$, or $\|$ (not comparable) between each pair of elements:

(*a*) 1 ___ 6, (*b*) 4 ___ 5, (*c*) 5 ___ 1, (*d*) 4 ___ 2

(2) Construct a diagram of the elements in W which defines the inverse order.

(3) Find all the totally ordered subsets of W, each of which contains at least three elements.

(4) Find all the totally ordered subsets of W with the inverse order, each of which contains at least three elements.

22. Let $A = (N, \leqq)$, the natural numbers with the natural order, let $B = (N, \geqq)$, the natural numbers with the inverse order, and let $A \times B$ be ordered lexicographically. Insert the correct symbol, $\prec$ or $\succ$, between each of the following pairs of elements of $A \times B$.

(*a*) $(1, 3)$ ___ $(1, 5)$, (*b*) $(4, 1)$ ___ $(2, 18)$, (*c*) $(4, 30)$ ___ $(4, 4)$, (*d*) $(2, 2)$ ___ $(15, 15)$

23. Let $D = \{1, 2, 3, 4\}$ be ordered as follows:

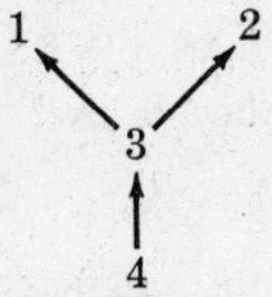

Let $\mathcal{B}$ be the family of all non-empty totally ordered subsets of D ordered by set inclusion. Construct a diagram of $\mathcal{B}$.

MINIMAL, MAXIMAL, FIRST AND LAST ELEMENTS

24. Let $B = \{a, b, c, d, e, f\}$ be ordered as follows:

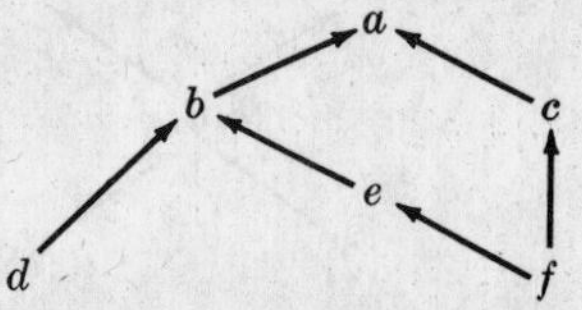

(1) (a) Find all the minimal elements of B. (c) Does B have a first element?
(b) Find all the maximal elements of B. (d) Does B have a last element?

(2) Let $\mathcal{B}$ be the family of all non-empty totally ordered subsets, and let $\mathcal{B}$ be ordered by set inclusion.
(a) Find all maximal elements of $\mathcal{B}$. (c) Does $\mathcal{B}$ have a first element?
(b) Find all minimal elements of $\mathcal{B}$. (d) Does $\mathcal{B}$ have a last element?

25. Let $M = \{2, 3, 4, \ldots\}$ and let $M \times M$ be ordered as follows:

$$(a, b) \precsim (c, d) \quad \text{if } a \text{ divides } c \text{ and if } b \text{ is less than or equal to } d.$$

(1) Find all the minimal elements. (2) Find all the maximal elements.

26. Let $M = \{2, 3, 4, \ldots\}$ be ordered by "x divides y". Furthermore, let $\mathcal{M}$ be the family of all non-empty totally ordered subsets of M and let $\mathcal{M}$ be partially ordered by set inclusion. (1) Find all minimal elements of $\mathcal{M}$. (2) Find all maximal elements of $\mathcal{M}$.

27. State whether each of the following statements is true or false and, if it false, give a counter-example.
(1) If a partially ordered set A has only one maximal element a, then a is also a last element.
(2) If a finite partially ordered set A has only one maximal element a, then a is also a last element.
(3) If a totally ordered set A has only one maximal element a, then a is also a last element.

UPPER AND LOWER BOUNDS

28. Let $W = \{1, 2, \ldots, 7, 8\}$ be ordered as follows:

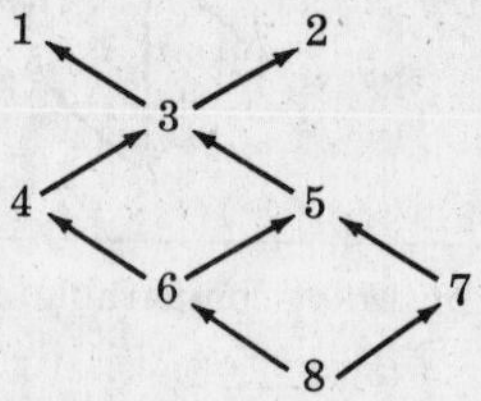

(1) Consider the subset $A = \{4, 5, 7\}$ of W.
(a) Find the set of upper bounds of A. (c) Does $\sup(A)$ exist?
(b) Find the set of lower bounds of A. (d) Does $\inf(A)$ exist?

(2) Consider the subset $B = \{2, 3, 6\}$ of W.
(a) Find the set of upper bounds of B. (c) Does $\sup(B)$ exist?
(b) Find the set of lower bounds of B. (d) Does $\inf(B)$ exist?

(3) Consider the subset $C = \{1, 2, 4, 7\}$ of W.
(a) Find the set of upper bounds of C. (c) Does $\sup(C)$ exist?
(b) Find the set of lower bounds of C. (d) Does $\inf(C)$ exist?

29. Consider Q, the set of rational numbers with the natural order, and its subset A:

$$A = \{x \mid x \in Q,\ 8 < x^3 < 15\}$$

(1) Is A bounded above? (2) Is A bounded below? (3) Does $\sup(A)$ exist? (4) Does $\inf(A)$ exist?

SIMILAR SETS

30. Find the maximum number of pairwise non-similar partially ordered sets with three elements, and construct a diagram of each.

31. Prove Theorem 10.2: The relation in sets defined by $A \simeq B$ is an equivalence relation.

Answers to Supplementary Problems

20. (1) (*a*) $3 \parallel 7$, (*b*) $2 \succ 8$, (*c*) $6 \prec 1$, (*d*) $3 \succ 33$
(2) (*a*) Yes, (*b*) Yes, (*c*) No, (*d*) Yes, (*e*) No, (*f*) No

21. (1) (*a*) $1 \succ 6$, (*b*) $4 \parallel 5$, (*c*) $5 \parallel 1$, (*d*) $4 \prec 2$
(2)

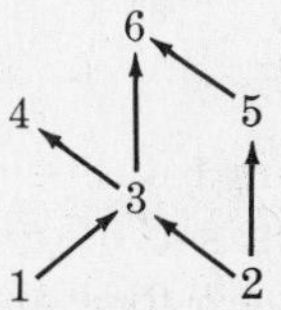

(3) $\{1,3,4\}$, $\{1,3,6\}$, $\{2,3,4\}$, $\{2,3,6\}$, $\{2,5,6\}$
(4) The same sets as in (3).

22. (*a*) $(1,3) \succ (1,5)$, (*b*) $(4,1) \succ (2,18)$, (*c*) $(4,30) \prec (4,4)$, (*d*) $(2,2) \prec (15,15)$

23.

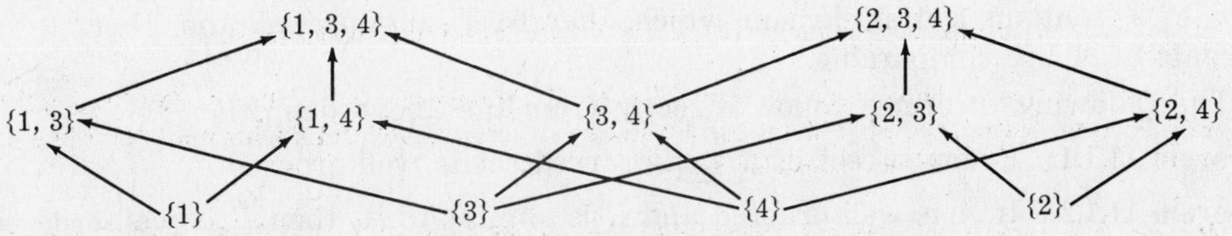

24. (1) (*a*) d and f, (*b*) a, (*c*) No, (*d*) Yes, a is a last element.
(2) (*a*) $\{a,b,d\}$, $\{a,b,e,f\}$, $\{a,c,f\}$
(*b*) The subsets consisting of one element: $\{a\}$, $\{b\}$, $\{c\}$, $\{d\}$, $\{e\}$, $\{f\}$.
(*c*) No, (*d*) No

25. (1) Any ordered pair $(p, 2)$, where p is a prime, is a minimal element.
(2) There is no maximal element.

26. (1) Each subset consisting of one element is a minimal element.
(2) Each set of the form $\{p_1, p_1p_2, p_1p_2p_3, \ldots\}$ where $p_1, p_2, \ldots$ is any sequence of primes, is a maximal element.

27. (1) False. Consider, for example, the set $\{a, 1, 2, 3, \ldots\}$ ordered as follows:

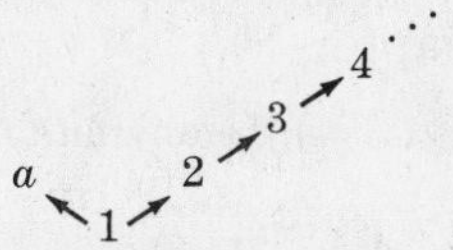

Note that the subset $\{1,2,3,\ldots\}$ has the natural order. Then a is a maximal element, the only one, and it is not a last element.
(2) True. (3) True. In fact, a totally ordered set can have at most one maximal element and it will always be a last element.

28. (1) (*a*) $\{1,2,3\}$, (*b*) $\{8\}$, (*c*) $\sup(A) = 3$, (*d*) $\inf(A) = 8$
(2) (*a*) $\{2\}$, (*b*) $\{6,8\}$, (*c*) $\sup(B) = 2$, (*d*) $\inf(B) = 6$
(3) (*a*) $\emptyset$. There are no upper bounds. (*b*) $\{8\}$, (*c*) No, (*d*) $\inf(C) = 8$.

29. (1) Yes, (2) Yes, (3) No, (4) $\inf(A) = 2$

30. There are five non-similar ways to order three elements, i.e. a set $A = \{a, b, c\}$:

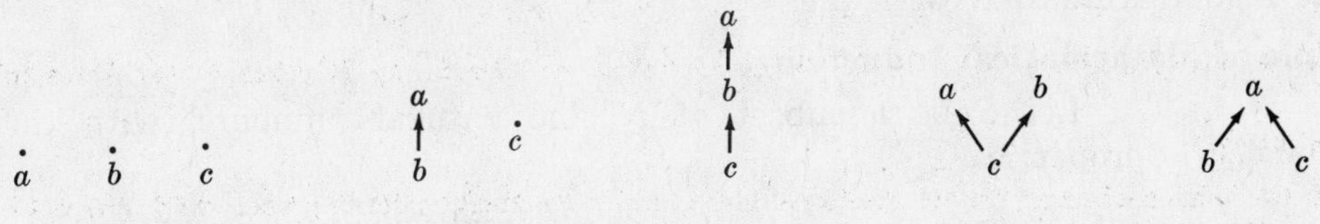

(1) (2) (3) (4) (5)

Chapter 11

Well-ordered Sets. Ordinal Numbers

WELL-ORDERED SETS

Not every ordered set, even if it is totally ordered, need have a first element. One of the fundamental properties of N, the set of natural numbers with the natural order, is that N and every subset of N does have a first element. We call an ordered set *well-ordered* if it has this property. Specifically,

Definition 11.1: Let A be an ordered set with the property that every subset of A contains a first element. Then A is called a *well-ordered* set.

In particular, any well-ordered set A is totally ordered. For if $a, b \in A$, then the subset $\{a, b\}$ of A contains a first element which, therefore, must precede the other; hence any two elements of A are comparable.

The following theorems follow directly from the above definition.

Theorem 11.1.1: Every subset of a well-ordered set is well-ordered.

Theorem 11.1.2: If A is well-ordered and B is similar to A, then B is well-ordered.

Example 1.1: Consider the ordered subsets

$$A_1 = \{1, 3, 5, \ldots\} \quad \text{and} \quad A_2 = \{2, 4, 6, \ldots\}$$

of N which are also well-ordered. Then the union (ordered from left to right)

$$A_1 \cup A_2 = \{1, 3, 5, \ldots;\ 2, 4, 6, \ldots\}$$

is also well-ordered. This example shows that it is possible for a set, such as $N = A_1 \cup A_2$ to be well-ordered in more than one way.

The preceding example can be generalized as follows:

Theorem 11.2: Let $\{A_i\}_{i \in I}$ be a well-ordered family of pairwise disjoint well-ordered sets. Then the union of the sets $\cup_{i \in I} A_i$ is well-ordered. (The order of the union of a totally ordered family of totally ordered sets is defined in Example 2.4 of Chapter 10.)

Example 1.2: Let $V = \{a_1, a_2, \ldots, a_n\}$ be any finite totally ordered set. Then V can be written as

$$V = \{a_{i_1}, a_{i_2}, \ldots, a_{i_n}\}$$

where the a_{i_j} are the original elements rearranged according to the order. Notice that V is well-ordered. Furthermore, notice that any other totally ordered set of n elements

$$W = \{b_{i_1}, b_{i_2}, \ldots, b_{i_n}\}$$

is similar to V.

In view of Example 1.2, we state

Theorem 11.3: All finite totally ordered sets with the same number of elements are well-ordered and are similar to each other.

TRANSFINITE INDUCTION

The reader is familiar with the

Principle of Mathematical Induction

Let S be a subset of N, the natural numbers, with the following properties:

(1) $1 \in S$

(2) $n \in S$ implies $n+1 \in S$.

Then S is the set of natural numbers (i.e. $S = N$).

The above principle is one of Peano's Axioms for the natural numbers. The principle can be shown to be a consequence of the fact that N is well-ordered. In fact, there is a somewhat similar statement which is true for every well-ordered set.

Principle of Transfinite Induction

Let S be a subset of a well-ordered set A with the following properties:

(1) $a_0 \in S$

(2) $s(a) \subset S$ implies $a \in S$.

Then $S = A$.

Here, a_0 is the first element of A and $s(a)$, called the *initial segment* of a, is defined to be the set of elements in A which strictly precede a.

LIMIT ELEMENTS

An element b in an ordered set A is called an *immediate successor* of an element $a \in A$, and a is called the *immediate predecessor* of b if $a \prec b$ and there does not exist an element $c \in A$ such that

$$a \prec c \prec b$$

Example 2.1: Let $A = \{a, b, c, d, e\}$ be ordered as follows:

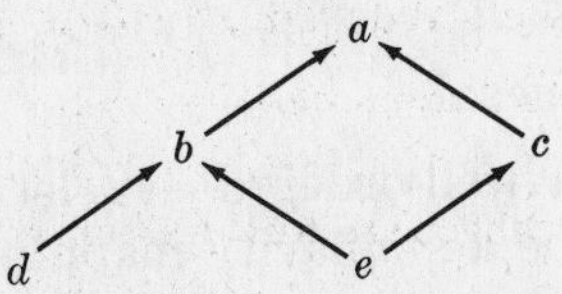

Then b is an immediate successor of both d and e, and e is an immediate predecessor of both b and c.

Example 2.2: Consider the set Q of rational numbers. No element in Q has an immediate successor or an immediate predecessor. For if $a, b \in Q$, say $a < b$, then $(a+b)/2 \in Q$ and

$$a < (a+b)/2 < b$$

Example 2.3: Let A be a well-ordered set and let $M(a)$ denote the set of elements which strictly dominate $a \in A$. If $M(a) \neq \emptyset$, i.e. if a is not a last element, then $M(a)$ has a first element b. Then b is the immediate successor of a.

In view of Example 2.3, we state

Theorem 11.4: Every element in a well-ordered set has an immediate successor except the last element.

There is no analogous statement to Theorem 11.4 about immediate predecessors, that is, there do exist elements in well-ordered sets, besides the first element, which do not have immediate predecessors.

Example 2.4: Let $D = \{1, 3, 5, \ldots\}$ and $E = \{2, 4, 6, \ldots\}$. Then, in the well-ordered set

$$\{D;\ E\} = \{1, 3, 5, \ldots;\ 2, 4, 6, \ldots\}$$

both 1 and 2 do not have immediate predecessors.

Remark 11.1: Here, and subsequently, $\{D; E\}$ means the set $D \cup E$ which is ordered positionwise from left to right, i.e., any element in D precedes any element in E, and elements in the same set keep the same order.

In view of the preceding example, we introduce

Definition 11.2: An element in a well-ordered set is called a *limit element* if it does not have an immediate predecessor and if it is not the first element.

INITIAL SEGMENTS

Let A be a well-ordered set. The *initial segment* $s(a)$ of an element $a \in A$ consists of all elements in A which strictly precede a. In other words,

$$s(a) = \{x \mid x \in A,\ x \prec a\}$$

Notice that $s(a)$ is a subset of A.

Example 3.1: Let $D = \{1, 3, 5, \ldots\}$ and $E = \{2, 4, 6, \ldots\}$. Consider the well-ordered set

$$\{D;\ E\} = \{1, 3, 5, \ldots;\ 2, 4, 6, \ldots\}$$

Then $s(1) = \emptyset$, $s(5) = \{1, 3\}$, $s(2) = \{1, 3, 5, \ldots\}$, and $s(8) = \{1, 3, 5, \ldots;\ 2, 4, 6\}$.

One basic property of initial segments is contained in the next theorem.

Theorem 11.5: Let $S(A)$ denote the family of all initial segments of elements in a well-ordered set A, and let $S(A)$ be ordered by set inclusion. Then A is similar to $S(A)$ and, in particular, the function $f: A \rightarrow S(A)$ defined by $f: x \rightarrow s(x)$ is a similarity mapping of A into $S(A)$.

SIMILARITY BETWEEN A WELL-ORDERED SET AND ITS SUBSET

Consider N, the natural numbers, and the subset $E = \{2, 4, 6, \ldots\}$ of N. The function $f: N \rightarrow E$ defined by $f(x) = 2x$ is a similarity mapping of N into its subset E. Notice that, for every $x \in N$,

$$x \precsim f(x)$$

This property is true in general.

Theorem 11.6: Let A be a well-ordered set, let B be a subset of A, and let the function $f: A \rightarrow B$ be a similarity mapping of A into B. Then, for every $a \in A$,

$$a \precsim f(a)$$

The following important properties of well-ordered sets are consequences of the preceding theorem.

Theorem 11.7: Let A and B be similar well-ordered sets. Then there exists only one similarity mapping of A into B.

Theorem 11.8: A well-ordered set cannot be similar to one of its initial segments.

COMPARISON OF WELL-ORDERED SETS

The next theorem gives an important relationship between any two well-ordered sets.

Theorem 11.9: Given any two well-ordered sets, either they are similar to each other or one of them is similar to an initial segment of the other.

If a well-ordered set A is equivalent to an initial segment of a well-ordered set B, then A is said to be *shorter* than B or B is said to be *longer* than A. With these definitions, Theorem 11.9 can be restated as follows:

Theorem 11.9′: Let A and B be well-ordered sets. Then A is shorter than B, A is similar to B or A is longer than B.

The preceding theorem can be strengthened as follows:

Theorem 11.10: Let $\mathcal{A}$ be any family of pairwise non-similar well-ordered sets. Then there exists a set $A \in \mathcal{A}$ such that A is shorter than every other set in $\mathcal{A}$.

Example 4.1: Consider two finite well-ordered sets

$$A = \{a_1, \ldots, a_n\} \quad \text{and} \quad B = \{b_1, \ldots, b_m\}$$

Then if $n < m$, A is similar to the initial segment $\{b_1, \ldots, b_n\}$ of B; hence A would be shorter than B. Similarly, if $n > m$ then A would be longer than B.

Example 4.2: Note that $N = \{1, 2, 3, \ldots\}$ is shorter than the well-ordered set

$$\{1, 3, 5, \ldots ;\ 2, 4, 6, \ldots\}$$

since it is similar to the initial segment $\{1, 3, 5, \ldots\}$.

ORDINAL NUMBERS

Notice once again, by Theorem 10.2, that the relation in ordered sets defined by

$$A \simeq B$$

i.e. A is similar to B, is an equivalence relation. Hence by the Fundamental Theorem on Equivalence Relations, all ordered sets, and in particular all well-ordered sets, are partitioned into disjoint classes of similar sets.

Definition 11.3: Let A be any well-ordered set and let λ denote the family of well-ordered sets which are similar to A. Then λ is called an *ordinal number* and it is denoted by

$$\lambda = \text{ord}(A)$$

Definition 11.4: The ordinal number of each of the well-ordered sets

$$\emptyset,\ \{1\},\ \{1, 2\},\ \{1, 2, 3\},\ \ldots$$

is denoted by $0, 1, 2, 3, \ldots$ respectively, and is called a *finite* ordinal number. All other ordinals are called *transfinite* numbers.

Definition 11.5: The ordinal number of N, the natural numbers, is denoted by

$$\omega = \text{ord}(N)$$

Remark 11.2: Although the symbols $0, 1, 2, 3, \ldots$ are used to denote natural numbers, cardinal numbers and, now, ordinal numbers, the context in which the symbols appear determines their particular meaning. Furthermore, since by Theorem 11.3 any two finite well-ordered sets with the same number of elements are similar, $0, 1, 2, \ldots$ are the only finite ordinal numbers.

In view of the definition of the order type of a totally ordered set in the preceding chapter, Definition 11.3 can be restated as follows:

Definition 11.3′: Let λ be the order type of a totally ordered set A. If A is well-ordered then λ is called an *ordinal number*.

INEQUALITIES AND ORDINAL NUMBERS

An inequality relation is defined for the ordinal numbers as follows:

Definition 11.6: Let λ and μ be two ordinal numbers and A and B be two well-ordered sets such that

$$\lambda = \text{ord}(A) \quad \text{and} \quad \mu = \text{ord}(B)$$

Then

$$\lambda < \mu$$

if A is similar to an initial segment of B.

In other words, for $\lambda = \text{ord}(A)$ and $\mu = \text{ord}(B)$,

$\lambda < \mu$ if A is shorter than B
$\lambda = \mu$ if A is similar to B
$\lambda > \mu$ if A is longer than B
$\lambda \leqq \mu$ if $\lambda < \mu$ or $\lambda = \mu$
$\lambda \geqq \mu$ if $\lambda > \mu$ or $\lambda = \mu$

Example 5.1: Consider two finite well-ordered sets

$$A = \{a_1, a_2, \ldots, a_n\} \quad \text{and} \quad B = \{b_1, \ldots, b_m\}$$

ordered positionwise from left to right. Say $n \leq m$. Then A is similar to the initial segment $\{b_1, \ldots, b_n\}$ of B. Hence $\text{ord}(A) \leq \text{ord}(B)$.

In other words, $n \leq m$ as ordinal numbers if and only if $n \leq m$ as natural numbers. Thus the inequality relation for ordinal numbers is an extension of the inequality relation in the set of natural numbers.

Example 5.2: Let $\lambda = \text{ord}(\{1, 3, 5, \ldots; 2, 4, 6, \ldots\})$. Since N, the natural numbers, is similar to the initial segment $\{1, 3, 5, \ldots\}$,

$$\omega < \lambda$$

The next theorem is a direct consequence of Theorem 11.9 and the above definition.

Theorem 11.11: Any set of ordinal numbers is totally ordered by the relation $\lambda \leq \mu$.

In view of Theorem 11.10, the preceding theorem can be strengthened as follows:

Theorem 11.12: Any set of ordinal numbers is well-ordered by the relation $\lambda \leq \mu$.

Now let λ be any ordinal number and $s(\lambda)$ denote the set of ordinal numbers less than λ. By the preceding theorem, $s(\lambda)$ is a well-ordered set and, therefore, $\text{ord}(s(\lambda))$ exists. Question: what is the relationship between λ and $\text{ord}(s(\lambda))$? The answer is given in the next theorem.

Theorem 11.13: Let $s(\lambda)$ be the set of ordinals less than the ordinal λ. Then $\lambda = \text{ord}(s(\lambda))$.

Remark 11.3: Since the ordinal numbers are themselves well-ordered, every ordinal has an immediate successor. Some non-zero ordinals, for example ω, do not have immediate predecessors; these are called *limit ordinal numbers* or, simply, *limit numbers*.

ORDINAL ADDITION

An operation of *addition* is defined for ordinal numbers as follows:

Definition 11.7: Let λ and μ be ordinal numbers such that $\lambda = \text{ord}(A)$ and $\mu = \text{ord}(B)$, where A and B are disjoint. Then

$$\lambda + \mu = \text{ord}(\{A; B\})$$

Example 6.1: Note $\omega = \text{ord}(\{1, 2, \ldots\})$ and $n = \text{ord}(\{a_1, \ldots, a_n\})$. Then

$$n + \omega = \text{ord}(\{a_1, \ldots, a_n; 1, 2, \ldots\}) = \omega$$

But

$$\omega + n = \text{ord}(\{1, 2, \ldots; a_1, \ldots, a_n\}) > \omega$$

since N is equivalent to $s(a_1)$, the initial segment of a_1.

Thus we see, by Example 6.1, that the operation of addition of ordinal numbers is not commutative. However the following conditions do hold.

Theorem 11.14: (1) Addition of ordinal numbers satisfies the associative law, i.e.,

$$(\lambda + \mu) + \eta = \lambda + (\mu + \eta)$$

(2) The ordinal 0 is an additive identity element, i.e.,

$$0 + \lambda = \lambda + 0 = \lambda$$

Example 6.2: In this example we will denote the finite ordinals by

$$0^*, 1^*, 2^*, \ldots$$

Consider, now, two finite well-ordered disjoint sets

$$A = \{a_1, \ldots, a_n\} \quad \text{and} \quad B = \{b_1, \ldots, b_m\}$$

Then

$$n^* + m^* = \operatorname{ord}(A) + \operatorname{ord}(B) = \operatorname{ord}(\{A;\ B\}) = (n+m)^*$$

Thus the operation of addition for finite ordinal numbers corresponds to the operation of addition for the set of natural numbers.

Note once again that the set of ordinal numbers is itself a well-ordered set; hence every ordinal has an immediate successor. For the finite ordinals, i.e. the natural numbers, it is easily seen that $n+1$ is the immediate successor to n. The next theorem states that this property is true in general.

Theorem 11.15: Let λ be any ordinal number. Then $\lambda + 1$ is the immediate successor of λ.

Addition of real numbers and, therefore, of natural numbers, is a binary operation and can be extended by induction to any finite sum

$$a_1 + a_2 + \cdots + a_n$$

of real numbers. The sum of an infinite number of real numbers, such as

$$1 + 2 + 3 + 4 + \cdots$$
$$1 + \tfrac{1}{2} + \tfrac{1}{4} + \tfrac{1}{8} + \cdots$$

has no meaning (unless one introduces the concepts of limits). On the other hand, it is possible to define the sum of an infinite number of ordinal numbers as follows:

Let $\{\lambda_i\}_{i \in I}$ be any well-ordered set, finite or infinite, of ordinal numbers. In other words, I is a well-ordered set and to each $i \in I$ there corresponds an ordinal number λ_i. Furthermore, let

$$\lambda_i = \operatorname{ord}(A_i)$$

Then the family of sets $\{A_i \times \{i\}\}_{i \in I}$ is a well-ordered family of pairwise disjoint well-ordered sets. By Theorem 11.2,

$$\cup_{i \in I}\{A_i \times \{i\}\}$$

is a well-ordered set. We therefore state

Definition 11.8: Let $\{\lambda_i\}_{i \in I}$ be a well-ordered set of ordinal numbers such that $\lambda_i = \operatorname{ord}(A_i)$. Then

$$\textstyle\sum_{i \in I} \lambda_i = \operatorname{ord}(\cup_{i \in I}\{A_i \times \{i\}\})$$

Example 6.3: By the above definition, $1 + 1 + 1 + \cdots = \omega$. In fact, if each λ_i is finite (and not 0) then

$$\lambda_1 + \lambda_2 + \lambda_3 + \cdots = \textstyle\sum_{i \in N} \lambda_i = \omega$$

ORDINAL MULTIPLICATION

An operation of multiplication is defined for ordinal numbers as follows:

Definition 11.9: Let λ and μ be ordinal numbers such that $\lambda = \operatorname{ord}(A)$ and $\mu = \operatorname{ord}(B)$. Then

$$\lambda\mu = \operatorname{ord}(\{A \times B\})$$

where $\{A \times B\}$ is ordered *reverse lexicographically.*

Note that $\{A \times B\}$ is ordered reverse lexicographically means that

$$(a, a') \prec (b, b') \quad \text{if} \quad a' \prec b' \text{ or } a' = b' \text{ but } a \prec b$$

Remark 11.4: Unless otherwise stated, the product set $\{A \times B\}$ of two well-ordered sets A and B is to be ordered reverse lexicographically.

Example 7.1: Note first that $2 = \text{ord}(\{a, b\})$ and $\omega = \text{ord}(\{1, 2, 3, \ldots\})$. Then

$$2\omega = \text{ord}(\{(a, 1), (b, 1), (a, 2), (b, 2), \ldots, (a, n), (b, n), \ldots\}) = \omega$$

But

$$\omega 2 = \text{ord}(\{(1, a), (2, a), \ldots; (1, b), (2, b), \ldots\}) > \omega$$

since N is similar to the initial segment $\{(1, a), (2, a), \ldots\}$.

Thus we see that the operation of multiplication of ordinal numbers is not commutative. However, the following conditions do hold.

Theorem 11.16: (1) The associative law for multiplication holds, i.e.,

$$\lambda(\mu\eta) = (\lambda\mu)\eta$$

(2) The left distributive law of multiplication over addition holds, i.e.,

$$\lambda(\mu + \eta) = \lambda\mu + \lambda\eta$$

(3) The ordinal 1 is a multiplicative identity element, i.e.,

$$1\lambda = \lambda 1 = \lambda$$

STRUCTURE OF ORDINAL NUMBERS

We now write down many of the ordinal numbers according to their order. First come the finite ordinals

$$0, 1, 2, 3, \ldots$$

and then comes the first limit ordinal ω and its successors

$$\omega, \omega + 1, \omega + 2, \ldots$$

Note (see Example 7.1) that $\text{ord}(\{0, 1, 2, \ldots; \omega, \omega + 1, \omega + 2, \ldots\}) = \omega 2$. Hence next comes the second limit number $\omega 2$ and its successors

$$\omega 2, \omega 2 + 1, \omega 2 + 2, \omega 2 + 3, \ldots$$

The next limit number is $\omega 3$. We proceed as follows:

$$\omega 3, \omega 3 + 1, \ldots, \omega 4, \ldots, \omega 5, \ldots, \ldots, \omega\omega = \omega^2$$

Here $\omega\omega = \omega^2$ is the limit number following the limit numbers ωn, where $n \in N$. We continue:

$$\omega^2, \omega^2 + 1, \ldots, \omega^2 + \omega, \omega^2 + \omega + 1, \ldots, \omega^2 + \omega 2, \ldots, \omega^2 + \omega 3, \ldots, \ldots, \omega^2 + \omega^2 = \omega^2 2$$

Then

$$\omega^2 2, \ldots, \omega^2 3, \ldots, \omega^2 4, \ldots, \omega^2\omega = \omega^3$$

Then we have the powers of ω:

$$\omega^3, \omega^3 + 1, \ldots, \omega^4, \ldots, \omega^5, \ldots, \ldots, \omega^\omega$$

Here ω^ω is the limit number after the limit numbers ω^n, where $n \in N$. We proceed:

$$\omega^\omega, \ldots, (\omega^\omega)^\omega, \ldots, ((\omega^\omega)^\omega)^\omega, \ldots, \ldots$$

After all these ordinals we have the ordinal ϵ_0. We can continue:

$$\epsilon_0, \epsilon_0 + 1, \ldots$$

Each of the ordinal numbers we have enumerated is still the ordinal number of a denumerable set.

AUXILIARY CONSTRUCTION OF ORDINAL NUMBERS

Note once again

Theorem 11.13: Let $s(\lambda)$ be the set of ordinal numbers which precede λ. Then

$$\lambda = \text{ord}(s(\lambda))$$

Some authors use this property of ordinal numbers to actually define the ordinal numbers. Roughly speaking, an ordinal number is defined to be the set of ordinal numbers which precede it. Specifically,

Definition:

$$0 \equiv \emptyset$$
$$1 \equiv \{0\}$$
$$2 \equiv \{0, 1\}$$
$$3 \equiv \{0, 1, 2\}$$
$$\vdots$$
$$\omega \equiv \{0, 1, 2, \ldots\}$$
$$\omega+1 \equiv \{0, 1, 2, \ldots, \omega\}$$
$$\omega+2 \equiv \{0, 1, 2, \ldots, \omega, \omega+1\}$$
$$\vdots$$
$$\omega 2 \equiv \{0, 1, \ldots, \omega, \omega+1, \ldots\}$$
$$\omega 2+1 \equiv \{0, 1, \ldots, \omega, \omega+1, \ldots, \omega 2\}$$
$$\vdots$$

One main reason the ordinal numbers are developed as above is in order to avoid certain inherent contradictions which appear in the preceding development of the ordinal numbers (see Chapter 13).

Solved Problems

1. Prove the Principle of Transfinite Induction: Let S be a subset of a well-ordered set A with the following properties: (1) $a_0 \in S$, (2) $s(a) \subset S$ implies $a \in S$. Then $S = A$.

Solution:

Suppose $S \neq A$, i.e. suppose $A - S = T$ is not empty. Since A is well-ordered, T has a first element t_0. Each element $x \in s(t_0)$ precedes t_0 and, therefore, cannot belong to T, i.e. belongs to S; hence $s(t_0) \subset S$. By (2), $t_0 \in S$. This contradicts the fact that $t_0 \in A - S$. Hence the original assumption that $S \neq A$ is not true; in other words, $S = A$. (Note that (1) is in fact a consequence of (2) since $\emptyset = s(a_0)$ is a subset of S and, therefore, implies $a_0 \in S$.)

2. Prove Theorem 11.5: Let $S(A)$ denote the family of all initial segments of elements in a well-ordered set A, and let $S(A)$ be ordered by set inclusion. Then A is similar to $S(A)$ and, in particular, the function $f: A \to S(A)$ defined by $f: x \to s(x)$ is a similarity mapping of A into $S(A)$.

Solution:

By definition f is onto. We show that f is one-one. Suppose $x \neq y$. Then one of them, say x, strictly precedes the other; hence $x \in s(y)$. But, by definition of initial segment, $x \notin s(x)$. Thus $s(x) \neq s(y)$, and hence f is one-one.

We show that f preserves order, that is,

$$x \precsim y \text{ if and only if } s(x) \subset s(y)$$

Let $x \precsim y$. If $a \in s(x)$ then $a \prec x$ and hence $a \prec y$; thus $a \in s(y)$. Since $a \in s(x)$ implies $a \in s(y)$, $s(x)$ is a subset of $s(y)$. Now suppose $x \not\precsim y$, that is, $x \succ y$; then $y \in s(x)$. But, by definition of initial segment, $y \notin s(y)$; hence $s(x) \not\subset s(y)$. In other words, $x \precsim y$ if and only if $s(x) \subset s(y)$.

3. Prove Theorem 11.6: Let A be a well-ordered set, let B be a subset of A, and let $f: A \to B$ be a similarity mapping of A into B. Then, for every $a \in A$, $a \precsim f(a)$.

Solution:

Let $D = \{x \mid f(x) \prec x\}$. If D is empty the theorem is true. Suppose $D \neq \emptyset$. Then, since A is well-ordered, D has a first element d_0. Note $d_0 \in D$ implies $f(d_0) \prec d_0$. Since f is a similarity mapping,

$$f(d_0) \prec d_0 \quad \text{implies} \quad f(f(d_0)) \prec f(d_0)$$

Consequently, $f(d_0)$ also belongs to D. But $f(d_0) \prec d_0$ and $f(d_0) \in D$ contradicts the fact that d_0 is the first element of D. Hence the original assumption that $D \neq \emptyset$ leads to a contradiction. Therefore D is empty and the theorem is true.

4. Prove Theorem 11.7: Let A and B be similar well-ordered sets. Then there exists only one similarity mapping of A into B.

Solution:

Let $f: A \to B$ and $g: A \to B$ be similarity mappings. Suppose $f \neq g$. Then there exists an element $x \in A$ such that $f(x) \neq g(x)$. Consequently, either $f(x) \prec g(x)$ or $g(x) \prec f(x)$. Say $f(x) \prec g(x)$.

Since $g: A \to B$ is a similarity mapping, $g^{-1}: B \to A$ is also a similarity mapping. Furthermore, $g^{-1} \circ f: A \to A$, the product of two similarity mappings, is a similarity mapping. But

$$f(x) \prec g(x) \quad \text{implies} \quad (g^{-1} \circ f)(x) \prec (g^{-1} \circ g)(x) = x$$

We have $g^{-1} \circ f$ is a similarity mapping and $(g^{-1} \circ f)(x) \prec x$. These facts contradict Theorem 11.6. Hence the assumption that $f \neq g$ leads to a contradiction. Accordingly, there can be only one similarity mapping of A into B.

5. Prove Theorem 11.8: A well-ordered set cannot be similar to one of its initial segments.

Solution:

Let A be a well-ordered set and let $f: A \to s(a)$ be a similarity mapping of A into one of its initial segments. Then $f(a) \in s(a)$. Therefore

$$f(a) \prec a$$

This last fact contradicts Theorem 11.6. Therefore A cannot be similar to one of its initial segments.

6. Prove: Let A be a well-ordered set and let S be a subset of A with the property that

$$a \precsim b \text{ and } b \in S \quad \text{implies} \quad a \in S$$

Then $S = A$ or S is an initial segment of A.

Solution:

Suppose $S \neq A$. Then $A - S$ has a first element a_0 where $a_0 \notin S$. We show that $S = s(a_0)$. If $x \prec a_0$ then $x \notin A - S$, i.e., $x \in S$; hence $s(a_0) \subset S$.

Now suppose $y \notin s(a_0)$, that is, suppose $a_0 \precsim y$. But

$$y \in S \text{ and } a_0 \precsim y \quad \text{implies} \quad a_0 \in S$$

which contradicts the fact that $a_0 \notin S$. Hence $y \notin S$.

In other words, $y \notin s(a_0)$ implies $y \notin S$, which means $S \subset s(a_0)$. Therefore $S = s(a_0)$.

7. Prove: Two different initial segments of a well-ordered set cannot be similar.

Solution:

Let $s(a)$ and $s(b)$ be two different initial segments, that is, $a \neq b$. Either $a \prec b$ or $b \prec a$; say $a \prec b$. Then $s(a)$ is an initial segment of the well-ordered set $s(b)$. Hence, by Theorem 11.8, $s(b)$ is not similar to $s(a)$.

8. Prove: Let A and B be well-ordered sets, and let an initial segment $s(a)$ of A be similar to an initial segment of B. Then $s(a)$ is similar to a unique initial segment $s(b)$ of B.

Solution:

Let $s(a) \simeq s(b)$ and $s(a) \simeq s(b')$ where $b, b' \in B$. Then $s(b) \simeq s(b')$. By Problem 7, $s(b) = s(b')$ and, therefore, $b = b'$.

9. Prove: Let A and B be well-ordered sets such that an initial segment $s(a)$ of A is similar to an initial segment $s(b)$ of B. Then each initial segment of $s(a)$ is similar to an initial segment of $s(b)$, that is,

$$a' \precsim a \quad \text{implies} \quad s(a') \simeq s(b') \text{ where } b' \precsim b$$

Furthermore, if $f: s(a) \to s(b)$ is the similarity mapping of $s(a)$ into $s(b)$, then f restricted to $s(a')$ is the similarity mapping of $s(a')$ into $s(b') = f(s(a'))$.

Solution:

Let $f(a') = b'$. Note that f restricted to $s(a')$ is one-one and preserves order; hence $s(a') \simeq f(s(a'))$. Furthermore, since f is a similarity mapping,

$$a^* \prec a' \quad \text{if and only if} \quad f(a^*) \prec b'$$

Then $f(s(a')) = s(b')$, and therefore $s(a') \simeq s(b')$.

10. Prove: Let A and B be well-ordered sets and let

$$S = \{x \mid x \in A,\ s(x) \simeq s(y) \text{ where } y \in B\}$$

(In other words, each element $x \in S$ has the property that its initial segment $s(x)$ is similar to an initial segment $s(y)$ of B.) Then $S = A$ or S is an initial segment of A.

Solution:

Let $x \in S$ and $y \precsim x$. By Problem 9, $s(y)$ is similar to an initial segment of B; hence $y \in S$. In other words,

$$y \precsim x \text{ and } x \in S \quad \text{implies} \quad y \in S$$

By Problem 6, $S = A$ or S is an initial segment of A.

11. Prove: Let A and B be well-ordered sets and let

$$S = \{x \mid x \in A,\ s(x) \simeq s(y) \text{ where } y \in B\}$$
$$T = \{y \mid y \in B,\ s(y) \simeq s(x) \text{ where } x \in A\}$$

Then S is similar to T.

Solution:

Let $x \in S$. Then, by Problem 8, $s(x)$ is similar to a unique segment $s(y)$ of B. Thus to each $x \in S$ there corresponds a unique $y \in Y$ such that $s(x) \simeq s(y)$, and vice versa. Hence the function $f: S \to T$ defined by

$$f(x) = y \quad \text{if} \quad s(x) \simeq s(y)$$

is one-one and onto.

Now let $x', x \in S$, $f(x) = y$, $f(x') = y'$ and $x' \prec x$. The theorem is proven if we can show that $y' \prec y$, that is, that f preserves order.

Let $\emptyset: s(x) \to s(y)$ be the similarity mapping of $s(x)$ into $s(f(x)) = s(y)$. By Problem 9, $\emptyset$ restricted to $s(x')$ is a similarity mapping of $s(x')$ into the initial segment $s(\emptyset(x'))$ of B. But, by Problem 8, there exists only one similarity mapping of $s(x')$ into B. Consequently, $\emptyset(x') = f(x') = y'$. Since $\emptyset(x') \in s(y)$,

$$\emptyset(x') = y' \prec y$$

Therefore, S is similar to T.

12. Prove Theorem 11.9: Let A and B be well-ordered sets. Then A is shorter than B, A is similar to B or A is longer than B.

Solution:

Let S and T be defined as in the preceding problem. Note $S \simeq T$. By Problem 10, there are four possibilities:

Case I. $S = A$ and $T = B$. Then A is similar to B.

Case II. $S = A$ and $T = s(b)$, an initial segment of B. Then A is shorter than B.

Case III. $T = B$ and $S = s(a)$, an initial segment of A. Then A is longer than B.

Case IV. $S = s(a)$ and $T = s(b)$. Then $a \in S$ since its initial segment $s(a)$ is similar to an initial segment $s(b)$ of B. But a cannot belong to its own initial segment; hence this case is impossible.

Thus the theorem is true.

13. Prove: Let $\mathcal{A}$ be a family of initial segments of a well-ordered set A. Then there is an initial segment $s(a) \in \mathcal{A}$ such that $s(a) \subset s(x)$ for any other initial segment $s(x)$ in $\mathcal{A}$, that is, there is an initial segment $s(a) \in \mathcal{A}$ which is shorter than every other initial segment in $\mathcal{A}$.

Solution:

By Theorem 11.5, A is similar to $S(A)$, the family of all initial segments of elements in A, ordered by set inclusion. Since A is well-ordered, $S(A)$ is also well-ordered. Consequently, $\mathcal{A}$, a subset of $S(A)$, has a first element $s(a)$. Therefore $s(a) \subset s(x)$ for any other initial segment $s(x) \in \mathcal{A}$.

14. Prove Theorem 11.10: Let $\mathcal{A}$ be any family of pairwise non-similar well-ordered sets. Then there exists a set $A_0 \in \mathcal{A}$ such that A_0 is shorter than every other set in $\mathcal{A}$.

Solution:

Let B be any set in $\mathcal{A}$. Define

$$\mathcal{B} = \{X \mid X \in \mathcal{A},\ X \text{ is shorter than } B\}$$

If $\mathcal{B}$ is empty, then B satisfies the requirements of the theorem. Suppose $\mathcal{B} \neq \emptyset$. If we show that $\mathcal{B}$ has a shortest set A_0 then, considering the way $\mathcal{B}$ was defined, A_0 will also be the shortest set in $\mathcal{A}$.

Now, by Theorem 11.9, every set $A \in \mathcal{B}$ is similar to an initial segment $s(a)$ of B. Let $\mathcal{B}'$ be the family of those initial segments of B each of which is similar to a set in $\mathcal{B}$. By Problem 13, $\mathcal{B}'$ contains an initial segment $s(a_0)$ which is shorter than every other initial segment in $\mathcal{B}'$. Consequently, the set $A_0 \in \mathcal{B}$, which is similar to $s(a_0)$, is shorter than every other set in $\mathcal{B}$.

Hence A_0 satisfies the requirements of the theorem.

ORDINAL NUMBERS

15. Prove: Let $\lambda = \text{ord}(A)$ and let $\mu < \lambda$. Then there is a unique initial segment $s(a)$ of A such that $\mu = \text{ord}(s(a))$.

Solution:

Let $\mu = \text{ord}(B)$. Since $\mu < \lambda$, B is shorter than A, that is, B is similar to an initial segment $s(a)$ of A. Therefore $\mu = \text{ord}(s(a))$. Furthermore, $s(a)$ is the only initial segment whose ordinal number is μ since, by Problem 7, two different initial segments of A cannot be similar.

16. Prove Theorem 11.13: Let $s(\lambda)$ be the set of ordinals less than the ordinal λ. Then $\lambda = \text{ord}(s(\lambda))$.

Solution:

Let $\lambda = \text{ord}(A)$ and let $S(A)$ denote the family of all initial segments of A, ordered by set inclusion. By Theorem 11.5, $A \simeq S(A)$; hence $\lambda = \text{ord}(S(A))$. If we show that $s(\lambda)$ is similar to $S(A)$, the theorem will follow.

Let $\mu \in s(\lambda)$; then $\mu < \lambda$. By Problem 15, there is a unique initial segment $s(a)$ of A such that $\mu = \text{ord}(s(a))$. Hence the function $f: s(\lambda) \to S(A)$ defined by

$$f(\mu) = s(a) \quad \text{if} \quad \mu = \text{ord}(s(a))$$

is one-one. Furthermore, f is onto; for if $s(b) \in S(A)$, then $s(b)$ is shorter than A and therefore $\text{ord}(s(b)) = \eta < \text{ord}(A) = \lambda$, i.e. $\eta \in s(\lambda)$; hence $f(\eta) = s(b)$.

To complete the proof of the theorem, it is only necessary to show that f preserves order; then f is a similarity mapping and $s(\lambda) \simeq S(A)$. Let $\mu < \eta$, where $\mu, \eta \in s(\lambda)$. Then $\mu = \text{ord}(s(a))$ and $\eta = \text{ord}(s(b))$, that is, $f(\mu) = s(a)$ and $f(\eta) = s(b)$. Since $\mu < \eta$, $s(a)$ is an initial segment of $s(b)$; hence $s(a)$ is a proper subset of $s(b)$. In other words, under the ordering of $S(A)$, $s(a) \prec s(b)$. Thus f preserves order.

17. Prove Theorem 11.15: Let λ be any ordinal number. Then $\lambda + 1$ is the immediate successor of λ.

Solution:

Let μ be the immediate successor of λ. Then, by definition of $s(\mu)$,

$$s(\mu) = s(\lambda) \cup \{\lambda\}$$

Hence,

$$\text{ord}(s(\mu)) = \text{ord}(s(\lambda)) + \text{ord}(\{\lambda\})$$

that is, $\mu = \lambda + 1$.

18. Prove, by giving a counter-example, that the right distributive law of multiplication over addition (for the ordinal numbers) is not true in general. In other words, exhibit three ordinal numbers λ, μ and η such that

$$(\lambda + \mu)\eta \neq \lambda\eta + \mu\eta$$

Solution:

Note, by Example 7.1, $(1+1)\omega = 2\omega = \omega$, and (using the left distributive law)

$$1\omega + 1\omega = \omega + \omega = \omega 1 + \omega 1 = \omega(1+1) = \omega 2 > \omega$$

Hence $(1+1)\omega \neq 1\omega + 1\omega$.

19. Let $\{A_i\}_{i \varepsilon I}$ be a well-ordered family of pairwise disjoint well-ordered sets, and let $\text{ord}(I) = \omega$ and $\text{ord}(A_i) = \omega$ for every $i \varepsilon I$. Find $\text{ord}(\cup_{i \varepsilon I} A_i)$.

Solution:

$$\text{ord}(\cup_{i \varepsilon I} A_i) = \omega + \omega + \omega + \cdots = \omega(1+1+1+\cdots) = \omega\omega \equiv \omega^2$$

20. Prove: $\omega + \omega = \omega 2$.

Solution:

Method 1. Note that

$$\omega + \omega = \omega 1 + \omega 1 = \omega(1+1) = \omega 2$$

Here, the left distributive law is used.

Method 2. Consider the well-ordered sets

$$A = \{a_1, a_2, \ldots\}, \quad B = \{b_1, b_2, \ldots\}, \quad C = \{c_1, c_2, \ldots\}, \quad D = \{r, s\}$$

Note that

$$\omega = \text{ord}(A) = \text{ord}(B) = \text{ord}(C) \quad \text{and} \quad 2 = \text{ord}(D)$$

Then

$$\omega + \omega = \text{ord}(\{A;\ B\}) = \text{ord}(\{a_1, a_2, \ldots;\ b_1, b_2, \ldots\})$$

$$\omega 2 = \text{ord}(\{C \times D\}) = \text{ord}(\{(c_1, r), (c_2, r), \ldots;\ (c_1, s), (c_2, s), \ldots\})$$

But the function $f: \{A;\ B\} \rightarrow \{C \times D\}$ defined by

$$f(x) = \begin{cases} (c_i, r) & \text{if } x = a_i \\ (c_i, s) & \text{if } x = b_i \end{cases}$$

is a similarity mapping of $\{A;\ B\}$ into $C \times D$. Hence

$$\omega + \omega = \text{ord}(\{A;\ B\}) = \text{ord}(\{C \times D\}) = \omega 2$$

Supplementary Problems

21. Prove Theorem 11.1.2: If A is a well-ordered set and B is similar to A, then B is well-ordered.

22. Prove Theorem 11.2: Let $\{A_i\}_{i \varepsilon I}$ be a well-ordered family of pairwise disjoint well-ordered sets. Then the union of the sets $\cup_{i \varepsilon I} A_i$ is well-ordered.

23. Assume that N, the set of natural numbers with the natural order, is well-ordered. Prove the Principle of Mathematical Induction: Let S be a subset of N with the properties (1) $1 \varepsilon S$ and (2) $n \varepsilon S$ implies $n+1 \varepsilon S$; then $S = N$.

24. Prove that 0 is the identity element for addition of ordinal numbers, that is, for any ordinal λ, $0 + \lambda = \lambda + 0 = \lambda$.

25. Prove that 1 is the identity element for multiplication of ordinal numbers, that is, for any ordinal λ, $1\lambda = \lambda 1 = \lambda$.

26. Prove: If each λ_i, $i \in N$, is a finite ordinal, then $\lambda_1 + \lambda_2 + \lambda_3 + \cdots = \sum_{i \in I} \lambda_i = \omega$.

27. Prove: Let λ be any infinite ordinal number. Then $\lambda = \mu + n$, where μ is a limit number and n is a finite ordinal.

28. State whether each of the following statements about ordinals is true or false; if it is true prove it, and if it is false give a counter-example:

(1) If $\lambda \neq 0$, then $\mu < \lambda + \mu$.

(2) If $\lambda \neq 0$, then $\mu < \mu + \lambda$.

29. State whether each of the following statements concerning ordinals is true or false; if it is true prove it, and if it is false give a counter-example:

(1) If $\lambda \neq 0$ and $\mu < \eta$, then $\lambda + \mu < \lambda + \eta$.

(2) If $\lambda \neq 0$ and $\mu < \eta$, then $\mu + \lambda < \eta + \lambda$.

30. Prove: The left distributive law of multiplication over addition holds for ordinal numbers, i.e., $\lambda(\mu + \eta) = \lambda\mu + \lambda\eta$.

Answers to Supplementary Problems

27. *Hint.* Note that a well-ordered set cannot contain an ordered subset $A = \{\cdots \prec a_3 \prec a_2 \prec a_1\}$, since A is not well-ordered.

28. (1) False, (2) True

29. (1) True, (2) False

Chapter 12

Axiom of Choice. Zorn's Lemma. Well-ordering Theorem

CARTESIAN PRODUCTS AND CHOICE FUNCTIONS

Definition 12.1: Let $\{A_i\}_{i \in I}$ be a non-empty family of non-empty sets. Then the Cartesian product of $\{A_i\}_{i \in I}$, denoted by

$$\prod_{i \in I} A_i$$

is the set of all choice functions defined on $\{A_i\}_{i \in I}$.

Recall that a function $f: \{A_i\}_{i \in I} \to X$, where $\{A_i\}_{i \in I}$ is a family of subsets of X, is called a choice function if $f(A_i) = a_i \in A_i$, for every $i \in I$. In other words, f "chooses" a point $a_i \in A_i$ for each set A_i.

Example 1.1: Let $\{A_1, A_2, \ldots, A_n\}$ be a finite family of sets. In Chapter 5, we defined the Cartesian product of the n sets,

$$A_1 \times A_2 \times \cdots \times A_n \equiv \prod_{i=1}^{n} A_i$$

to be the set of n-tuples

$$(a_1, a_2, \ldots, a_n)$$

where $a_i \in A_i$ for $i = 1, \ldots, n$. But to each choice function f defined on $\{A_1, \ldots, A_n\}$ there corresponds the unique n-tuple

$$(f(A_1), f(A_2), \ldots, f(A_n))$$

and vice versa. Accordingly, in the finite case, Definition 12.1 agrees with the previous definition of the Cartesian product.

The main reason for introducing Definition 12.1 is that it applies to any family of sets: finite, denumerable or even non-denumerable. The previous definition, which used the concept of n-tuples, applied only to a finite family of sets.

Remark 12.1: Although a choice function is defined for a family of subsets, any family of sets $\{A_i\}_{i \in I}$ can be considered to be a family of subsets of their union $\bigcup_{i \in I} A_i$.

AXIOM OF CHOICE

The axiom of choice lies at the foundations of mathematics and, in particular, the theory of sets. This "innocent looking" axiom, which follows, has as a consequence some of the most powerful and important results in mathematics.

Axiom of Choice: The Cartesian product of a non-empty family of non-empty sets is non-empty.

In view of Definition 12.1, the Axiom of Choice can be stated as follows:

Axiom of Choice: There exists a choice function for any non-empty family of non-empty sets.

The axiom of choice is equivalent to the following postulate:

Zermelo's Postulate: Let $\{A_i\}_{i \in I}$ be any non-empty family of disjoint non-empty sets. Then there exists a subset B of $\cup_{i \in I} A_i$ such that the intersection of B and each set A_i consists of exactly one element.

Observe that in Zermelo's Postulate the sets are disjoint whereas in the Axiom of Choice may not be disjoint.

WELL-ORDERING THEOREM, ZORN'S LEMMA

The following theorem is attributed to Zermelo who proved the theorem directly from the axiom of choice.

Well-ordering Theorem: Every set can be well-ordered.

Zorn's Lemma, which follows, is one of the most important tools in mathematics; it establishes the existence of certain types of elements although no constructive process is given to find these elements.

Zorn's Lemma: Let X be a non-empty partially ordered set in which every totally ordered subset has an upper bound in X. Then X contains at least one maximal element.

We formally state and prove (Problem 4) the following basic result of set theory:

Theorem 12.1: The following are equivalent: (i) Axiom of Choice, (ii) Well-ordering Theorem, (iii) Zorn's Lemma.

CARDINAL AND ORDINAL NUMBERS

To each ordinal number $\lambda = \operatorname{ord}(A)$ we can associate a unique cardinal number $\alpha = \#(A)$. We call α the cardinal number of λ and denote it by

$$\alpha = \bar{\lambda}$$

This function from the ordinal numbers to the cardinal numbers is not one-one, that is, there are different ordinal numbers with the same cardinal number. For example,

$$\omega = \operatorname{ord}(\{1, 2, 3, \ldots\})$$

$$\omega 2 = \operatorname{ord}(\{a_1, a_2, \ldots; b_1, b_2, \ldots\})$$

are both ordinal numbers of denumerable sets, i.e. sets with the same cardinal number $\mathfrak{a}$. In other words,

$$\bar{\omega} = \mathfrak{a} = \overline{\omega 2}$$

The well-ordering theorem implies that the above function from the ordinal numbers to the cardinal numbers is onto. For, suppose $\alpha = \#(A)$ is any cardinal number. By the well-ordering theorem, A can be well-ordered; say $\lambda = \operatorname{ord}(A)$. Then $\alpha = \bar{\lambda}$. Hence α is the cardinal number of at least one ordinal number λ. (Here, A is used both as the original set and then as the well-ordered set.)

The following correspondence between the ordinal and cardinal numbers is easily established.

Theorem 12.2: Let $\alpha = \bar{\lambda}$ and $\beta = \bar{\mu}$ be cardinal numbers. Then

(1) $\alpha < \beta$ implies $\lambda < \mu$

(2) $\lambda < \mu$ implies $\alpha \leq \beta$

The next result, mentioned previously, is a direct consequence of the Well-ordering Theorem.

Theorem 9.11 (Law of Trichotomy): Let α and β be any cardinal numbers. Then one of the following holds:

$$\alpha < \beta, \quad \alpha = \beta \quad \text{or} \quad \alpha > \beta$$

In other words, the inequality relation defined for the cardinal numbers is a total order and not only a partial order. Since the ordinal numbers are themselves well-ordered, we can make an even stronger statement.

Theorem 12.3: Any set of cardinal numbers is well-ordered by the relation $\alpha \leqq \beta$.

ALEPHS

We mentioned previously that $\mathfrak{a}$, the cardinal number of denumerable sets, is also denoted by

$$\aleph_0$$

(Here aleph $\aleph$ is the first letter of the Hebrew alphabet.) Since the cardinal numbers are well-ordered, the following system of notation is used to denote cardinal numbers. The immediate successor of $\aleph_0$ is denoted by $\aleph_1$, and its immediate successor by $\aleph_2$, and so on. The cardinal number which succeeds all the $\aleph_n$ is denoted by $\aleph_\omega$. In fact every infinite cardinal can be uniquely denoted by an $\aleph$ with an ordinal number as a subscript as follows:

Notation: Let α be any infinite cardinal number. Let $s(\alpha)$ be the set of infinite cardinal numbers less than α. Note that $s(\alpha)$ is well-ordered; say $\lambda = \text{ord}\,(s(\alpha))$. Then α is denoted by

$$\aleph_\lambda$$

The continuum hypothesis can now be reformulated as follows:

Continuum Hypothesis: $\aleph_1 = \mathfrak{c}$.

Solved Problems

AXIOM OF CHOICE

1. Show that the Axiom of Choice is equivalent to Zermelo's Postulate.

Solution:

Let $\{A_i\}_{i \in I}$ be a non-empty family of disjoint non-empty sets and let f be a choice function on $\{A_i\}_{i \in I}$. Set $B = \{f(A_i) : i \in I\}$. Then

$$A_i \cap B = \{f(A_i)\}$$

consists of exactly one element since the A_i are disjoint and f is a choice function. Accordingly, the Axiom of Choice implies Zermelo's Postulate.

Now let $\{A_i\}_{i \in I}$ be any non-empty family of non-empty sets which may or may not be disjoint. Set

$$A_i^* = \{A_i\} \times \{i\}, \qquad \text{for every } i \in I$$

Then certainly $\{A_i^*\}$ is a disjoint family of sets since $i \neq j$ implies $A_i \times \{i\} \neq A_j \times \{j\}$, even if $A_i = A_j$. By Zermelo's Postulate, there exists a subset B of $\cup_i A_i^*$ such that

$$B \cap A_i^* = \{(a_i, i)\}$$

consists of exactly one element. Then $a_i \in A_i$, and so the function f on $\{A_i\}_{i \in I}$ defined by $f(A_i) = a_i$ is a choice function. Accordingly, Zermelo's Postulate implies the Axiom of Choice.

2. Prove the Well-ordering Theorem (Zermelo): Every non-empty set X can be well-ordered.

Solution:

Let f be a choice function on the collection $\mathcal{P}(X)$ of all subsets of X, i.e.,

$$f : \mathcal{P}(X) \to X \quad \text{with} \quad f(A) \in A, \quad \text{for every } A \subset X$$

A subset A of X will be called *normal* if it has a well-ordering with the additional property that, for every $a \in A$,

$$f(X - s_A(a)) = a \quad \text{where} \quad s_A(a) = \{x \in A : x \prec a\}$$

i.e. $s_A(a)$ is the initial segment of a in the ordering of A. We show that normal sets exist. Set

$$x_0 = f(X), \quad x_1 = f(X - \{x_0\}) \quad \text{and} \quad x_2 = f(X - \{x_0, x_1\})$$

Then $A = \{x_0, x_1, x_2\}$ is normal. We claim that if A and B are normal subsets of X, then either $A = B$ or one is an initial segment of the other. Since A and B are well-ordered, one of them, say A, is similar to B or to an initial segment of B (Theorem 11.9). Thus there exists a similarity mapping $\alpha : A \to B$. Set

$$A^* = \{x \in A : \alpha(x) \neq x\}$$

If A^* is empty, then $A = B$ or A is an initial segment of B. Suppose $A^* \neq \emptyset$, and let a_0 be the first element of A^*. Then $s_A(a_0) = s_B(\alpha(a_0))$. But A and B are normal, and so

$$a_0 = f(X - s_A(a_0)) = f(X - s_B(\alpha(a_0))) = \alpha(a_0)$$

But this contradicts the definition of A^*, and so $A = B$ or A is an initial segment of B. In particular, if $a \in A$ and $b \in B$, then either $a, b \in A$ or $a, b \in B$. Furthermore, if $a, b \in A$ and $a, b \in B$, then $a \precsim b$ as elements of A if and only if $a \precsim b$ as elements of B.

Now let Y consist of all those elements in X which belong to at least one normal set. If $a, b \in Y$, then $a \in A$ and $b \in B$ where A and B are normal and so, as noted above, $a, b \in A$ or $a, b \in B$. We define an order in Y as follows: $a \precsim b$ as elements of Y iff $a \precsim b$ as elements of A or of B. This order is well-defined, i.e. independent of the particular choice of A and B, and, furthermore, it is a total order. Now let Z be any non-empty subset of Y and let a be any arbitrary element in Z. Then a belongs to a normal set A. Hence $A \cap Z$ is a non-empty subset of the well-ordered set A and so contains a first element a_0. Furthermore, a_0 is a first element of Z (Problem 12); thus Y is, in fact, well-ordered.

We next show that Y is normal. If $a \in Y$, then a belongs to a normal set A. Furthermore, $s_A(a) = s_Y(a)$, (Problem 12), and so

$$f(X - s_Y(a)) = f(X - s_A(a)) = a$$

that is, Y is normal. Lastly, we claim that $Y = X$. Suppose not, i.e. suppose $X - Y \neq \emptyset$ and, say, $a = f(X - Y)$. Set $Y^* = Y \cup \{a\}$ and let Y^* be ordered by the order in Y and with a dominating every element in Y. Then $f(X - s_Y(a)) = f(X - Y) = a$ and so Y^* is normal. Thus $a \in Y$. But this contradicts the fact that f is a choice function, i.e. $f(X - Y) = a \in X - Y$ which is disjoint from Y. Hence $Y = X$, and so X is well-ordered.

3. Prove (using the Well-ordering Theorem): Let X be a partially ordered set. Then X contains a maximal totally ordered subset, i.e. a totally ordered subset which is not a proper subset of any other totally ordered subset.

Solution:

The result clearly holds if X is empty (or even finite); hence we can assume that X is not empty and that X can be well-ordered with, say, first element x_0. (Observe that X now has both a partial ordering and a well-ordering; the terms initial segment of X and first element of a subset of X will only be used with respect to the well-ordering, and the term comparable will only be used with respect to the partial ordering.)

Let A be an initial segment of X, (we also allow $A = X$). A function $f : A \to A$ will be called *special* if

$$f(x) = \begin{cases} x, & \text{if } x \text{ is comparable to every element of } f[s(x)] \\ x_0, & \text{otherwise} \end{cases}$$

Here $s(x)$ denotes the initial segment of x. We claim that if a special function exists then it is unique. If not, then there exist special functions f and f' on A and a first element a_0 for which $f(a_0) \neq f'(a_0)$; hence f and f' agree on $s(a_0)$ which implies $f(a_0) = f'(a_0)$, a contradiction.

Remark: If A and A' are initial segments with special functions f and f' respectively and if $A \subset A'$, then the uniqueness of f on A implies that f' restricted to A equals f, i.e. $f'(a) = f(a)$ for every $a \in A$.

Now let B be the union of those A_i which admit a special function f_i. Since the A_i are initial segments, so is B. Furthermore, B admits the special function $g: B \to B$ defined by $g(b) = f_i(b)$ where $b \in A_i$. By the above remark, g is well defined. We next show that $B = X$. Let $y \in X$ be the first element for which $y \notin B$. Then $C = B \cup \{y\}$ is an initial segment. Moreover, C admits the special function $h: C \to C$ defined as follows: $h(c) = g(c)$ if $c \in B$, and $h(y) = y$ or x_0 according as y is or is not comparable to every element in $h[B]$. It now follows that $y \in B$, a contradiction. Thus no such y exists and so $B = X$.

Lastly, we claim that $g[B]$, i.e. $g[X]$, is a maximal totally ordered subset of X. If not, then there exists an element $z \in X$ such that $z \notin g[X]$ but is comparable to every element of $g[X]$. Thus, in particular, z is comparable to every element of $g[s(z)]$. By definition of a special function, $g(z) = z$ which implies $z \in g[X]$, a contradiction. Thus $g[X]$ is a maximal totally ordered subset of X, and the theorem is proved.

4. Prove Theorem 12.1: The following are equivalent: (i) Axiom of Choice, (ii) Well-ordering Theorem, (iii) Zorn's Lemma.

Solution:

By Problem 2, (i) implies (ii). We use Problem 3 to prove that (ii) implies (iii). Let X be a partially ordered subset in which every totally ordered subset has an upper bound. We need to show that X has a maximal element. By Problem 3, X has a maximal totally ordered subset Y. By hypothesis, Y has an upper bound m in X. We claim that m is a maximal element of X. If not, then there exists $z \in X$ such that z dominates m. It follows that $z \notin Y$ since m is an upper bound for Y, and that $Y \cup \{z\}$ is totally ordered. This contradicts the maximality of Y. Thus m is a maximal element of X and, consequently, (ii) implies (iii).

It remains to show that (iii) implies (i). By Problem 1, it suffices to prove that (iii) implies Zermelo's Postulate (page 180). Let $\{A_i\}$ be a non-empty family of disjoint non-empty sets. Let $\mathcal{B}$ be the class of all subsets of $\cup_i A_i$ which intersect each A_i in at most one element. We partially order $\mathcal{B}$ by set inclusion. Let $\{B_j\}$ be a totally ordered subset of $\mathcal{B}$. We claim that $B = \cup_j B_j$ belongs to $\mathcal{B}$. If not, then B intersects some A_{i_0} in more than one element; say, $a, b \in B \cap A_{i_0}$ where $a \neq b$. Since $a, b \in B$, there exist sets B_{j_1} and B_{j_2} such that $a \in B_{j_1}$ and $b \in B_{j_2}$. But $\{B_j\}$ is totally ordered by set inclusion; hence a and b both belong to B_{j_1} or B_{j_2}. This implies that B_{j_1} or B_{j_2} intersects A_{i_0} in more than one element, a contradiction. Accordingly, B belongs to $\mathcal{B}$, and so B is an upper bound for $\{B_j\}$.

We have shown that every totally ordered subset of $\mathcal{B}$ has an upper bound. By Zorn's Lemma, $\mathcal{B}$ has a maximal element M. If M does not intersect each A_i in exactly one element, then M and some A_{i_k} are disjoint. Say $c \in A_{i_k}$. Then $M \cup \{c\}$ belongs to $\mathcal{B}$, which contradicts the maximality of M. Thus M intersects each A_i in exactly one element, and therefore (iii) implies Zermelo's Postulate. Thus the theorem is proved.

APPLICATIONS OF ZORN'S LEMMA

5. Prove: Let R be a relation from A to B where the domain of R is A. Note R is a subset of $A \times B$. Then there exists a subset f^* of R such that f^* is a function from A into B.

Solution:

Let $\mathcal{A}$ be the family of subsets of R in which each $f \in \mathcal{A}$ is a function from a subset of A into B. Partially order $\mathcal{A}$ by set inclusion. Note that if $f: A_1 \to B$ is a subset of $g: A_2 \to B$ then $A_1 \subset A_2$.

Now suppose $\{f_i: A_i \to B\}_{i \in I}$ is a totally ordered subset of $\mathcal{A}$. Then (see Problem 12) $f = \cup_{i \in I} f_i$ is a function from $\cup_{i \in I} A_i$ into B and, therefore, f is an upper bound of $\{f_i\}_{i \in I}$. By Zorn's Lemma, $\mathcal{A}$ has a maximal element $f^*: A^* \to B$. If we show that $A^* = A$, then the theorem is proven.

Suppose $A^* \neq A$. Then there exists an element $a \in A$ such that $a \notin A^*$. Furthermore, since the domain of R is A, there exists an ordered pair $(a, b) \in R$. Then $f^* \cup \{(a, b)\}$ is a function from $A^* \cup \{a\}$ into B. But this contradicts the fact that f^*, which would be a proper subset of $f^* \cup \{(a, b)\}$, is a maximal element of $\mathcal{A}$. Therefore $A^* = A$, and the theorem is proven.

6. (Application to Linear Algebra.) Prove: Let V be a vector space; then V has a basis.

Solution:

If V consists of the zero vector alone then, by definition, the empty set is a basis for V; hence we assume V contains a non-zero vector a. Let $\mathcal{B}$ be the family of independent sets of vectors in V.

In other words, each element $B \varepsilon \mathcal{B}$ is an independent set of vectors. Note that $\mathcal{B}$ is non-empty since, e.g., $\{a\}$ belongs to $\mathcal{B}$. Partial order $\mathcal{B}$ by set inclusion.

Now suppose $\{B_i\}_{i \varepsilon I}$ is a totally ordered subset of $\mathcal{B}$. If we show that $A = \cup_{i \varepsilon I} B_i$ belongs to $\mathcal{B}$, i.e. is an independent set of vectors, then A would be an upper bound of $\{B_i\}_{i \varepsilon I}$. Assume that A is dependent. Then there exist vectors $\mathbf{a}_1, \ldots, \mathbf{a}_n \,\varepsilon\, A$ such that

$$c_1 \mathbf{a}_1 + \cdots + c_n \mathbf{a}_n = 0 \qquad (1)$$

where at least one $c_i \neq 0$. Note that there also exist elements $i_1, \ldots, i_n \,\varepsilon\, I$ such that $\mathbf{a}_i \,\varepsilon\, B_{i_1}, \ldots, \mathbf{a}_n \,\varepsilon\, B_{i_n}$. Since $\{B_i\}_{i \varepsilon I}$ is totally ordered, one of the sets, say B_{i_1}, is a superset of the others; hence $\mathbf{a}_1, \ldots, \mathbf{a}_n \,\varepsilon\, B_{i_1}$. In view of (1), B_{i_1} would be dependent, which is a contradiction. Thus A is independent, belongs to $\mathcal{B}$, and is an upper bound of $\{B_i\}_{i \varepsilon I}$.

By Zorn's Lemma, $\mathcal{B}$ has an upper bound B^*. B^* can then be shown to be a basis for V.

Note that the main part of the proof consists in showing that $A = \cup_{i \varepsilon I} B_i$ does belong to $\mathcal{B}$. This is a typical example of how Zorn's Lemma is used.

Supplementary Problems

7. State whether each of the following statements about cardinal numbers is true or false; give reasons for your answer:

   (1) $\aleph_0 + \aleph_\lambda = \aleph_\lambda$ (2) $\aleph_\lambda + \aleph_\mu = \aleph_{\lambda+\mu}$

8. Prove Theorem 12.2: Let $\alpha = \lambda$ and $\beta = \mu$ be cardinal numbers. Then

   (1) $\alpha < \beta$ implies $\lambda < \mu$ (2) $\lambda < \mu$ implies $\alpha \leqq \beta$

9. Prove Theorem 9.11: For any cardinal numbers α and β, one of the following holds: $\alpha < \beta$, $\alpha = \beta$ or $\alpha > \beta$.

10. Prove Theorem 12.3: Any set of cardinal numbers is well-ordered by the relation $\alpha \leqq \beta$.

11. Consider the proof of the following statement: There exists a finite set of natural numbers which is not a proper subset of another finite set of natural numbers.

    Proof. Let $\mathcal{B}$ be the family of all finite sets of natural numbers. Partially order $\mathcal{B}$ by set inclusion. Now let $\{B_i\}_{i \varepsilon I}$ be a totally ordered subset of $\mathcal{B}$. Consider the set $A = \cup_{i \varepsilon I} B_i$. Note that, for every $i \,\varepsilon\, I$, $B_i \subset A$; hence A is an upper bound of $\{B_i\}_{i \varepsilon I}$.

    Since every totally ordered subset of $\mathcal{B}$ has an upper bound, by Zorn's Lemma, $\mathcal{B}$ has a maximal element, a finite set which is not a proper subset of another finite set.

    Question. Since the statement is obviously false, which step in the proof is incorrect?

12. Prove the following two statements which were assumed in the proof in Problem 2:

    (i) The first element a_0 of the set $A \cap Z$ is a first element of the set Z.

    (ii) $s_A(a) = s_Y(a)$.

13. Prove the following statement which was assumed in the proof in Problem 5: Let $\{f_i : A_i \to B\}_{i \varepsilon I}$ be a collection of functions which is totally ordered by set inclusion. Then $\cup_{i \varepsilon I} f_i$ is a function from $\cup_{i \varepsilon I} A_i$ into B.

Answers to Supplementary Problems

7. (1) True. For $\aleph_0$ is the cardinal number of a denumerable set and, as proven previously, the union of a denumerable set and an infinite set does not change the cardinality of the infinite set.

   (2) False. For, since the addition of cardinals is commutative,

   $$\aleph_{\lambda+\mu} = \aleph_\lambda + \aleph_\mu = \aleph_\mu + \aleph_\lambda = \aleph_{\mu+\lambda}$$

   would imply that the addition of ordinal numbers is commutative, which is not true.

Chapter 13

Paradoxes in Set Theory

INTRODUCTION

The theory of sets was first studied as a mathematical discipline by Cantor (1845-1918) in the latter part of the nineteenth century. Today, the theory of sets lies at the foundations of mathematics and has revolutionized almost every branch of mathematics. At about the same time that set theory began to influence other branches of mathematics, various contradictions, called paradoxes, were discovered, the first by Burali-Forti in 1897. In this chapter, some of these paradoxes are presented. Although it is possible to eliminate these known contradictions by a strict axiomatic development of set theory, there are still many questions which are unanswered.

SET OF ALL SETS (CANTOR'S PARADOX)

Let $\mathcal{C}$ be the set of all sets. Then every subset of $\mathcal{C}$ is also a member of $\mathcal{C}$; hence the power set of $\mathcal{C}$ is a subset of $\mathcal{C}$, i.e.,

$$2^{\mathcal{C}} \subset \mathcal{C}$$

But $2^{\mathcal{C}} \subset \mathcal{C}$ implies that

$$\#(2^{\mathcal{C}}) \leq \#(\mathcal{C})$$

However, according to Cantor's theorem,

$$\#(\mathcal{C}) < \#(2^{\mathcal{C}})$$

Thus the concept of the set of all sets leads to a contradiction.

RUSSELL'S PARADOX

Let Z be the set of all sets which do not contain themselves as members, that is,

$$Z = \{X \mid X \notin X\}$$

Question: Does Z belong to itself or not? If Z does not belong to Z then, by definition of Z, Z does belong to itself. Furthermore, if Z does belong to Z then, by definition of Z, Z does not belong to itself. In either case we are led to a contradiction.

The above paradox is somewhat analogous to the following popular paradox: In a certain town, there is a barber who shaves only and all those men who do not shave themselves. Question: Who shaves the barber?

SET OF ALL ORDINAL NUMBERS (BURALI-FORTI PARADOX)

Let Δ be the set of all ordinal numbers. By a previous theorem Δ is a well-ordered set, say $\alpha = \text{ord}(\Delta)$. Now consider $s(\alpha)$, the set of all ordinal numbers less than α. Note:

(1) Since $s(\alpha)$ consists of all elements in Δ which precede α, $s(\alpha)$ is an initial segment of Δ.

(2) By a previous theorem $\alpha = \text{ord}\,(s(\alpha))$; hence

$$\text{ord}\,(s(\alpha)) = \alpha = \text{ord}\,(\Delta)$$

Therefore Δ is similar to one of its initial segments. Thus the concept of the set of all ordinal numbers leads to a contradiction of Theorem 11.8.

SET OF ALL CARDINAL NUMBERS

Let $\mathcal{A}$ be the set of all cardinal numbers. Then for each cardinal $\alpha \in \mathcal{A}$ there is a set A_α such that $\alpha = \#(A_\alpha)$. Let

$$A = \cup_{\alpha \in \mathcal{A}} A_\alpha$$

Consider the power set 2^A of A. Note $2^A \sim A_{\#(2^A)}$, which is a subset of A. Hence $2^A \precsim A$ and, in particular,

$$\#(2^A) \leq \#(A)$$

But by Cantor's Theorem,

$$\#(A) < \#(2^A)$$

Thus the concept of the set of all cardinal numbers leads to a contradiction.

FAMILY OF ALL SETS EQUIVALENT TO A SET

Let $A = \{a, b, \ldots\}$ be any set (not necessarily countable) and let $\mathcal{A} = \{i, j, \ldots\}$ be any other set. Consider the sets

$$A_i = \{(a, i), (b, i), \ldots\}$$
$$A_j = \{(a, j), (b, j), \ldots\}$$
$$\ldots\ldots\ldots\ldots\ldots\ldots$$
$$\ldots\ldots\ldots\ldots\ldots\ldots$$
$$\ldots\ldots\ldots\ldots\ldots\ldots$$

that is, the family of sets $\{A_i\}_{i \in \mathcal{A}}$. Note that

$$\#(\{A_i\}_{i \in \mathcal{A}}) = \#(\mathcal{A})$$

and $A_i \sim A$, for every $i \in \mathcal{A}$.

Now let α be the family of all sets equivalent to A. Consider the power set 2^α of α, and define the family of sets $\{A_i\}_{i \in 2^\alpha}$ as above. Since each $A_i \sim A$,

$$\{A_i\}_{i \in 2^\alpha} \subset \alpha$$

Hence

$$\#(2^\alpha) = \#(\{A_i\}_{i \in 2^\alpha}) \leq \#(\alpha)$$

But by Cantor's Theorem, $\#(\alpha) < \#(2^\alpha)$. Thus the concept of the family of all sets equivalent to a set (our definition of cardinal number) leads to a contradiction.

FAMILY OF ALL SETS SIMILAR TO A WELL-ORDERED SET

Let A be any well-ordered set. Then the set A_i, defined as above and ordered by

$$(a, i) \precsim (b, i) \quad \text{if } a \precsim b$$

is well-ordered and is similar to A, that is, $A_i \simeq A$.

Now let λ be the family of all sets similar to the well-ordered set A. Consider the power set 2^λ of λ, and define the family of sets $\{A_i\}_{i \in 2^\lambda}$ as above. Since each set A_i is similar to A,

$$\{A_i\}_{i \in 2^\lambda} \subset \lambda$$

Hence

$$\#(2^\lambda) = \#(\{A_i\}_{i \in 2^\lambda}) \leq \#(\lambda)$$

Since, by Cantor's Theorem, $\#(\lambda) < \#(2^\lambda)$, the concept of the family of all sets similar to a well-ordered set (our definition of ordinal number) leads to a contradiction.

Part III – Related Topics

Chapter 14

Algebra of Propositions

STATEMENTS

Statements (or *verbal assertions*) will be denoted by the letters

$$p, q, r$$

(with or without subscripts). The fundamental property of a statement is that it is either *true* or *false*, but not both. The truthfulness or falsity of a statement is called its *truth value*. Some statements are *composite*, that is, composed of *substatements* and various connectives which will be discussed subsequently.

Example 1.1: "Roses are red and violets are blue," is a composite statement with substatements "Roses are red" and "Violets are blue".

Example 1.2: "Where are you going?" is not a statement since it is neither true nor false.

Example 1.3: "John is sick or old" is, implicitly, a composite statement with substatements "John is sick" and "John is old".

A fundamental property of a composite statement is that its truth value is completely determined by the truth value of each of its substatements and the way they are connected to form the composite statement.

CONJUNCTION, $p \wedge q$

Any two statements can be combined by the word "and" to form a composite statement which is called the *conjunction* of the original statements. Symbolically, the conjunction of the two statements p and q is denoted by

$$p \wedge q$$

Example 2.1: Let p be "It is raining", and let q be "The sun is shining". Then $p \wedge q$ denotes the statement "It is raining and the sun is shining".

Example 2.2: The symbol $\wedge$ can be used to define the intersection of two sets; specifically,

$$A \cap B \;=\; \{x \mid x \varepsilon A \wedge x \varepsilon B\}$$

The truth value of the composite statement $p \wedge q$ satisfies the following property:

$\mathbf{T}_1$: If p is true and q is true, then $p \wedge q$ is true; otherwise $p \wedge q$ is false. In other words, the conjunction of two statements is true only if each component is true.

Example 2.3: Consider the following four statements:

(1) Paris is in France and $2+2 = 5$.
(2) Paris is in England and $2+2 = 4$.
(3) Paris is in England and $2+2 = 5$.
(4) Paris is in France and $2+2 = 4$.

By $\mathbf{T}_1$, only (4) is true. Each of the other statements is false since at least one of its substatements is false.

A convenient way to state $\mathbf{T}_1$ is by means of a table as follows:

p	q	$p \wedge q$
T	T	T
T	F	F
F	T	F
F	F	F

Note that the first line is a short way of saying that if p is true and q is true then $p \wedge q$ is true. The other lines have analogous meaning.

DISJUNCTION, $p \vee q$

Any two statements can be combined by the word "or" (in the sense of "and/or") to form a new statement which is called the *disjunction* of the original two statements. Symbolically, the disjunction of statements p and q is denoted by

$$p \vee q$$

Example 3.1: Let p be "He studied French at the university", and let q be "He lived in France". Then $p \vee q$ is the statement "He studied French at the university or he lived in France".

Example 3.2: The symbol $\vee$ can be used to define the union of two sets; specifically,

$$A \cup B = \{x \mid x \varepsilon A \vee x \varepsilon B\}$$

The truth value of the composite statement $p \vee q$ satisfies the following property:

T_2: If p is true or q is true or both p and q are true, then $p \vee q$ is true; otherwise, $p \vee q$ is false. In other words, the disjunction of two statements is false only if each component is false.

T_2 can also be written in the form of a table as follows:

p	q	$p \vee q$
T	T	T
T	F	T
F	T	T
F	F	F

Example 3.3: Consider the following four statements:

(1) Paris is in France or $2+2=5$.
(2) Paris is in England or $2+2=4$.
(3) Paris is in France or $2+2=4$.
(4) Paris is in England or $2+2=5$.

Only (4) is false. Each of the other statements is true since at least one of its components is true.

NEGATION, $\sim p$

Given any statement p, another statement, called the *negation* of p, can be formed by writing "It is false that . . ." before p or, if possible, by inserting in p the word "not". Symbolically, the negation of p is denoted by

$$\sim p$$

Example 4.1: Consider the following three statements:

(1) Paris is in France.
(2) It is false that Paris is in France.
(3) Paris is not in France.

Then (2) and (3) are each the negation of (1).

Example 4.2: Consider the following statements:

(1) $2+2=5$
(2) It is false that $2+2=5$.
(3) $2+2 \neq 5$

Then (2) and (3) are each the negation of (1).

The truth value of the negation of a statement satisfies the following property:

T_3: If p is true, then $\sim p$ is false; if p is false, then $\sim p$ is true. In other words, the truth value of the negation of a statement is always the opposite of the truth value of the original statement.

Example 4.3: Consider the statements in Example 4.1. Notice that (1) is true and (2) and (3), its negations, are false.

Example 4.4: Consider the statements in Example 4.2. Notice that (1) is false and (2) and (3) are true.

$\mathbf{T}_3$ can also be written in the form of a table as follows:

p	$\sim p$
T	F
F	T

CONDITIONAL, $p \to q$

Many statements, especially in mathematics, are of the form "If p then q". Such statements are called *conditional* statements and are denoted by

$$p \to q$$

The conditional $p \to q$ can also be read:

(*a*) p implies q (*c*) p is sufficient for q

(*b*) p only if q (*d*) q is necessary for p

The truth value of the conditional statement $p \to q$ satisfies the following property:

$\mathbf{T}_4$: The conditional $p \to q$ is true unless p is true and q is false. In other words, $\mathbf{T}_4$ states that a true statement cannot imply a false statement.

$\mathbf{T}_4$ can be written in the form of a table as follows:

p	q	$p \to q$
T	T	T
T	F	F
F	T	T
F	F	T

Example 5.1: Consider the following statements:

(1) If Paris is in France then $2+2 = 5$.
(2) If Paris is in England then $2+2 = 4$.
(3) If Paris is in France then $2+2 = 4$.
(4) If Paris is in England then $2+2 = 5$.

By $\mathbf{T}_4$, only (1) is a false statement; the others are true.

BICONDITIONAL, $p \leftrightarrow q$

Another common statement is of the form "p if and only if q" or, simply, "p iff q". Such statements are called *biconditional* statements and are denoted by

$$p \leftrightarrow q$$

The truth value of the biconditional statement $p \leftrightarrow q$ satisfies the following property:

$\mathbf{T}_5$: If p and q have the same truth value, then $p \leftrightarrow q$ is true; if p and q have opposite truth values, then $p \leftrightarrow q$ is false.

Example 6.1: Consider the following statements:

(1) Paris is in France if and only if $2+2 = 5$.
(2) Paris is in England if and only if $2+2 = 4$.
(3) Paris is in France if and only if $2+2 = 4$.
(4) Paris is in England if and only if $2+2 = 5$.

According to $\mathbf{T}_5$, (3) and (4) are true and (1) and (2) are false.

$\mathbf{T}_5$ can be written in the form of a table as follows:

p	q	$p \leftrightarrow q$
T	T	T
T	F	F
F	T	F
F	F	T

POLYNOMIALS AND BOOLEAN POLYNOMIALS

Forming finite sums (+), products (·) and differences (−) of the *indeterminants* (or, simply, *variables*)

$$x, y, \ldots$$

subject to the usual rules of ordinary algebra, leads to the construction of polynomials in the above variables.

Example 7.1: The following are polynomials in two indeterminants:

$$f(x, y) \;=\; x \cdot x - x \cdot y + y \cdot y \cdot y + x \cdot x \;=\; 2x^2 - xy + y^3$$
$$g(x, y) \;=\; (x - y) \cdot (x + y) \;=\; x^2 - y^2$$

Now suppose each of the indeterminants $x, y, \ldots$ in the polynomial $f(x, y, \ldots)$ is replaced by specific real numbers $x_0, y_0, \ldots$. Then the expression

$$f(x_0, y_0, \ldots)$$

which denotes sums, products and differences of real numbers, is itself a number. In other words, if $x, y, \ldots$ are considered as *real variables* then the polynomial $f(x, y, \ldots)$ defines a function in the sense that it assigns a specific image value $f(x_0, y_0, \ldots)$ to the real numbers $x_0, y_0, \ldots$.

Example 7.2: Consider the polynomials in Example 7.1. Then

$$f(2, 3) \;=\; 2 \cdot 2 - 2 \cdot 3 + 3 \cdot 3 \cdot 3 + 2 \cdot 2 \;=\; 4 - 6 + 27 + 4 \;=\; 29$$
$$g(3, 1) \;=\; (3 - 1) \cdot (3 + 1) \;=\; 2 \cdot 4 \;=\; 8$$

Furthermore, the operations sum, product and difference defined for the real numbers, induces similar operations, also called sum, product and difference, on the polynomials.

Example 7.3: Consider the polynomials in Example 7.1. Then

$$f(x, y) - g(x, y) \;=\; (2x^2 - xy + y^3) - (x^2 - y^2)$$
$$f(x, y) \cdot g(x, y) \;=\; (2x^2 - xy + y^3) \cdot (x^2 - y^2)$$

Now let the letters

$$p, q, \ldots$$

which previously denoted statements, also be indeterminants, i.e. variables. Combining these variables by the connectives $\wedge$, $\vee$ and $\sim$, or more generally by the connectives $\wedge$, $\vee$, $\sim$, $\rightarrow$ and $\leftrightarrow$, leads to the construction of expressions which we shall call *Boolean polynomials*.

Example 7.4: The following are Boolean polynomials in two variables.

$$f(p, q) \;=\; \sim p \vee (p \rightarrow q)$$
$$g(p, q) \;=\; (p \leftrightarrow \sim q) \wedge q$$

Furthermore, the symbols $\wedge$, $\vee$, $\sim$, $\rightarrow$ and $\leftrightarrow$ can now be used as connectives for the Boolean polynomials; hence we can speak of the conjunction, disjunction and negation of Boolean polynomials.

Example 7.5: Consider the Boolean polynomials in Example 7.4. Then

$$f(p, q) \wedge g(p, q) \;=\; [\sim p \vee (p \rightarrow q)] \wedge [(p \leftrightarrow \sim q) \wedge q]$$
$$f(p, q) \rightarrow g(p, q) \;=\; [\sim p \vee (p \rightarrow q)] \rightarrow [(p \leftrightarrow \sim q) \wedge q]$$

Now suppose each of the variables $p, q, \ldots$ in a Boolean polynomial $f(p, q, \ldots)$ is replaced, respectively, by specific statements, denoted by $p_0, q_0, \ldots$. Then the expression

$$f(p_0, q_0, \ldots)$$

is also a statement and, furthermore, has a truth value.

Example 7.6: Let $f(p, q) = \sim p \wedge (p \rightarrow q)$, and let p_0 be "$2+2 = 5$" and q_0 be "$1+1 = 2$". Then $f(p_0, q_0)$ reads

"$2+2 \neq 5$, and if $2+2 = 5$ then $1+1 = 2$"

By $\mathbf{T}_4$, $r_0 = p_0 \rightarrow q_0$ is true. Note that $s_0 = \sim p_0$ is true. Therefore, by $\mathbf{T}_1$, $f(p_0, q_0) = s_0 \wedge r_0$ is also true.

Remark 14.1: Let $f(p, q, \ldots)$ be a Boolean polynomial, and let statements $p_0', q_0', \ldots$ have, respectively, the same truth value as statements $p_0, q_0, \ldots$. Then $f(p_0', q_0', \ldots)$ has the same truth value as $f(p_0, q_0, \ldots)$.

PROPOSITIONS AND TRUTH TABLES

Definition 14.1: A proposition, denoted by

$$P(p, q, \ldots),\ Q(p, q, \ldots),\ \ldots$$

or, simply, $P, Q, \ldots$, is a Boolean polynomial in the variables $p, q, \ldots$.

By Remark 14.1, the truth value of a proposition $P(p, q, \ldots)$ evaluated on any statements is a function only of the truth values of the statements, and not the particular statements themselves. Hence we speak of the "truth value" of each of the variables $p, q, \ldots$, and the "truth value" of the proposition $P(p, q, \ldots)$.

A simple concise way to show the relationship between the truth value of a proposition $P(p, q, \ldots)$ and the truth values of its variables $p, q, \ldots$ is through a *truth table.* The truth table, for example, of the proposition $\sim(p \wedge \sim q)$ is constructed as follows:

p	q	$\sim q$	$p \wedge \sim q$	$\sim(p \wedge \sim q)$
T	T	F	F	T
T	F	T	T	F
F	T	F	F	T
F	F	T	F	T

Note that the first columns of the table are for the variables $p, q, \ldots$. Note also that there are enough rows in the table to allow for all combinations of T and F for these variables. (For 2 variables, as above, 4 rows are necessary; for 3 variables, 8 rows are necessary; and, in general, for n variables, 2^n rows are necessary.) Then there is an additional column for each step in the computation of the desired truth value of the proposition which appears in the last column.

The truth table of the above proposition $\sim(p \wedge \sim q)$ consists only of the columns under the variables and the column under the proposition, that is, the following table:

p	q	$\sim(p \wedge \sim q)$
T	T	T
T	F	F
F	T	T
F	F	T

Another way to construct the above truth table for $\sim(p \wedge \sim q)$ is as follows. First draw the following table.

p	q	$\sim$	$(p$	$\wedge$	$\sim$	$q)$
T	T					
T	F					
F	T					
F	F					
Step						

Notice that the proposition is written on the top row to the right of the variables in the proposition. Also notice that there is a column under each variable or connective in the proposition. Truth values are then entered into the truth table in various steps as follows:

p	q	$\sim$	$(p$	$\wedge$	$\sim$	$q)$
T	T		T			T
T	F		T			F
F	T		F			T
F	F		F			F
Step			1			1

p	q	$\sim$	$(p$	$\wedge$	$\sim$	$q)$
T	T		T		F	T
T	F		T		T	F
F	T		F		F	T
F	F		F		T	F
Step			1		2	1

p	q	$\sim$	$(p$	$\wedge$	$\sim$	$q)$
T	T		T	F	F	T
T	F		T	T	T	F
F	T		F	F	F	T
F	F		F	F	T	F
Step			1	3	2	1

p	q	$\sim$	$(p$	$\wedge$	$\sim$	$q)$
T	T	T	T	F	F	T
T	F	F	T	T	T	F
F	T	T	F	F	F	T
F	F	T	F	F	T	F
Step		4	1	3	2	1

The truth table of the proposition then consists of the original columns under the variables and the last column entered into the table, i.e. the last step.

TAUTOLOGIES AND CONTRADICTION

Some propositions $P(p, q, \ldots)$ contain only T in the last column of their truth tables. In other words the proposition $P(p, q, \ldots)$ will always become a true statement no matter which statements $p_0, q_0, \ldots$, true or false, are substituted for the variables. Such propositions are called *tautologies*. Specifically,

Definition 14.2: A proposition $P(p, q, \ldots)$ is a *tautology* if $P(p_0, q_0, \ldots)$ is true for any statements $p_0, q_0, \ldots$.

Similarly,

Definition 14.3: A proposition $P(p, q, \ldots)$ is a *contradiction* if $P(p_0, q_0, \ldots)$ is false for any statements $p_0, q_0, \ldots$. In other words, a contradiction will contain only F in the last column of its truth table.

Example 8.1: The proposition "p or not p", i.e. $p \vee \sim p$, is a tautology. This fact is verified by constructing a truth table.

p	$\sim p$	$p \vee \sim p$
T	F	T
F	T	T

Example 8.2: The proposition "p and not p", i.e. $p \wedge \sim p$, is a contradiction. This fact is verified by the following table.

p	$\sim p$	$p \wedge \sim p$
T	F	F
F	T	F

Example 8.3: A fundamental principle of logical reasoning, called the "Law of Syllogism", states: "If p implies q and q implies r, then p implies r." In other words, the proposition

$$[(p \rightarrow q) \wedge (q \rightarrow r)] \rightarrow (p \rightarrow r)$$

is a tautology. This fact is verified by a truth table.

p	q	r	$[(p$	$\rightarrow$	$q)$	$\wedge$	$(q$	$\rightarrow$	$r)]$	$\rightarrow$	$(p$	$\rightarrow$	$r)$
T	T	T	T	T	T	T	T	T	T	T	T	T	T
T	T	F	T	T	T	F	T	F	F	T	T	F	F
T	F	T	T	F	F	F	F	T	T	T	T	T	T
T	F	F	T	F	F	F	F	T	F	T	T	F	F
F	T	T	F	T	T	T	T	T	T	T	F	T	T
F	T	F	F	T	T	F	T	F	F	T	F	T	F
F	F	T	F	T	F	T	F	T	T	T	F	T	T
F	F	F	F	T	F	T	F	T	F	T	F	T	F
Step			1	2	1	3	1	2	1	4	1	2	1

Note that eight rows are required in the above truth table in order to allow for all combinations of T and F for the variables p, q and r.

Since a tautology is always true, the negation of a tautology is always false, i.e. is a contradiction; and vice versa. In other words

Remark 14.2: If $P(p, q, \ldots)$ is a tautology then $\sim P(p, q, \ldots)$ is a contradiction, and vice versa.

Now let $P_1(p, q, \ldots), P_2(p, q, \ldots), \ldots$ be any propositions, and let $P(p, q, \ldots)$ be a tautology. Then $P(p, q, \ldots)$ does not depend upon the particular truth values of $p, q, \ldots$. Therefore if we substitute P_1 for p, P_2 for q, $\ldots$ in $P(p, q, \ldots)$, we still would have a tautology. In other words:

Theorem (Principle of Substitution): If $P(p, q, \ldots)$ is a tautology, then

$$P(P_1, P_2, \ldots)$$

is also a tautology for any propositions $P_1, P_2, \ldots$.

LOGICAL EQUIVALENCE

Two propositions $P(p, q, \ldots)$ and $Q(p, q, \ldots)$ are said to be logically equivalent if their truth tables are identical. We denote the logical equivalence of $P(p, q, \ldots)$ and $Q(p, q, \ldots)$ by

$$P(p, q, \ldots) \equiv Q(p, q, \ldots)$$

Example 9.1: The truth tables of $(p \rightarrow q) \wedge (q \rightarrow p)$ and $p \leftrightarrow q$ are as follows:

p	q	$p \to q$	$q \to p$	$p \to q \wedge q \to p$
T	T	T	T	T
T	F	F	T	F
F	T	T	F	F
F	F	T	T	T

p	q	$p \leftrightarrow q$
T	T	T
T	F	F
F	T	F
F	F	T

Hence $(p \to q) \wedge (q \to p) \equiv p \leftrightarrow q$.

Example 9.2: The truth tables below show that $p \to q$ and $\sim p \vee q$ are logically equivalent, i.e., $p \to q \equiv \sim p \vee q$.

p	q	$p \to q$
T	T	T
T	F	F
F	T	T
F	F	T

p	q	$\sim p$	$\sim p \vee q$
T	T	F	T
T	F	F	F
F	T	T	T
F	F	T	T

The following theorems are direct consequences of our previous definitions.

Theorem 14.1: The relation in propositions defined by

$$P(p, q, \ldots) \equiv Q(p, q, \ldots)$$

is an equivalence relation. In other words:

(1) for every $P(p, q, \ldots)$, $P(p, q, \ldots) \equiv P(p, q, \ldots)$;

(2) if $P(p, q, \ldots) \equiv Q(p, q, \ldots)$, then $Q(p, q, \ldots) \equiv P(p, q, \ldots)$;

(3) if $P(p, q, \ldots) \equiv Q(p, q, \ldots)$ and $Q(p, q, \ldots) \equiv R(p, q, \ldots)$ then $P(p, q, \ldots) \equiv R(p, q, \ldots)$.

Theorem 14.2: $P(p, q, \ldots) \equiv Q(p, q, \ldots)$ if and only if the proposition

$$P(p, q, \ldots) \leftrightarrow Q(p, q, \ldots)$$

is a tautology.

Theorem 14.3: If $P(p, q, \ldots)$ and $Q(p, q, \ldots)$ are either both tautologies or both contradictions, then

$$P(p, q, \ldots) \equiv Q(p, q, \ldots)$$

The following corollary is a consequence of the Principle of Substitution in tautologies and of Theorem 14.2 above.

Corollary 14.1: If $P(p, q, \ldots) \equiv Q(p, q, \ldots)$, then

$$P(P_1, P_2, \ldots) \equiv Q(P_1, P_2, \ldots)$$

for any propositions $P_1, P_2, \ldots$.

In other words, if propositions are substituted for the variables in equivalent propositions, then the resulting propositions are also equivalent.

ALGEBRA OF PROPOSITIONS

The following propositions are logically equivalent:

Theorem 14.4:

(1a) $p \vee p \equiv p$	(1b) $p \wedge p \equiv p$
(2a) $(p \vee q) \vee r \equiv p \vee (q \vee r)$	(2b) $(p \wedge q) \wedge r \equiv p \wedge (q \wedge r)$
(3a) $p \vee q \equiv q \vee p$	(3b) $p \wedge q \equiv q \wedge p$
(4a) $p \vee (q \wedge r) \equiv (p \vee q) \wedge (p \vee r)$	(4b) $p \wedge (q \vee r) \equiv (p \wedge q) \vee (p \wedge r)$
(5a) $p \vee f \equiv p$	(5b) $p \wedge t \equiv p$
(6a) $p \vee t \equiv t$	(6b) $p \wedge f \equiv f$
(7a) $p \vee \sim p \equiv t$	(7b) $p \wedge \sim p \equiv f$
(8a) $\sim\sim p \equiv p$	(8b) $\sim t \equiv f,\ \sim f \equiv t$
(9a) $\sim(p \vee q) \equiv \sim p \wedge \sim q$	(9b) $\sim(p \wedge q) \equiv \sim p \vee \sim q$

The above theorem can be proven by constructing the necessary truth tables. Here t and f denote variables which are restricted, respectively, to true and false statements.

In view of Corollary 14.1, any proposition can be substituted for the variables in Theorem 14.4 (except for t and f which can only be replaced, respectively, by a tautology T or a contradiction F). Hence propositions satisfy the laws in Table 14.1 below. Notice the similarity between the laws of the algebra of propositions in Table 14.1 and the laws of the algebra of sets on Page 104.

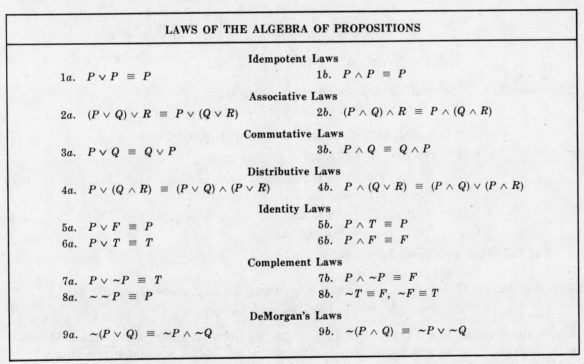

LAWS OF THE ALGEBRA OF PROPOSITIONS	
Idempotent Laws	
1a. $P \vee P \equiv P$	1b. $P \wedge P \equiv P$
Associative Laws	
2a. $(P \vee Q) \vee R \equiv P \vee (Q \vee R)$	2b. $(P \wedge Q) \wedge R \equiv P \wedge (Q \wedge R)$
Commutative Laws	
3a. $P \vee Q \equiv Q \vee P$	3b. $P \wedge Q \equiv Q \wedge P$
Distributive Laws	
4a. $P \vee (Q \wedge R) \equiv (P \vee Q) \wedge (P \vee R)$	4b. $P \wedge (Q \vee R) \equiv (P \wedge Q) \vee (P \wedge R)$
Identity Laws	
5a. $P \vee F \equiv P$	5b. $P \wedge T \equiv P$
6a. $P \vee T \equiv T$	6b. $P \wedge F \equiv F$
Complement Laws	
7a. $P \vee \sim P \equiv T$	7b. $P \wedge \sim P \equiv F$
8a. $\sim\sim P \equiv P$	8b. $\sim T \equiv F,\ \sim F \equiv T$
DeMorgan's Laws	
9a. $\sim(P \vee Q) \equiv \sim P \wedge \sim Q$	9b. $\sim(P \wedge Q) \equiv \sim P \vee \sim Q$

Table 14.1

LOGICAL IMPLICATION

Consider the following theorem.

Theorem 14.5: Let $P(p, q, \ldots)$ and $Q(p, q, \ldots)$ be propositions. Then the following three conditions are equivalent:

(1) $\sim P(p, q, \ldots) \vee Q(p, q, \ldots)$ is a tautology.

(2) $P(p, q, \ldots) \wedge \sim Q(p, q, \ldots)$ is a contradiction.

(3) $P(p, q, \ldots) \rightarrow Q(p, q, \ldots)$ is a tautology.

In view of the above theorem, we can now introduce

Definition 14.4: A proposition $P(p,q,\ldots)$ is said to *logically imply* a proposition $Q(p,q,\ldots)$, denoted by

$$P(p,q,\ldots) \Rightarrow Q(p,q,\ldots)$$

if one of the conditions in Theorem 14.5 holds.

Example 10.1: The proposition "p implies q and q implies r" logically implies the proposition "p implies r" since, as shown in Example 8.3,

$$[(p \to q) \wedge (q \to r)] \to (p \to r)$$

is a tautology. In other words,

$$(p \to q) \wedge (q \to r) \Rightarrow p \to r$$

Example 10.2: Consider the truth table of $(p \wedge q) \wedge \sim(p \vee q)$.

p	q	$(p$	$\wedge$	$q)$	$\wedge$	$\sim$	$(p$	$\vee$	$q)$
T	T	T	T	T	F	F	T	T	T
T	F	T	F	F	F	F	T	T	F
F	T	F	F	T	F	F	F	T	T
F	F	F	F	F	F	T	F	F	F
Step		1	2	1	4	3	1	2	1

Note that $(p \wedge q) \wedge \sim(p \vee q)$ is a contradiction; hence $p \wedge q \Rightarrow p \vee q$.

Theorem 14.6: The relation in propositions defined by

$$P(p,q,\ldots) \Rightarrow Q(p,q,\ldots)$$

is reflexive, anti-symmetric and transitive, that is:

(1) $P(p,q,\ldots) \Rightarrow P(p,q,\ldots)$.

(2) If $P(p,q,\ldots) \Rightarrow Q(p,q,\ldots)$ and $Q(p,q,\ldots) \Rightarrow P(p,q,\ldots)$, then $P(p,q,\ldots) \equiv Q(p,q,\ldots)$.

(3) If $P(p,q,\ldots) \Rightarrow Q(p,q,\ldots)$ and $Q(p,q,\ldots) \Rightarrow R(p,q,\ldots)$, then $P(p,q,\ldots) \Rightarrow R(p,q,\ldots)$.

The following condition also holds:

Theorem 14.7: If $P(p,q,\ldots) \Rightarrow Q(p,q,\ldots)$, then, for any propositions $P_1, P_2, \ldots$,

$$P(P_1, P_2, \ldots) \Rightarrow Q(P_1, P_2, \ldots)$$

In other words, if a proposition logically implies another, the relation still holds when arbitrary propositions are substituted for the variables in the original propositions.

Remark 14.3: Consider the symbols

$$\to \text{ and } \Rightarrow$$

Note that

$$P(p,q,\ldots) \to Q(p,q,\ldots)$$

is just a proposition and that its truth table could contain either T or F in its last column. But

$$P(p,q,\ldots) \Rightarrow Q(p,q,\ldots)$$

defines a relation in composite propositions which states that the composite proposition

$$P(p,q,\ldots) \to Q(p,q,\ldots)$$

contains only T in the last column of its truth table, i.e. is a tautology.

Remark 14.4: Similarly, consider the symbols

$$\leftrightarrow \text{ and } \equiv$$

Note that

$$P(p, q, \ldots) \leftrightarrow Q(p, q, \ldots)$$

is also just a composite proposition and that its truth table can contain both T's or F's in its last column. But

$$P(p, q, \ldots) \equiv Q(p, q, \ldots)$$

defines a relation in composite propositions which states that $P(p, q, \ldots)$ and $Q(p, q, \ldots)$ have identical truth tables or, equivalently, that

$$P(p, q, \ldots) \leftrightarrow Q(p, q, \ldots)$$

contains only T in the last column of its truth table. Furthermore, since, logically speaking, we do not distinguish between two equivalent propositions, some authors use the equal sign $=$ for $\equiv$.

LOGICALLY TRUE AND LOGICALLY EQUIVALENT STATEMENTS

A statement is said to be *logically true* if it is derivable from a tautology, that is, if the statement is of the form $P(p_0, q_0, \ldots)$ where $P(p, q, \ldots)$ is a tautology.

Example 11.1: Consider the following two statements:

(1) It is raining.

(2) It is raining or it is not raining.

The first statement may be true; its truth value depends on conditions outside of the sentence itself, i.e. the weather outside. The second statement is logically true since it is derivable from the tautology $p \vee \sim p$. Notice that its truth value does not depend upon any conditions outside of the sentence itself.

Similarly, statements of the form $P(p_0, q_0, \ldots)$ and $Q(p_0, q_0, \ldots)$ are said to be logically equivalent if the propositions $P(p, q, \ldots)$ and $Q(p, q, \ldots)$ are logically equivalent.

Example 11.2: Since $\sim(p \wedge q) \equiv \sim p \vee \sim q$, the statement "It is not true that roses are red and violets are blue" is logically equivalent to the statement "Roses are not red or violets are not blue."

Remark 14.5: We emphasize that the main purpose of this chapter has been to show that the Boolean polynomials, i.e. propositions, and their truth tables satisfy certain algebraic properties. No attempt has been made to analyze the logical foundations of our basic assumptions.

Solved Problems

STATEMENTS

1. Let p be "It is cold" and let q be "It is raining." Give a simple verbal sentence which describes each of the following statements.

(1) $\sim p$	(5) $p \rightarrow \sim q$	(9) $\sim\sim q$
(2) $p \wedge q$	(6) $q \vee \sim p$	(10) $(p \wedge \sim q) \rightarrow p$
(3) $p \vee q$	(7) $\sim p \wedge \sim q$	
(4) $q \leftrightarrow p$	(8) $p \leftrightarrow \sim q$	

Solution:

In each case, translate $\wedge$, $\vee$, $\sim$, $\rightarrow$ and $\leftrightarrow$ to read "and", "or", "it is false that" or "not", "if ... then" and "if and only if", respectively, and then simplify the English sentence.

(1) It is not cold.
(2) It is cold and raining.
(3) It is cold or it is raining.
(4) It is raining if and only if it is cold.
(5) If it is cold, then it is not raining.
(6) It is raining or it is not cold.
(7) It is not cold and it is not raining.
(8) It is cold if and only if it is not raining.
(9) It is not true that it is not raining.
(10) If it is cold and not raining, then it is cold.

2. Let p be "He is tall" and let q be "He is handsome." Write each of the following statements in symbolic form using p and q.

(1) He is tall and handsome.
(2) He is tall but not handsome.
(3) It is false that he is short or handsome.
(4) He is neither tall nor handsome.
(5) He is tall, or he is short and handsome.
(6) It is not true that he is short or not handsome.

Solution:

(1) $p \wedge q$ (3) $\sim(\sim p \vee q)$ (5) $p \vee (\sim p \wedge q)$

(2) $p \wedge \sim q$ (4) $\sim p \wedge \sim q$ (6) $\sim(\sim p \vee \sim q)$

3. Determine the truth value of each of the following composite statements.

(1) If $3+2 = 7$, then $4+4 = 8$.
(2) It is not true that $2+2 = 5$ if and only if $4+4 = 10$.
(3) Paris is in England or London is in France.
(4) It is not true that $1+1 = 3$ or $2+1 = 3$.
(5) It is false that if Paris is in England then London is in France.

Solution:

(1) Let p be "$3+2 = 7$" and let q be "$4+4 = 8$". Note p is false and q is true. By $\mathbf{T}_4$, $p \to q$ is true. In other words, the given statement is true.

(2) Let p be "$2+2 = 5$", let q be "$4+4 = 10$", and let r be "p iff q". Note p and q are each false; hence by $\mathbf{T}_5$, $p \leftrightarrow q$ is true, i.e. r is true. Since r is true, the given statement, which is the negation of r, is false.

(3) Let p be "Paris is in England" and let q be "London is in France". Note p and q are each false; hence by $\mathbf{T}_2$ the given statement, which is $p \vee q$, is false.

(4) Let p be "$1+1 = 3$", let q be "$2+1 = 3$", and let r be "p or q". Note that p is false and q is true; hence, by $\mathbf{T}_2$, $p \vee q$, which is r, is true. Since the given statement is $\sim r$, it is false.

(5) Let p be "Paris is in England", let q be "London is in France", and let r be "If p then q". Note p and q are each false; hence, by $\mathbf{T}_4$, $p \to q$ is true, i.e. r is true. Consequently the given statement, $\sim r$, is false.

TRUTH TABLES OF PROPOSITIONS

4. Find the truth table of each proposition.

(1) $\sim p \wedge q$ (3) $(p \wedge q) \to (p \vee q)$

(2) $\sim(p \to \sim q)$ (4) $\sim(p \wedge q) \vee \sim(q \leftrightarrow p)$

Solution:

(1)

p	q	$\sim p$	$\sim p \wedge q$
T	T	F	F
T	F	F	F
F	T	T	T
F	F	T	F

Method 1

p	q	$\sim$	p	$\wedge$	q
T	T	F	T	F	T
T	F	F	T	F	F
F	T	T	F	T	T
F	F	T	F	F	F
Step		2	1	3	1

Method 2

(2)

p	q	$\sim q$	$p \to \sim q$	$\sim(p \to \sim q)$
T	T	F	F	T
T	F	T	T	F
F	T	F	T	F
F	F	T	T	F

Method 1

p	q	$\sim$	$(p$	$\to$	$\sim$	$q)$
T	T	T	T	F	F	T
T	F	F	T	T	T	F
F	T	F	F	T	F	T
F	F	F	F	T	T	F
Step		4	1	3	2	1

Method 2

(3)

p	q	$p \wedge q$	$p \vee q$	$(p \wedge q) \to (p \vee q)$
T	T	T	T	T
T	F	F	T	T
F	T	F	T	T
F	F	F	F	T

Method 1

p	q	$(p$	$\wedge$	$q)$	$\to$	$(p$	$\vee$	$q)$
T	T	T	T	T	T	T	T	T
T	F	T	F	F	T	T	T	F
F	T	F	F	T	T	F	T	T
F	F	F	F	F	T	F	F	F
Step		1	2	1	3	1	2	1

Method 2

Note that $(p \wedge q) \to (p \vee q)$ is a tautology.

(4)

p	q	$p \wedge q$	$q \leftrightarrow p$	$\sim(p \wedge q)$	$\sim(q \leftrightarrow p)$	$\sim(p \wedge q) \vee \sim(q \leftrightarrow p)$
T	T	T	T	F	F	F
T	F	F	F	T	T	T
F	T	F	F	T	T	T
F	F	F	T	T	F	T

Method 1

p	q	$\sim$	$(p$	$\wedge$	$q)$	$\vee$	$\sim$	$(q$	$\leftrightarrow$	$p)$
T	T	F	T	T	T	F	F	T	T	T
T	F	T	T	F	F	T	T	F	F	T
F	T	T	F	F	T	T	T	T	F	F
F	F	T	F	F	F	T	F	F	T	F
Step		3	1	2	1	4	3	1	2	1

Method 2

Usually if a proposition is very involved, the second method takes less time and less space.

5. Find the truth table of each proposition.

(1) $(p \to q) \vee \sim(p \leftrightarrow \sim q)$ (2) $[p \to (\sim q \vee r)] \wedge \sim[q \vee (p \leftrightarrow \sim r)]$

Solution:

(1)

p	q	$(p$	$\to$	$q)$	$\vee$	$\sim$	$(p$	$\leftrightarrow$	$\sim$	$q)$
T	T	T	T	T	T	T	T	F	F	T
T	F	T	F	F	F	F	T	T	T	F
F	T	F	T	T	T	F	F	T	F	T
F	F	F	T	F	T	T	F	F	T	F
Step		1	2	1	5	4	1	3	2	1

(2)

p	q	r	$[p$	$\to$	$(\sim$	q	$\vee$	$r)]$	$\wedge$	$\sim$	$[q$	$\vee$	$(p$	$\leftrightarrow$	$\sim$	$r)]$
T	T	T	T	T	F	T	T	T	F	F	T	T	T	F	F	T
T	T	F	T	F	F	T	F	F	F	F	T	T	T	T	T	F
T	F	T	T	T	T	F	T	T	T	T	F	F	T	F	F	T
T	F	F	T	T	T	F	T	F	F	F	F	T	T	T	T	F
F	T	T	F	T	F	T	T	T	F	F	T	T	F	T	F	T
F	T	F	F	T	F	T	F	F	F	F	T	T	F	F	T	F
F	F	T	F	T	T	F	T	T	F	F	F	T	F	T	F	T
F	F	F	F	T	T	F	T	F	T	T	F	F	F	F	T	F
Step			1	4	2	1	3	1	6	5	1	4	1	3	2	1

NEGATION

6. Verify, by truth tables, that the negation of $p \wedge q$, $p \vee q$, $p \to q$, and $p \leftrightarrow q$ is logically equivalent to, respectively, $\sim p \vee \sim q$, $\sim p \wedge \sim q$, $p \wedge \sim q$, and $p \leftrightarrow \sim q$ or $\sim p \leftrightarrow q$. In other words, verify that:

(1) $\sim(p \wedge q) \equiv \sim p \vee \sim q$ (DeMorgan's Law)
(3) $\sim(p \to q) \equiv p \wedge \sim q$

(2) $\sim(p \vee q) \equiv \sim p \wedge \sim q$ (DeMorgan's Law)
(4) $\sim(p \leftrightarrow q) \equiv p \leftrightarrow \sim q \equiv \sim p \leftrightarrow q$

Solution:

(1)

p	q	$p \wedge q$	$\sim(p \wedge q)$	$\sim p$	$\sim q$	$\sim p \vee \sim q$
T	T	T	F	F	F	F
T	F	F	T	F	T	T
F	T	F	T	T	F	T
F	F	F	T	T	T	T

(2)

p	q	$p \vee q$	$\sim(p \vee q)$	$\sim p$	$\sim q$	$\sim p \wedge \sim q$
T	T	T	F	F	F	F
T	F	T	F	F	T	F
F	T	T	F	T	F	F
F	F	F	T	T	T	T

(3)

p	q	$p \to q$	$\sim(p \to q)$	$\sim q$	$p \wedge \sim q$
T	T	T	F	F	F
T	F	F	T	T	T
F	T	T	F	F	F
F	F	T	F	T	F

(4)

p	q	$p \leftrightarrow q$	$\sim(p \leftrightarrow q)$	$\sim p$	$\sim p \leftrightarrow q$	$\sim q$	$p \leftrightarrow \sim q$
T	T	T	F	F	F	F	F
T	F	F	T	F	T	T	T
F	T	F	T	T	T	F	T
F	F	T	F	T	F	T	F

7. Verify: $\sim\sim p \equiv p$.

p	$\sim p$	$\sim\sim p$
T	F	T
F	T	F

8. Use the results of Problems 6 and 7 to simplify each of the following propositions:

(1) $\sim(p \vee \sim q)$ (3) $\sim(p \wedge \sim q)$ (5) $\sim(\sim p \leftrightarrow q)$

(2) $\sim(\sim p \rightarrow q)$ (4) $\sim(\sim p \wedge \sim q)$ (6) $\sim(\sim p \rightarrow \sim q)$

Solution:

(1) $\sim(p \vee \sim q) \equiv \sim p \wedge \sim\sim q \equiv \sim p \wedge q$

(2) $\sim(\sim p \rightarrow q) \equiv \sim p \wedge \sim q$

(3) $\sim(p \wedge \sim q) \equiv \sim p \vee \sim\sim q \equiv \sim p \vee q$

(4) $\sim(\sim p \wedge \sim q) \equiv \sim\sim p \vee \sim\sim q \equiv p \vee q$

(5) $\sim(\sim p \leftrightarrow q) \equiv \sim\sim p \leftrightarrow q \equiv p \leftrightarrow q$

(6) $\sim(\sim p \rightarrow \sim q) \equiv \sim p \wedge \sim\sim q \equiv \sim p \wedge q$

9. Simplify each of the following statements.

(1) It is not true that roses are red implies violets are blue.

(2) It is not true that it is cold and raining.

(3) It is not true that he is short or handsome.

(4) It is not true that it is not cold or it is raining.

(5) It is not true that if it is raining then it is cold.

(6) It is not true that, roses are red iff violets are blue.

Solution:

(1) Let p be "Roses are red" and let q be "Violets are blue". Then the given statement can be denoted by $\sim(p \rightarrow q)$. By Problem 6, $\sim(p \rightarrow q) \equiv p \wedge \sim q$. Hence the given statement is logically equivalent to "Roses are red and violets are not blue".

(2) Since $\sim(p \wedge q) \equiv \sim p \vee \sim q$, the given statement is logically equivalent to "It is not cold or it is not raining".

(3) Since $\sim(p \vee q) \equiv \sim p \wedge \sim q$, the given statement is logically equivalent to "He is not short and not handsome".

(4) Note that $\sim(\sim p \vee q) \equiv \sim\sim p \wedge \sim q \equiv p \wedge \sim q$. Hence the given statement, which can be denoted by $\sim(\sim p \vee q)$ where p is "It is cold" and q is "It is raining", can be rewritten "It is cold and it is not raining".

(5) Since $\sim(p \rightarrow q) \equiv p \wedge \sim q$, the given statement can be rewritten "It is raining and it is not cold".

(6) Since $\sim(p \leftrightarrow q) \equiv p \leftrightarrow \sim q$, the given statement is logically equivalent to "Roses are red iff violets are not blue".

LOGICAL EQUIVALENCE

10. (1) Prove the Associative Law: $(p \wedge q) \wedge r \;\equiv\; p \wedge (q \wedge r)$.

(2) Prove the Distributive Law: $p \vee (q \wedge r) \;\equiv\; (p \vee q) \wedge (p \vee r)$.

Solution: In each case construct a truth table.

(1)

p	q	r	$p \wedge q$	$(p \wedge q) \wedge r$	$q \wedge r$	$p \wedge (q \wedge r)$
T	T	T	T	T	T	T
T	T	F	T	F	F	F
T	F	T	F	F	F	F
T	F	F	F	F	F	F
F	T	T	F	F	T	F
F	T	F	F	F	F	F
F	F	T	F	F	F	F
F	F	F	F	F	F	F

(2)

p	q	r	$q \wedge r$	$p \vee (q \wedge r)$	$p \vee q$	$p \vee r$	$(p \vee q) \wedge (p \vee r)$
T	T	T	T	T	T	T	T
T	T	F	F	T	T	T	T
T	F	T	F	T	T	T	T
T	F	F	F	T	T	T	T
F	T	T	T	T	T	T	T
F	T	F	F	F	T	F	F
F	F	T	F	F	F	T	F
F	F	F	F	F	F	F	F

11. Prove that the operation of disjunction can be written in terms of the operations of conjunction and negation. Specifically, $p \vee q \equiv \sim(\sim p \wedge \sim q)$.

Solution:

p	q	$p \vee q$	$\sim p$	$\sim q$	$\sim p \wedge \sim q$	$\sim(\sim p \wedge \sim q)$
T	T	T	F	F	F	T
T	F	T	F	T	F	T
F	T	T	T	F	F	T
F	F	F	T	T	T	F

12. Prove that the conditional operation distributes over the operation of conjunction: $p \rightarrow (q \wedge r) \;\equiv\; (p \rightarrow q) \wedge (p \rightarrow r)$.

Solution:

p	q	r	$q \wedge r$	$p \rightarrow (q \wedge r)$	$p \rightarrow q$	$p \rightarrow r$	$(p \rightarrow q) \wedge (p \rightarrow r)$
T	T	T	T	T	T	T	T
T	T	F	F	F	T	F	F
T	F	T	F	F	F	T	F
T	F	F	F	F	F	F	F
F	T	T	T	T	T	T	T
F	T	F	F	T	T	T	T
F	F	T	F	T	T	T	T
F	F	F	F	T	T	T	T

ALGEBRA OF PROPOSITIONS

13. Use the laws of the algebra of sets to simplify each of the following:
(1) $(P \vee Q) \wedge \sim P$, (2) $P \vee (P \wedge Q)$, (3) $\sim(P \vee Q) \vee (\sim P \wedge Q)$.

Solution:

(1) $(P \vee Q) \wedge \sim P \;\equiv\; \sim P \wedge (P \vee Q) \;\equiv\; (\sim P \wedge P) \vee (\sim P \wedge Q) \;\equiv\; F \vee (\sim P \wedge Q) \;\equiv\; \sim P \wedge Q$

(2) $P \vee (P \wedge Q) \;\equiv\; (P \wedge T) \vee (P \wedge Q) \;\equiv\; P \wedge (T \vee Q) \;\equiv\; P \wedge T \;\equiv\; P$

(3) $\sim(P \vee Q) \vee (\sim P \wedge Q) \;\equiv\; (\sim P \wedge \sim Q) \vee (\sim P \wedge Q) \;\equiv\; \sim P \wedge (\sim Q \vee Q) \;\equiv\; \sim P \wedge T \;\equiv\; \sim P$

LOGICAL IMPLICATION

14. Decide whether each of the following is true or false: (1) $p \Rightarrow p \wedge q$, (2) $p \Rightarrow p \vee q$.

Solution:

Construct the truth tables of $p \rightarrow (p \wedge q)$ and $p \rightarrow (p \vee q)$.

p	q	$p \wedge q$	$p \rightarrow (p \wedge q)$	$p \vee q$	$p \rightarrow (p \vee q)$
T	T	T	T	T	T
T	F	F	F	T	T
F	T	F	T	T	T
F	F	F	T	F	T

Note that $p \rightarrow (p \wedge q)$ is not a tautology; hence (1) is false.

Note that $p \rightarrow (p \vee q)$ is a tautology; hence (2) is true.

15. Prove $p \wedge q$ logically implies $p \leftrightarrow q$.

Solution:

Construct the truth table for $(p \wedge q) \rightarrow (p \leftrightarrow q)$.

p	q	$p \wedge q$	$p \leftrightarrow q$	$(p \wedge q) \rightarrow (p \leftrightarrow q)$
T	T	T	T	T
T	F	F	F	T
F	T	F	F	T
F	F	F	T	T

Since $(p \wedge q) \rightarrow (p \leftrightarrow q)$ is a tautology, $p \wedge q \Rightarrow p \leftrightarrow q$.

16. Prove Theorem 14.7: If $P(p,q,\ldots) \Rightarrow Q(p,q,\ldots)$, then

$$P(P_1, P_2, \ldots) \Rightarrow Q(P_1, P_2, \ldots)$$

Solution:

Note that $P(p,q,\ldots) \Rightarrow Q(p,q,\ldots)$ if and only if

$$P(p,q,\ldots) \rightarrow Q(p,q,\ldots)$$

is a tautology. By the Principle of Substitution,

$$P(P_1,P_2,\ldots) \rightarrow Q(P_1,P_2,\ldots)$$

is also a tautology. In other words,

$$P(P_1,P_2,\ldots) \Rightarrow Q(P_1,P_2,\ldots)$$

17. Prove: Let $P(p,q,\ldots)$ be any proposition; then $p \Rightarrow p \vee P(p,q,\ldots)$.

Solution:

By Problem 14, $p \Rightarrow p \vee q$. By Theorem 14.7, $P(p,q,\ldots)$ can be substituted for q, i.e.

$$p \Rightarrow p \vee P(p,q,\ldots)$$

18. Prove: Let $P(p, q, \ldots) \Rightarrow Q(p, q, \ldots)$. Then Q is true whenever P is true.

Solution:

Note that $P(p, q, \ldots) \Rightarrow Q(p, q, \ldots)$ if and only if $P(p, q, \ldots) \rightarrow Q(p, q, \ldots)$ is a tautology, that is, is always true. By $\mathbf{T}_4$, $p_0 \rightarrow q_0$ is false if p_0 is true and q_0 is false; hence, if $P(p, q, \ldots)$ is true, then $Q(p, q, \ldots)$ must also be true.

MISCELLANEOUS PROBLEMS

19. The propositional connective $\underline{\vee}$ is called the *exclusive disjunction*; $p \underline{\vee} q$ is read "p or q but not both".

(1) Construct a truth table for $p \underline{\vee} q$.

(2) Prove: $p \underline{\vee} q \equiv (p \vee q) \wedge \sim(p \wedge q)$. Hence $\underline{\vee}$ can be written in terms of the original three connectives $\vee$, $\wedge$ and $\sim$.

Solution:

(1) Note that $p \underline{\vee} q$ is true if p is true or q is true but not if both p and q are true; hence the truth table of $p \underline{\vee} q$ is as follows:

p	q	$p \underline{\vee} q$
T	T	F
T	F	T
F	T	T
F	F	F

(2) Consider the following truth table.

p	q	$(p$	$\vee$	$q)$	$\wedge$	$\sim$	$(p$	$\wedge$	$q)$
T	T	T	T	T	F	F	T	T	T
T	F	T	T	F	T	T	T	F	F
F	T	F	T	T	T	T	F	F	T
F	F	F	F	F	F	T	F	F	F
Step		1	2	1	4	3	1	2	1

Since the truth tables of $p \underline{\vee} q$ and $(p \vee q) \wedge \sim(p \wedge q)$ are identical, $p \underline{\vee} q \equiv (p \vee q) \wedge \sim(p \wedge q)$.

20. The propositional connective $\downarrow$ is called the *joint denial*; $p \downarrow q$ is read "Neither p nor q".

(1) Construct a truth table for $p \downarrow q$.

(2) Prove: The three connectives $\vee$, $\wedge$ and $\sim$ may be expressed in terms of the connective $\downarrow$ as follows:

$$(a)\ \sim p \equiv p \downarrow p, \quad (b)\ p \wedge q \equiv (p \downarrow p) \downarrow (q \downarrow q), \quad (c)\ p \vee q \equiv (p \downarrow q) \downarrow (p \downarrow q).$$

Solution:

(1) Note $p \downarrow q$ is true if neither p is true nor q is true; hence the truth table of $p \downarrow q$ is as follows:

p	q	$p \downarrow q$
T	T	F
T	F	F
F	T	F
F	F	T

(2) (*a*)

p	$\sim p$	$p \downarrow p$
T	F	F
F	T	T

(*b*)

p	q	$p \wedge q$	$p \downarrow p$	$q \downarrow q$	$(p \downarrow p) \downarrow (q \downarrow q)$
T	T	T	F	F	T
T	F	F	F	T	F
F	T	F	T	F	F
F	F	F	T	T	F

(*c*)

p	q	$p \vee q$	$p \downarrow q$	$(p \downarrow q) \downarrow (p \downarrow q)$
T	T	T	F	T
T	F	T	F	T
F	T	T	F	T
F	F	F	T	F

21. There are at most four different non-equivalent propositions of one variable. The truth tables of such propositions are as follows:

p	$P_1(p)$	$P_2(p)$	$P_3(p)$	$P_4(p)$
T	T	T	F	F
F	T	F	T	F

Find four such propositions.

Solution:

Note

p	$\sim p$	$p \vee \sim p$	$p \wedge \sim p$
T	F	T	F
F	T	T	F

Hence $P_1(p) \equiv p \vee \sim p$, $P_2(p) \equiv p$, $P_3(p) \equiv \sim p$, $P_4(p) \equiv p \wedge \sim p$.

22. Find the number of different non-equivalent propositions of (1) two variables p and q, (2) three variables p, q and r, (3) n variables $p_1, p_2, \ldots, p_n$.

Solution:

(1) The truth table of a proposition $P(p, q)$ will contain $2^2 = 4$ lines. In each line T or F can appear; hence there are $2^{2^2} = 2^4 = 16$ different non-equivalent propositions $P(p, q)$.

(2) The truth table of a proposition $P(p, q, r)$ will contain $2^3 = 8$ lines. In each line T or F can appear; hence there are $2^{2^3} = 2^8 = 256$ different non-equivalent propositions $P(p, q, r)$.

(3) The truth table of a proposition $P(p_1, \ldots, p_n)$ will contain 2^n lines; hence, as above, there are 2^{2^n} different non-equivalent propositions $P(p_1, \ldots, p_n)$.

23. Let Apq denote $p \wedge q$ and let Np denote $\sim p$. Rewrite the following propositions using A and N instead of $\wedge$ and $\sim$.

(1) $p \wedge \sim q$ (3) $\sim p \wedge (\sim q \wedge r)$

(2) $\sim(\sim p \wedge q)$ (4) $\sim(p \wedge \sim q) \wedge (\sim q \wedge \sim r)$

Solution:

(1) $p \wedge \sim q = p \wedge Nq = ApNq$

(2) $\sim(\sim p \wedge q) = \sim(Np \wedge q) = \sim(ANpq) = NANpq$

(3) $\sim p \wedge (\sim q \wedge r) = Np \wedge (Nq \wedge r) = Np \wedge (ANqr) = ANpANqr$

(4) $\sim(p \wedge \sim q) \wedge (\sim q \wedge \sim r) = \sim(ApNq) \wedge (ANqNr) = (NApNq) \wedge (ANqNr) = ANApNqANqNr$

Notice that there are no parenthesis in the final answer when A and N are used instead of $\wedge$ and $\sim$. It has been proven that they are not needed. Furthermore, since every connective is logically equivalent to A and N, i.e. $\wedge$ and $\sim$, the above notation suffices for any development of the algebra of propositions.

24. Rewrite the following propositions using $\wedge$ and $\sim$ instead of A and N.

(1) $NApq$ (3) $ApNq$ (5) $AApqr$ (7) $NAANpqr$

(2) $ANpq$ (4) $ApAqr$ (6) $ANpAqNr$ (8) $ANApAqpAANqrp$

Solution:

(1) $NApq = N(p \wedge q) = \sim(p \wedge q)$

(2) $ANpq = A(\sim p)q = \sim p \wedge q$

(3) $ApNq = Ap(\sim q) = p \wedge \sim q$

(4) $ApAqr = Ap(q \wedge r) = p \wedge (q \wedge r)$

(5) $AApqr = A(p \wedge q)r = (p \wedge q) \wedge r$

(6) $ANpAqNr = ANpAq(\sim r) = ANp(q \wedge \sim r) = A(\sim p)(q \wedge \sim r) = \sim p \wedge (q \wedge \sim r)$

(7) $NAANpqr = NAA(\sim p)qr = NA(\sim p \wedge q)r = N[(\sim p \wedge q) \wedge r] = \sim[(\sim p \wedge q) \wedge r]$

(8) $ANApAqpAANqrp = ANApAqpAA(\sim q)rp = ANApAqpA(\sim q \wedge r)p = ANApAqp[(\sim q \wedge r) \wedge p]$
$= ANAp[q \wedge p][(\sim q \wedge r) \wedge p] = AN[p \wedge (q \wedge p)][(\sim q \wedge r) \wedge p]$
$= A \sim[p \wedge (q \wedge p)][(\sim q \wedge r) \wedge p] = \sim[p \wedge (q \wedge p)] \wedge [(\sim q \wedge r) \wedge p]$

Supplementary Problems

STATEMENTS

25. Let p be "He is rich" and let q be "He is happy". Give a simple verbal sentence which describes each of the following statements.

(1) $p \vee q$ (3) $q \rightarrow p$ (5) $q \leftrightarrow \sim p$ (7) $\sim \sim p$

(2) $p \wedge q$ (4) $p \vee \sim q$ (6) $\sim p \rightarrow q$ (8) $(\sim p \wedge q) \rightarrow p$

26. Let p be "He is rich" and let q be "He is happy". Write each of the following statements in symbolic form using p and q.

(1) He is neither rich nor happy.
(2) To be poor is to be unhappy.
(3) One is never happy if he is rich.
(4) He is poor but happy.
(5) He cannot be both rich and happy.
(6) If he is unhappy he is poor.
(7) If he is not poor and happy, then he is rich.
(8) To be rich means the same as to be happy.
(9) He is poor or else he is both rich and unhappy.
(10) If he is not poor, then he is happy.

27. Let p be "He is rich" and let q be "He is happy". Write each of the following conditional statements in symbolic form using p and q.

(1) If he is poor, he is happy.
(2) Being poor implies being happy.
(3) It is necessary to be poor in order to be happy.
(4) Being rich is a sufficient condition to being happy.
(5) Being rich is a necessary condition to being happy.
(6) He is poor only if he is unhappy.

28. Determine the truth value of each of the following statements.

(1) If $5 < 3$, then $-3 < -5$.
(2) It is not true that $2+2=4$ or $3+5=6$.
(3) It is true that $2+2 \neq 4$ and $3+3=6$.
(4) If $3 < 5$, then $-3 < -5$.

29. Determine the truth value of each of the following statements.

(1) It is not true that if $2+2=4$ then $3+3=5$ or $1+1=2$.
(2) If $2+2=4$, then it is not true that $2+1=3$ and $5+5=10$.
(3) It is not true that $2+7=9$ if and only if $2+1=5$ implies $5+5=8$.
(4) If $2+2 \neq 4$, then it is not true that $3+3=7$ iff $1+1=2$.

30. Write the negation of each of the following statements in as simple a sentence as possible.

(1) He is tall but handsome.
(2) He is neither rich nor happy.
(3) If stock prices fall, then unemployment rises.
(4) Neither Marc nor Erik is rich.
(5) He has blond hair or blue eyes.
(6) He has blond hair if and only if he has blue eyes.
(7) Both Marc and Erik are intelligent.
(8) If Marc is rich, then both Erik and Audrey are happy.
(9) Marc or Erik is intelligent, and Audrey is pretty.

TRUTH TABLES

31. Find the truth table of each of the following propositions.

(1) $\sim p \wedge \sim q$
(2) $\sim(\sim p \leftrightarrow q)$
(3) $p \rightarrow (\sim p \vee q)$
(4) $(p \wedge \sim q) \rightarrow (\sim p \vee q)$

32. Find a truth table of each proposition.

(1) $[p \wedge (\sim q \rightarrow p)] \wedge \sim[(p \leftrightarrow \sim q) \rightarrow (q \vee \sim p)]$
(2) $[p \vee (q \rightarrow \sim r)] \wedge [(\sim p \vee r) \leftrightarrow \sim q]$

LOGICAL EQUIVALENCE AND LOGICAL IMPLICATION

33. Prove: (1) $p \rightarrow \sim q \;\equiv\; q \rightarrow \sim p$
(2) $p \wedge (q \vee r) \;\equiv\; (p \wedge q) \vee (p \wedge r)$
(3) $(p \wedge q) \rightarrow r \;\equiv\; (p \rightarrow r) \vee (q \rightarrow r)$
(4) $[(p \rightarrow q) \rightarrow r] \;\equiv\; [(p \wedge \sim r) \rightarrow \sim q]$

34. State whether each of the following is true or false.

(1) $p \wedge q \Rightarrow p$ (2) $p \vee q \Rightarrow p$ (3) $q \Rightarrow p \rightarrow q$

35. Prove (by constructing necessary truth tables):

(1) $p \leftrightarrow q \Rightarrow q \rightarrow p$
(2) $\sim p \Rightarrow p \rightarrow q$
(3) $p \wedge (q \vee r) \Rightarrow (p \wedge q) \vee r$
(4) $q \Rightarrow [(p \wedge q) \leftrightarrow p]$

36. Prove: For any proposition $P(p, q, \ldots)$, $p \wedge P(p, q, \ldots) \Rightarrow p$.

37. Let Apq denote $p \wedge q$ and let Np denote $\sim p$. (See Problem 23.) Rewrite the following propositions using A and N instead of $\wedge$ and $\sim$.

(1) $\sim p \wedge q$
(2) $p \wedge \sim(p \wedge q)$
(3) $\sim(p \wedge q) \wedge (p \wedge \sim q)$
(4) $\sim(p \wedge q) \wedge \sim(\sim p \wedge \sim q)$
(5) $[p \wedge \sim(p \wedge \sim q)] \wedge \sim(p \wedge q)$
(6) $(\sim p \wedge \sim q) \wedge \sim[(p \wedge q) \wedge (\sim q \wedge p)]$

38. Rewrite the following propositions using $\wedge$ and $\sim$ instead of A and N.

(1) $NApNq$
(2) $ANApqNp$
(3) $AApNrAqNp$
(4) $ANANqANpqNp$
(5) $ANAApAqNrpAqr$
(6) $ANANpNAqNrApNAqNr$

Answers to Supplementary Problems

26. (2) $\sim p \leftrightarrow \sim q$, (5) $\sim(p \wedge q)$, (9) $\sim p \vee (p \wedge \sim q)$

27. (3) $q \rightarrow \sim p$, (4) $p \rightarrow q$, (6) $\sim p \rightarrow \sim q$

29. (1) F, (2) F, (3) Ambiguous, (4) T

30. (2) He is rich or happy. (8) Marc is rich, and Erik or Audrey is unhappy.

31. (2) T F F T, (4) T F T T

32. (1) F T F F

33. *Hint.* Construct truth tables.

34. (1) True, (2) False, (3) True

37. (2) $ApNApq$, (4) $ANApqNANpNq$, (6) $AANpNqNAApqANqp$

38. (2) $\sim(p \wedge q) \wedge \sim p$, (4) $\sim[\sim q \wedge (\sim p \wedge q)] \wedge \sim p$, (6) $\sim[\sim p \wedge \sim(q \wedge \sim r)] \wedge [p \wedge \sim(q \wedge \sim r)]$

Chapter 15

Quantifiers

PROPOSITIONAL FUNCTIONS AND TRUTH SETS

Let a set A be given, explicitly or implicitly. A *propositional function* or, simply, an *open-sentence* (or *condition*) on A is an expression denoted by

$$p(x)$$

which has the property that $p(a)$ is true or false for each $a \varepsilon A$. In other words, $p(x)$ is a propositional function on A if $p(x)$ becomes a statement whenever any element $a \varepsilon A$ is substituted for the variable x.

Example 1.1: Let $p(x)$ be "$x+2>7$". Then $p(x)$ is a propositional function on N, the set of natural numbers.

Example 1.2: Let $p(x)$ be "$x+2>7$". Then $p(x)$ is not a propositional function on C, the set of complex numbers, since inequalities are not defined for all the complex numbers.

Moreover, if $p(x)$ is a propositional function on a set A, then the set of elements $a \varepsilon A$ with the property that $p(a)$ is true is called the *truth set* T_p of $p(x)$. In other words,

$$T_p \;=\; \{x \mid x \varepsilon A,\ p(x) \text{ is true}\}$$

or, simply,

$$T_p \;=\; \{x \mid p(x)\}$$

Example 1.3: Consider the propositional function "$x+2>7$" defined on N, the set of natural numbers. Then

$$\{x \mid x \varepsilon N,\ x+2>7\} \;=\; \{6, 7, 8, \ldots\}$$

is its truth set.

Example 1.4: Let $p(x)$ be "$x+5<3$". Then the truth set of $p(x)$ on N is

$$\{x \mid x \varepsilon N,\ x+5<3\} \;=\; \emptyset$$

the empty set.

Example 1.5: Let $p(x)$ be "$x+5>1$". Then the truth set of $p(x)$ on N is

$$\{x \mid x+5>1\} \;=\; N$$

Notice, by the preceding examples, that if $p(x)$ is a propositional function defined on a set A then $p(x)$ could be true for all $x \varepsilon A$, for some $x \varepsilon A$ or for no $x \varepsilon A$.

UNIVERSAL QUANTIFIER

Let $p(x)$ be a propositional function on a set A. Then

$$(\forall x \varepsilon A)\, p(x) \quad \text{or} \quad \forall_x\, p(x) \quad \text{or} \quad \forall x,\, p(x) \tag{1}$$

is a statement which reads "For every element x in A, $p(x)$ is a true statement", or, simply, "For all x, $p(x)$". The symbol

$$\forall$$

which reads "for all" or "for every" is called the *universal quantifier*. Notice that (*1*) is equivalent to the set-theoretic statement that the truth set of $p(x)$ is the entire set A, that is,

$$T_p \;=\; \{x \mid x \varepsilon A,\ p(x)\} \;=\; A \tag{2}$$

Example 2.1: Let M denote the set of men. Then "All men are mortal" can be written as

$$(\forall x \varepsilon M)(x \text{ is mortal})$$

Note that $p(x)$, by itself, is an open sentence and therefore does not have a truth value. But $p(x)$ with the quantifier $\forall$ in front of it, that is, $\forall x\, p(x)$ is a statement and does have a truth value. In view of the equivalence of statements (*1*) and (*2*), we state:

Q$_1$: If $\{x \mid x \in A,\ p(x)\} = A$, then $\forall x\, p(x)$ is true; if $\{x \mid x \in A,\ p(x)\} \neq A$, then $\forall x\, p(x)$ is false.

Example 2.2: The proposition $(\forall n \in N)(n+4>3)$, where N is the set of natural numbers, is true since

$$\{n \mid n+4>3\} = \{1, 2, 3, \ldots\} = N$$

Example 2.3: The proposition $\forall n\,(n+2>8)$ is false since

$$\{n \mid n+2>8\} = \{7, 8, 9, \ldots\} \neq N$$

Example 2.4: The symbol $\forall$ can be used in defining the intersection of a family of sets $\{A_i\}_{i \in I}$ as follows:

$$\cap_{i \in I} A_i = \{x \mid \forall i \in I,\ x \in A_i\}$$

EXISTENTIAL QUANTIFIER

Let $p(x)$ be a propositional function on a set A. Then

$$(\exists x \in A)\, p(x) \quad \text{or} \quad \exists_x\, p(x) \quad \text{or} \quad \exists x\, p(x) \qquad (1)$$

is a proposition which reads "There exists an $x \in A$ such that $p(x)$ is a true statement" or, simply, "For some $x, p(x)$". The symbol

$$\exists$$

which reads "there exists" or "for some" or "for at least one" is called the *existential quantifier*. Notice that (*1*) is equivalent to the set-theoretic statement that the truth set of $p(x)$ is not empty, that is,

$$T_p = \{x \mid x \in A,\ p(x)\} \neq \emptyset$$

Hence

Q$_2$: If $\{x \mid p(x)\} \neq \emptyset$, then $\exists x\, p(x)$ is true; if $\{x \mid p(x)\} = \emptyset$, then $\exists x\, p(x)$ is false.

Example 3.1: The statement

$$(\exists n \in N)(n+4<7)$$

is true since

$$\{n \mid n+4<7\} = \{1, 2\} \neq \emptyset$$

Example 3.2: The proposition $\exists n\,(n+6<4)$ is false since $\{n \mid n+6<4\} = \emptyset$.

Example 3.3: The symbol $\exists$ can be used in defining the union of a family of sets $\{A_i\}_{i \in I}$ as follows:

$$\cup_{i \in I} A_i = \{x \mid \exists i \in I,\ x \in A_i\}$$

Remark 15.1: The symbol $\ni$ is frequently used for the words "such that" in many sentences using the existential quantifier $\exists$. For example, the statement "There exists a natural number n such that $50 < n^2 < 100$" is written

$$\exists n \in N \ \ni\ 50 < n^2 < 100$$

NEGATION OF PROPOSITIONS WHICH CONTAIN QUANTIFIERS

The negation of the proposition "All men are mortal" reads "It is not true that all men are mortal"; in other words, there exists at least one man who is not mortal. Symbolically, then, if M denotes the set of men, then the above can be written as

$$\sim(\forall x \in M)(x \text{ is mortal}) \equiv (\exists x \in M)(x \text{ is not mortal})$$

Furthermore, if $p(x)$ denotes "x is mortal", then the above can be written

$$\sim(\forall x \varepsilon M)\, p(x) \equiv (\exists x \varepsilon M) \sim p(x) \quad \text{or} \quad \sim\forall_x\, p(x) \equiv \exists_x \sim p(x)$$

The above is true in general. Specifically,

Theorem (DeMorgan) 15.1: $\sim(\forall x \varepsilon A)\, p(x) \equiv (\exists x \varepsilon A) \sim p(x)$.

There is an analogous theorem for the negation of a proposition which contains the existential quantifier.

Theorem (DeMorgan) 15.2: $\sim(\exists x \varepsilon A)\, p(x) \equiv (\forall x \varepsilon A) \sim p(x)$.

In other words, the statement

"It is not true that, for every $a \varepsilon A$, $p(a)$ is true"

is equivalent to the statement

"There exists an $a \varepsilon A$ such that $p(a)$ is false".

Similarly, the statement

"It is not true that there exists an $a \varepsilon A$ such that $p(a)$ is true".

is equivalent to the statement

"For all $a \varepsilon A$, $p(a)$ is false".

Example 4.1: The negation of the proposition "For all natural numbers n, $n+2 > 8$" is equivalent to the proposition "There exists an n such that $n+2 \not> 8$". In other words,

$$\sim(\forall n \varepsilon N)(n+2 > 8) \equiv (\exists n \varepsilon N)(n+2 \leqq 8)$$

Example 4.2: The negation of the statement "There exists a habitable planet" is the statement "All planets are not habitable". In other words, if P is the set of planets, then

$$\sim(\exists x \varepsilon P)(x \text{ is habitable}) \equiv (\forall x \varepsilon P)(x \text{ is not habitable})$$

Remark 15.2: Here $\sim p(x)$ has the obvious meaning, i.e. it is the propositional function that can be gotten by writing "It is not true that ..." in front of $p(x)$. Note that the truth set of $\sim p(x)$ is the complement of the truth set of $p(x)$, for

if $p(a)$ is true then $\sim p(a)$ is false

and vice versa. Previously $\sim$ was used as an operation on statements; here it is used in a different sense, i.e. as an operation on propositional functions.

Furthermore, $p(x) \wedge q(x)$ reads "$p(x)$ and $q(x)$", and $p(x) \vee q(x)$ reads "$p(x)$ or $q(x)$". It can also be shown that the same laws for propositions hold for propositional functions, e.g. $\sim(p(x) \wedge q(x)) \equiv \sim p(x) \vee \sim q(x)$.

COUNTER-EXAMPLE

By Theorem 15.1, $\sim\forall x, p(x) \equiv \exists x \sim p(x)$. Therefore, to show that a statement $\forall x, p(x)$ is false, it is equivalent to showing that $\exists x \sim p(x)$ is true, that is, that there is an element x_0 with the property that $p(x_0)$ is false. Such an element x_0 is called a *counter-example* to the statement $\forall x, p(x)$.

Example 5.1: Consider the statement $\forall x, |x| \neq 0$. The statement is false since the number 0 is a counter-example, i.e. $|0| \neq 0$ is not true.

Example 5.2: Consider the statement $\forall x, x^2 > x$. The statement is not true since, for example, $\frac{1}{2}$ is a counter-example, that is, $(\frac{1}{2})^2 \not> \frac{1}{2}$.

NOTATION

Let $A = \{2, 3, 5\}$, and let $p(x)$ be "x is a prime number". Then the proposition

"Two is a prime number and three is a prime number and five is a prime number" *(1)*

can be denoted by

$$p(2) \wedge p(3) \wedge p(5) \equiv \bigwedge_{a \in A} p(a)$$

Similarly, the proposition

"Two is a prime number or three is a prime number or five is a prime number" (*2*)

can be denoted by

$$p(2) \vee p(3) \vee p(5) \equiv \bigvee_{a \in A} p(a)$$

But (*1*) is equivalent to the statement

"Every number in A is a prime" (*3*)

and (*2*) is equivalent to the statement

"At least one number in A is a prime"

In other words,

$$\bigwedge_{a \in A} p(a) \equiv \forall_{a \in A}\, p(a)$$
$$\bigvee_{a \in A} p(a) \equiv \exists_{a \in A}\, p(a)$$

Hence the symbols $\wedge$ and $\vee$ are sometimes used instead of $\forall$ and $\exists$.

Remark 15.3: If A were an infinite set then a statement of the form (*1*) cannot be made since the sentence would not end; but a statement of the form (*3*) can always be made, even if A is infinite.

PROPOSITIONAL FUNCTIONS CONTAINING MORE THAN ONE VARIABLE

Consider the sets $A_1, A_2, \ldots, A_n$. A propositional function (of n variables) on $A_1 \times A_2 \times \cdots \times A_n$ is an expression denoted by

$$p(x_1, \ldots, x_n)$$

which has the property that $p(a_1, \ldots, a_n)$ is true or false for any ordered n-tuple $(a_1, \ldots, a_n) \in A_1 \times A_2 \times \cdots \times A_n$.

Example 6.1: Let M be a set of men and let W be a set of women. Then "x is married to y" is a propositional function on $M \times W$.

Example 6.2: Let N denote the natural numbers. Then "$x + 2y + 3z < 18$" is a propositional function on $N \times N \times N$.

Basic Principle: A propositional function preceded by quantifiers for each variable, for example,

$$\forall x\, \exists y\, p(x, y) \quad \text{or} \quad \exists x\, \forall z\, \forall y\, p(x, y, z)$$

denotes a statement and has a truth value.

Example 6.3: Let $M = \{\text{Erik, Mark, Paul}\}$, let $W = \{\text{Karen, Audrey}\}$, and let $p(x, y)$ be "x is the brother of y". Then

$$\forall x \in M [\exists y \in W \;\; p(x, y)] \equiv \forall x \in M \;\; \exists y \in W \;\; p(x, y)$$

reads "For every x in M there exists a y in W such that x is the brother of y." In other words, each member of M is the brother of Karen or Audrey.

Example 6.4: Let M, W and $p(x, y)$ be as in Example 6.3. Then

$$\exists y \in W \;\; \forall x \in M \;\; p(x, y)$$

states that at least one of the women in W is the sister of all the men in M. Hence a different order of the quantifiers defines a different proposition.

The negation of a proposition which contains quantifiers can be found as follows:

$$\sim \forall x\, [\exists y\, p(x, y)] \equiv \exists x \sim [\exists y\, p(x, y)] \equiv \exists x\, \forall y \sim p(x, y)$$

Example 6.5: Let M, W and $p(x, y)$ be as in Example 6.3. Then

$$\sim \forall x \in M \;\; \exists y \in W \;\; p(x, y) \equiv \exists x \in M \;\; \forall y \in W \;\; \sim p(x, y)$$

In other words, the statement "It is false that each man is the brother of at least one woman" is equivalent to "At least one of the men is not the brother of any of the women".

Solved Problems

1. Let $p(x)$ denote the sentence "$x+2 > 5$". State whether or not $p(x)$ is a propositional function on each of the following sets: (1) N, the set of natural numbers; (2) $M = \{-1, -2, -3, \ldots\}$; (3) C, the set of complex numbers.

Solution:

(1) Yes. (2) Although $p(x)$ is false for every element in M, $p(x)$ is still a propositional function on M. (3) No. Note that $2i+2 > 5$ does not have any meaning. In other words, inequalities are not defined for complex numbers.

2. Determine the truth value of each of the following statements. (Here the universal set is the set of real numbers.)

(1) $\forall x,\ |x| = x$ (3) $\forall x,\ x+1 > x$ (5) $\exists x,\ |x| = 0$

(2) $\exists x,\ x^2 = x$ (4) $\exists x,\ x+2 = x$

Solution:

(1) False. Note that if $x_0 = -3$ then $|x_0| \neq x_0$.

(2) True. For if $x_0 = 1$ then $x_0^2 = x_0$.

(3) True. For every real number is a solution to $x+1 > x$.

(4) False. There is no solution to $x+2 = x$.

(5) True. For if $x_0 = 0$ then $|x_0| = 0$.

3. Negate each of the statements in Problem 2.

Solution:

(1) $\sim\forall x,\ |x| = x \ \equiv\ \exists x \sim(|x| = x) \ \equiv\ \exists x,\ |x| \neq x$

(2) $\sim\exists x,\ x^2 = x \ \equiv\ \forall x \sim(x^2 = x) \ \equiv\ \forall x,\ x^2 \neq x$

(3) $\sim\forall x,\ x+1 > x \ \equiv\ \exists x \sim(x+1 > x) \ \equiv\ \exists x,\ x+1 \leq x$

(4) $\sim\exists x,\ x+2 = x \ \equiv\ \forall x \sim(x+2 = x) \ \equiv\ \forall x,\ x+2 \neq x$

(5) $\sim\exists x,\ |x| = 0 \ \equiv\ \forall x \sim(|x| = 0) \ \equiv\ \forall x,\ |x| \neq 0$

4. Let $A = \{1, 2, 3, 4, 5\}$. Determine the truth value of each of the statements.

(1) $(\exists x \in A)(x+3 = 10)$ (3) $(\exists x \in A)(x+3 < 5)$

(2) $(\forall x \in A)(x+3 < 10)$ (4) $(\forall x \in A)(x+3 \leq 7)$

Solution:

(1) False. For no number in A is a solution to $x+3 = 10$.

(2) True. For every number in A satisfies $x+3 < 10$.

(3) True. For if $x_0 = 1$, then $x_0 + 3 < 5$, i.e. 1 is a solution.

(4) False. For if $x_0 = 5$, then $x_0 + 3 \not\leq 7$. In other words, 5 is not a solution to the given condition.

5. Negate each of the statements in Problem 4.

Solution:

(1) $\sim(\exists x \in A)(x+3 = 10) \ \equiv\ (\forall x \in A) \sim(x+3 = 10) \ \equiv\ (\forall x \in A)(x+3 \neq 10)$

(2) $\sim(\forall x \in A)(x+3 < 10) \ \equiv\ (\exists x \in A) \sim(x+3 < 10) \ \equiv\ (\exists x \in A)(x+3 \geq 10)$

(3) $\sim(\exists x \in A)(x+3 < 5) \ \equiv\ (\forall x \in A) \sim(x+3 < 5) \ \equiv\ (\forall x \in A)(x+3 \geq 5)$

(4) $\sim(\forall x \in A)(x+3 \leq 7) \ \equiv\ (\exists x \in A) \sim(x+3 \leq 7) \ \equiv\ (\exists x \in A)(x+3 > 7)$

6. Negate each of the statements: (1) $\forall x\, p(x) \wedge \exists y\, q(y)$, (2) $\exists x\, p(x) \vee \forall y\, q(y)$.

Solution:

(1) Note that $\sim(p \wedge q) \equiv \sim p \vee \sim q$; hence

$$\sim(\forall x\, p(x) \wedge \exists y\, q(y)) \ \equiv\ \sim\forall x\, p(x) \vee \sim\exists y\, q(y) \ \equiv\ \exists x \sim p(x) \vee \forall y \sim q(y)$$

(2) Note that $\sim(p \vee q) \equiv \sim p \wedge \sim q$; hence

$$\sim(\exists x\, p(x) \vee \forall y\, q(y)) \ \equiv\ \sim\exists x\, p(x) \wedge \sim\forall y\, q(y) \ \equiv\ \forall x \sim p(x) \wedge \exists y \sim q(y)$$

7. Negate each of the following statements.

(1) If there is a riot then someone is killed.

(2) It is daylight and all the people have arisen.

Solution:

(1) Note that $\sim(p \rightarrow q) \equiv p \wedge \sim q$. Hence:

"It is false that, if there is a riot then someone is killed"

$\equiv$ "There is a riot and it is false that someone is killed"

$\equiv$ "There is a riot and everyone is alive".

(2) Note that $\sim(p \wedge q) \equiv \sim p \vee \sim q$. Hence:

"It is false that it is daylight and all the people have arisen"

$\equiv$ "It is not daylight or it is false that all the people have arisen"

$\equiv$ "It is night or someone has not arisen".

8. Find a counter-example for each of the following statements. Here $B = \{2, 3, \ldots, 8, 9\}$.

(1) $\forall x \in B,\ x + 5 < 12$

(2) $\forall x \in B,\ x$ is prime

(3) $\forall x \in B,\ x^2 > 1$

(4) $\forall x \in B,\ x$ is even

Solution:

(1) If $x_0 = 7, 8$ or 9, then $x_0 + 5 < 12$ is not true; hence either 7, 8 or 9 is a counter-example.

(2) Note that 4 is not a prime; hence 4 is a counter-example.

(3) The statement is true; hence there is no counter-example.

(4) Note that 3 is odd; hence 3 is a counter-example.

9. Let $\{1, 2, 3\}$ be the universal set. Determine the truth value of each of the following statements.

(1) $\exists x\ \forall y,\ x^2 < y + 1$

(2) $\forall x\ \exists y,\ x^2 + y^2 < 12$

(3) $\forall x\ \forall y,\ x^2 + y^2 < 12$

(4) $\exists x\ \forall y\ \exists z,\ x^2 + y^2 < 2z^2$

(5) $\exists x\ \exists y\ \forall z,\ x^2 + y^2 < 2z^2$

Solution:

(1) True. For if $x = 1$, then $1 < y + 1$ has as solutions each of the numbers 1, 2 and 3.

(2) True. For each x_0, let $y = 1$; then $x_0^2 + 1 < 12$ is a true statement.

(3) False. For if $x_0 = 2$ and $y_0 = 3$, then $x_0^2 + y_0^2 < 12$ is not a true statement.

(4) True. For if $x_0 = 1$ and $z_0 = 3$, then the truth set of $x_0^2 + y^2 < 2z_0^2$, i.e. $1 + y^2 < 18$, is the universal set 1, 2, 3.

(5) False. For if $z_0 = 1$, then $x^2 + y^2 < 2z_0^2$ has no solution.

10. Let $A = \{1, 2, \ldots, 9, 10\}$. Consider each of the following sentences. If it is a statement, then determine its truth value. If it is a propositional function, determine its truth set.

(1) $(\forall x \in A)(\exists y \in A) \ni (x + y < 14)$

(2) $(\forall y \in A)(x + y < 14)$

(3) $(\forall x \in A)(\forall y \in A)(x + y < 14)$

(4) $(\exists y \in A)(x + y < 14)$

Solution:

(1) The open sentence in two variables is preceded by two quantifiers; hence it is a statement. Moreover, the statement is true.

(2) The open sentence is preceded by one quantifier; hence it is a propositional function of the other variable. Note that for every $y \in A$, $x_0 + y < 14$ if and only if $x_0 = 1$, 2 or 3. Hence the truth set is $\{1, 2, 3\}$.

(3) It is a statement and it is false. For if $x_0 = 8$ and $y_0 = 9$, then $x_0 + y_0 < 14$ is not true.

(4) It is an open sentence in x. The truth set is A itself.

11. Negate each of the following statements.

(1) $\exists x \, \forall y, \; p(x,y)$ (4) $\forall x \, \exists y \; (p(x) \vee q(y))$

(2) $\forall x \, \forall y, \; p(x,y)$ (5) $\exists x \, \forall y \; (p(x,y) \rightarrow q(x,y))$

(3) $\exists y \, \exists x \, \forall z, \; p(x,y,z)$ (6) $\exists y \, \exists x \; (p(x) \wedge \sim q(y))$

Solution:

(1) $\sim(\exists x \, \forall y, \; p(x,y)) \;\equiv\; \forall x \, \exists y \sim p(x,y)$

(2) $\sim(\forall x \, \forall y, \; p(x,y)) \;\equiv\; \exists x \, \exists y \sim p(x,y)$

(3) $\sim(\exists y \, \exists x \, \forall z, \; p(x,y,z)) \;\equiv\; \forall y \, \forall x \, \exists z \sim p(x,y,z)$

(4) $\sim[(\forall x \, \exists y \; (p(x) \vee q(y))] \;\equiv\; \exists x \, \forall y \sim(p(x) \vee q(y)) \;\equiv\; \exists x \, \forall y \; (\sim p(x) \wedge \sim q(y))$

(5) $\sim[\exists x \, \forall y \; (p(x,y) \rightarrow q(x,y))] \;\equiv\; \forall x \, \exists y \sim(p(x,y) \rightarrow q(x,y)) \;\equiv\; \forall x \, \exists y \; (p(x,y) \wedge \sim q(x,y))$

(6) $\sim[\exists y \, \exists x \; (p(x) \wedge \sim q(y))] \;\equiv\; \forall y \, \forall x \sim(p(x) \wedge \sim q(y)) \;\equiv\; \forall y \, \forall x \; (\sim p(x) \vee q(y))$

12. Consider the following sentence which is the definition that the sequence $a_1, a_2, \ldots$ has zero as a limit:

$$\forall \epsilon > 0 \; \exists n_0 \; \forall n \; (n > n_0 \rightarrow |a_n| < \epsilon)$$

Negate the sentence.

Solution:

$$\begin{aligned} \sim[\forall \epsilon > 0 \; \exists n_0 \; \forall n \; (n > n_0 \rightarrow |a_n| < \epsilon)] \;&\equiv\; \exists \epsilon > 0 \; \forall n_0 \; \exists n \sim(n > n_0 \rightarrow |a_n| < \epsilon) \\ &\equiv\; \exists \epsilon > 0 \; \forall n_0 \; \exists n \; (n > n_0 \wedge \sim(|a_n| < \epsilon)) \\ &\equiv\; \exists \epsilon > 0 \; \forall n_0 \; \exists n \; (n > n_0 \wedge |a_n| \geqq \epsilon) \end{aligned}$$

Supplementary Problems

13. Determine the truth value of each of the following statements. (Here the universal set is the set of real numbers.)

(1) $\forall x, \; x^2 = x$ (4) $\forall x, \; x - 3 < x$

(2) $\exists x, \; 2x = x$ (5) $\exists x, \; x^2 - 2x + 5 = 0$

(3) $\exists x, \; x^2 + 3x - 2 = 0$ (6) $\forall x, \; 2x + 3x = 5x$

14. Negate each statement in Problem 13.

15. Let {1, 2, 3, 4} be the universal set. Determine the truth value of each statement.

(1) $\forall x, \; x + 3 < 6$ (3) $\forall x, \; x^2 - 10 \leqq 8$

(2) $\exists x, \; x + 3 < 6$ (4) $\exists x, \; 2x^2 + x = 15$

16. Negate each statement in Problem 15.

17. Negate each statement: (1) $\forall x \; p(x) \wedge \exists x \; q(x)$, (2) $\exists y \; p(y) \rightarrow \forall x \sim q(x)$, (3) $\exists x \sim p(x) \vee \forall x \; q(x)$.

18. Negate each of the following statements.

(1) If the teacher is absent, then some students do not complete their homework.

(2) All the students completed their homework and the teacher is present.

(3) Some of the students did not complete their homework or the teacher is absent.

19. Find a counter-example for each statement which is false. Here {3, 5, 7, 9} is the universal set.

(1) $\forall x, \; x + 3 \geqq 7$ (3) $\forall x, \; x$ is prime

(2) $\forall x, \; x$ is odd (4) $\forall x, \; |x| = x$

20. Negate each statement in Problem 19.

21. Let {1, 2, 3} be the universal set. Determine the truth value of each statement.

(1) $\forall x\ \forall y,\ x^2 + 2y < 10$ (3) $\forall x\ \exists y,\ x^2 + 2y < 10$
(2) $\exists x\ \forall y,\ x^2 + 2y < 10$ (4) $\exists x\ \exists y,\ x^2 + 2y < 10$

22. Negate each statement in Problem 21.

23. Negate each of the following statements:

(1) $\forall x\ \exists y\ \forall z\ p(x, y, z)$ (3) $\forall x\ \exists y\ (p(x, y) \rightarrow q(y))$
(2) $\exists x\ \forall y\ (p(x) \vee \sim q(y))$ (4) $\exists x\ \exists y\ (p(x) \wedge q(y))$

24. Let {1, 2, 3, 4, 5} be the universal set. Find the truth set of each of the following propositional functions.

(1) $\exists x,\ 2x + y < 7$ (3) $\forall x,\ 2x + y < 10$
(2) $\exists y,\ 2x + y < 7$ (4) $\forall y,\ 2x + y < 10$

Answers to Supplementary Problems

13. (2) T, (5) F

14. (1) $\exists x,\ x^2 \neq x$ (4) $\exists x,\ x - 3 \geqq x$ (5) $\forall x,\ x^2 - 2x + 5 \neq 0$

15. (1) F, (3) T

16. (2) $\forall x,\ x + 3 \geqq 6$ (3) $\exists x,\ x^2 - 10 > 8$

17. (2) $\exists y\ p(y) \wedge \exists x\ q(x)$ (3) $\forall x\ p(x) \wedge \exists x \sim q(x)$

18. (1) The teacher is absent and all the students completed their homework.
(2) Some of the students did not complete their homework or the teacher is absent.
(3) All the students completed their homework and the teacher is present.

19. (1) 3, (3) 9

20. (1) $\exists x,\ x + 3 < 7$ (3) $\exists x$, x is not a prime
(2) $\exists x$, x is even (4) $\exists x,\ |x| \neq x$

21. (2) T, (3) F

23. (2) $\forall x\ \exists y\ (\sim p(x) \wedge q(y))$ (3) $\exists x\ \forall y\ (p(x, y) \wedge \sim q(y))$

24. (1) {1, 2, 3, 4}, (2) {1, 2}, (3) $\emptyset$, (4) {1, 2}

Chapter 16

Boolean Algebra

DEFINITION

Both sets and propositions were seen to have similar properties, that is, satisfy identical laws. These laws are those that are used to define an abstract mathematical structure called a Boolean algebra, which is named after the mathematician George Boole (1813-1864).

Definition 16.1: A Boolean algebra is a set B of elements $a, b, \ldots$ and two binary operations called the *sum* and *product*, denoted respectively by $+$ and $*$, such that:

B_0. **Closure Law:** For any $a, b \in B$, the sum $a+b$ and the product $a * b$ exist and are unique elements in B.

B_1. **Commutative Law:**
(1a) $a + b = b + a$ (1b) $a * b = b * a$

B_2. **Associative Law:**
(2a) $(a+b)+c = a+(b+c)$ (2b) $(a * b) * c = a * (b * c)$

B_3. **Distributive Law:**
(3a) $a + (b * c) = (a+b) * (a+c)$ (3b) $a * (b+c) = (a * b) + (a * c)$

B_4. **Identity:** An additive identity 0 and a multiplicative identity U exist such that, for any $a \in B$,
(4a) $a + 0 = a$ (4b) $a * U = a$

B_5. **Complement:** For any $a \in B$ there exists an $a' \in B$, called the *complement* of a, such that
(5a) $a + a' = U$ (5b) $a * a' = 0$

Remark 16.1: Note that, by definition, a binary operation satisfies the closure law; hence axiom B_0 need not have been explicitly stated.

Example 1.1: Let $B = \{1, 0\}$ and let two operations $+$ and $*$ be defined on B as follows:

+	1	0
1	1	1
0	1	0

*	1	0
1	1	0
0	0	0

Then B, or more precisely the triplet $(B, +, *)$, is a Boolean algebra.

Example 1.2: Let $\mathcal{A}$ be a family of sets which is closed under the operations of union, intersection and complement. Then $(\mathcal{A}, \cup, \cap)$ is a Boolean algebra. Note that the universal set is then the unit element and the null set $\emptyset$ is the zero element.

Example 1.3: Let $\mathcal{B}$ be the set of propositions generated by variables $p, q, \ldots$. Then $(\mathcal{B}, \vee, \wedge)$ is a Boolean algebra.

Remark 16.2: Since sets and propositions are classical examples of Boolean algebras, many texts denote the operations of a Boolean algebra by $\vee$ and $\wedge$ or by $\cup$ and $\cap$.

DUALITY IN A BOOLEAN ALGEBRA

By definition, the dual of any statement in a Boolean algebra $(B, +, *)$ is the statement that is derived by interchanging $+$ and $*$, and their identity elements U and 0, in the original statement; for example, the dual of

$$(U + a) * (b + 0) = b$$

is

$$(0 * a) + (b * U) = b$$

Notice that the dual of each axiom of a Boolean algebra is also an axiom. Accordingly, the Principle of Duality holds, that is,

Theorem (Principle of Duality): The dual of any theorem in a Boolean algebra is also a theorem.

In other words, if any statement is a consequence of the axioms of a Boolean algebra, then the dual is also a consequence of those axioms since the dual statement can be proven by using the dual of each step of the proof of the original statement.

BASIC THEOREMS

Although the five axioms B_1-B_5 do not include all the properties of sets and propositions listed on Pages 104 and 195, the other properties are a direct consequence of the axioms B_1-B_5. Specifically,

Theorem 16.1.1 (Idempotent Law): (i) $a + a = a$ (ii) $a * a = a$

Theorem 16.1.2: (i) $a + U = U$ (ii) $a * 0 = 0$

Theorem 16.1.3 (Involution Law): $(a')' = a$

Theorem 16.1.4: (i) $U' = 0$ (ii) $0' = U$

Theorem 16.1.5 (DeMorgan's Law): (i) $(a + b)' = a' * b'$ (ii) $(a * b)' = a' + b'$

ORDER IN A BOOLEAN ALGEBRA

Consider the following theorem.

Theorem 16.2: Let $a, b \in B$, a Boolean algebra. Then the following conditions are equivalent:

$$(1)\ a * b' = 0, \quad (2)\ a + b = b, \quad (3)\ a' + b = U, \quad (4)\ a * b = a$$

For a proof, see Problem 9.

In view of the preceding theorem, the following definition is introduced:

Definition 16.2: Let $a, b \in B$, a Boolean algebra. Then a is said to *precede* b, denoted by

$$a \precsim b,$$

if one of the properties in Theorem 16.2 holds.

Example 2.1: Consider a Boolean algebra of sets $(\mathcal{A}, \cup, \cap)$. Then A precedes B means that $A \subset B$. In other words, Theorem 16.2 states that if A is a subset of B, as illustrated in the adjoining Venn diagram, then the following conditions hold:

(1) $A \cap B' = \emptyset$ (3) $A' \cup B = U$
(2) $A \cup B = B$ (4) $A \cap B = A$

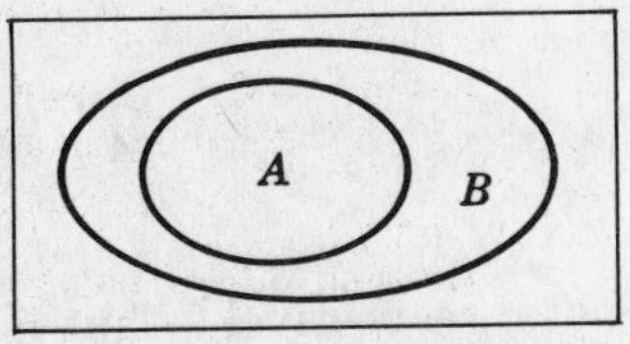

A is a subset of B

Example 2.2: Consider a Boolean algebra of proposition $(\mathcal{B}, \vee, \wedge)$. Then p precedes q means that p logically implies q, i.e. $p \Rightarrow q$.

Theorem 16.3: The relation in a Boolean algebra B defined by $a \precsim b$ is a partial order in B, i.e.,

(1) $a \precsim a$ for every $a \in B$ (Reflexive Law)
(2) $a \precsim b$ and $b \precsim a$ implies $a = b$ (Anti-symmetric Law)
(3) $a \precsim b$ and $b \precsim c$ implies $a \precsim c$ (Transitive Law)

Unless otherwise stated, a Boolean algebra is assumed to be partially ordered by the above definition.

The relationship between the properties of the partial order in a Boolean algebra B and the operations of B is presented in the next theorem.

Theorem 16.4: Let $a, b \in B$, a Boolean algebra. Then

(i) $a + b = \sup\{a, b\}$ **(ii)** $a * b = \inf\{a, b\}$

Remark 16.3: Any partially ordered set A such that $\inf\{a, b\}$ and $\sup\{a, b\}$ exist for any elements $a, b \in A$, is called a *lattice*. Hence a Boolean algebra is a special type of a lattice.

SWITCHING CIRCUIT DESIGNS

Let $A, B, \ldots$ denote electrical switches, and let A and A' denote switches with the property that if one is on then the other is off, and vice versa. Two switches, say A and B, can be connected by wire in a series or parallel combination as follows:

Series combination, $A \wedge B$

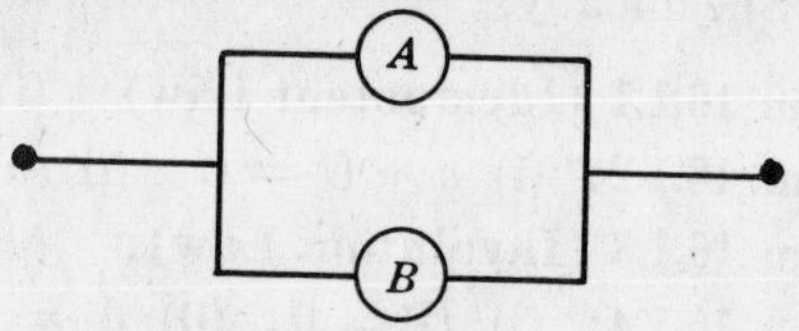

Parallel combination, $A \vee B$

Let

$$A \wedge B \quad \text{and} \quad A \vee B$$

denote respectively that A and B are connected in series and A and B are connected in parallel.

A Boolean switching circuit design means an arrangement of wires and switches that can be constructed by repeated use of series and parallel combinations; hence it can be described by the use of the connectives $\wedge$ and $\vee$.

Example 3.1:

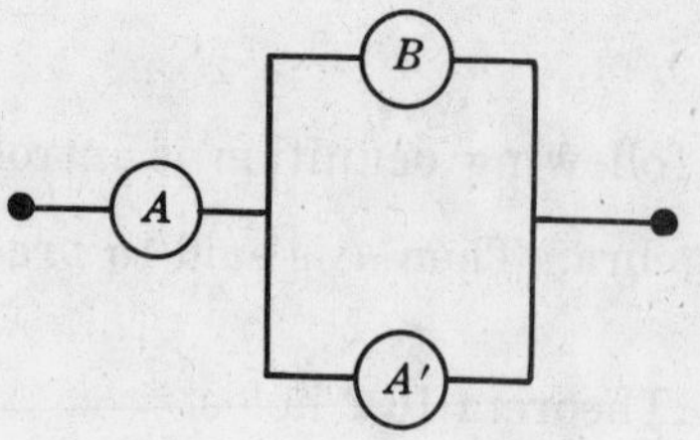

(1): $A \wedge (B \vee A')$

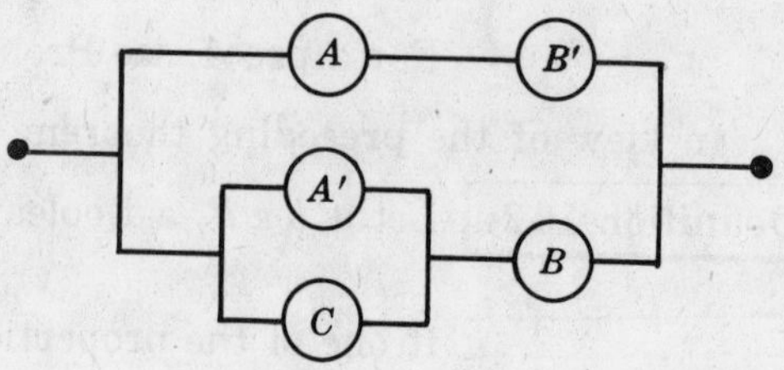

(2): $(A \wedge B') \vee [(A' \vee C) \wedge B]$

Circuit (1) can be described by $A \wedge (B \vee A')$, and circuit (2) can be described by $(A \wedge B') \vee [(A' \vee C) \wedge B]$.

Now let

$$1 \text{ and } 0$$

denote, respectively, that a switch or circuit is on and that a switch or circuit is off. The next two tables describe the behavior of a series circuit $A \wedge B$ and a parallel circuit $A \vee B$.

A	B	$A \wedge B$
1	1	1
1	0	0
0	1	0
0	0	0

A	B	$A \vee B$
1	1	1
1	0	1
0	1	1
0	0	0

The next table shows the relationship between a switch A and a switch A'.

A	A'
1	0
0	1

Notice that the above three tables are identical with the tables of conjunction, disjunction and negation for statements (and propositions). The only difference is that 1 and 0 are used here instead of T and F. Thus

Theorem 16.5: The algebra of Boolean switching circuits is a Boolean algebra.

In order to find the behavior of a Boolean switching circuit, a table is constructed which is analogous to the truth tables for propositions.

Example 3.2: Consider circuit (1) in Example 3.1. What is the behavior of the circuit, that is, when will the circuit be on (i.e. when will current flow) and when will the circuit be off? A "truth" table is constructed for $A \wedge (B \vee A')$ as follows:

A	B	A'	$B \vee A'$	$A \wedge (B \vee A')$
1	1	0	1	1
1	0	0	0	0
0	1	1	1	0
0	0	1	1	0

Thus current will flow only if both A and B are on.

Example 3.3: The behavior of circuit (2) in Example 3.1 is indicated by the following truth table for $(A \wedge B') \vee [(A' \vee C) \wedge B]$:

A	B	C	$(A$	$\wedge$	$B')$	$\vee$	$[(A'$	$\vee$	$C)$	$\wedge$	$B]$
1	1	1	1	0	0	1	0	1	1	1	1
1	1	0	1	0	0	0	0	0	0	0	1
1	0	1	1	1	1	1	0	1	1	0	0
1	0	0	1	1	1	1	0	0	0	0	0
0	1	1	0	0	0	1	1	1	1	1	1
0	1	0	0	0	0	1	1	1	0	1	1
0	0	1	0	0	1	0	1	1	1	0	0
0	0	0	0	0	1	0	1	1	0	0	0
Step			1	2	1	4	1	2	1	3	1

Remark 16.4: Any combination of switches using the connectives $\wedge$ and $\vee$, such as $(A \wedge B') \vee [(A' \vee C) \wedge B]$, will also be called a Boolean polynomial.

Solved Problems

BASIC THEOREMS

1. Prove Theorem 16.1.1 (Idempotent Law): **(i)** $a + a = a$, **(ii)** $a * a = a$.

Solution:

(ii)

Statement	**Reason**
(1) $a = a * U$	(1) B_4, Identity
(2) $= a * (a + a')$	(2) B_5, Complement
(3) $= (a * a) + (a * a')$	(3) B_3, Distributive Law
(4) $= (a * a) + 0$	(4) B_5, Complement
(5) $= a * a$	(5) B_4, Identity

(i) True by Principle of Duality.

2. Prove Theorem 16.1.2: **(i)** $a + U = U$, **(ii)** $a * 0 = 0$.

Solution:

(i)

Statement	**Reason**
(1) $U = a + a'$	(1) B_5, Complement
(2) $a + U = a + (a + a')$	(2) Substitution
(3) $= (a + a) + a'$	(3) B_2, Associative Law
(4) $= a + a'$	(4) Theorem 16.1.1, Idempotent Law
(5) $= U$	(5) B_5, Complement

(ii) True by Principle of Duality.

3. Prove Theorem 16.1.3 (Involution Law): $(a')' = a$; that is, if $(a)\ a + a' = U$, $(b)\ a * a' = 0$, $(c)\ a' + a'' = U$ and $(d)\ a' * a'' = 0$, then $a = a''$.

Solution:

Statement	**Reason**
(1) $a = a + 0$	(1) B_4, Identity
(2) $= a + (a' * a'')$	(2) Hypothesis (d)
(3) $= (a + a') * (a + a'')$	(3) B_3, Distributive Law
(4) $= U * (a + a'')$	(4) Hypothesis (a)
(5) $= (a' + a'') * (a + a'')$	(5) Hypothesis (c)
(6) $= (a'' + a') * (a'' + a)$	(6) B_1, Commutative Law
(7) $= a'' + (a' * a)$	(7) B_3, Distributive Law
(8) $= a'' + (a * a')$	(8) B_1, Commutative Law
(9) $= a'' + 0$	(9) Hypothesis (b)
(10) $= a''$	(10) B_4, Identity

4. Prove the uniqueness of the identity elements, that is:

(a) If 0_1 and 0_2 are additive identity elements, then $0_1 = 0_2$.

(b) If I_1 and I_2 are multiplicative identity elements, then $I_1 = I_2$.

Solution:

(a)

Statement	**Reason**
(1) $0_1 = 0_1 + 0_2$	(1) Hypothesis (0_2 is an additive identity)
(2) $= 0_2 + 0_1$	(2) B_1, Commutative Law
(3) $= 0_2$	(3) Hypothesis (0_1 is an additive identity)

(b) Principle of Duality

5. Prove Theorem 16.1.4: The identity elements are the complements of each other, i.e., (i) $U' = 0$ and (ii) $0' = U$.

Solution:

(i)

Statement	Reason
(1) $U' = U' * U$	(1) B_4, Identity
(2) $\quad = U * U'$	(2) B_1, Commutative Law
(3) $\quad = 0$	(3) B_5, Complement

(ii) Principle of Duality

6. Prove Theorem 16.1.5 (DeMorgan's Law):

(i) $(a + b)' = a' * b'$, that is, $(a + b) * (a' * b') = 0$ and $(a + b) + (a' * b') = U$.

(ii) $(a * b)' = a' + b'$, that is, $(a * b) * (a' + b') = 0$ and $(a * b) + (a' + b') = U$.

Solution:

(i)

Statement	Reason
(1) $(a + b) * (a' * b') = (a' * b') * (a + b)$	(1) B_1, Commutative Law
(2) $\quad = ((a' * b') * a) + ((a' * b') * b)$	(2) B_3, Distributive Law
(3) $\quad = ((b' * a') * a) + ((a' * b') * b)$	(3) B_1, Commutative Law
(4) $\quad = (b' * (a' * a)) + (a' * (b' * b))$	(4) B_2, Associative Law
(5) $\quad = (b' * (a * a')) + (a' * (b * b'))$	(5) B_1, Commutative Law
(6) $\quad = (b' * 0) + (a' * 0)$	(6) B_5, Complement
(7) $\quad = 0 + 0$	(7) Theorem 16.1.2
(8) $\quad = 0$	(8) Theorem 16.1.1
(9) $(a + b) + (a' * b') = U$	(9) Steps (1) through (8)

(ii) Principle of Duality

7. Prove (Uniqueness of complement): If a_1' and a_2' are complements of a, i.e. $a + a_1' = U$, $a + a_2' = U$, $a * a_1' = 0$ and $a * a_2' = 0$, then $a_1' = a_2'$.

Solution:

Statement	Reason
(1) $a_1' = a_1' + 0$	(1) B_4, Identity
(2) $\quad = a_1' + (a * a_2')$	(2) Hypothesis
(3) $\quad = (a_1' + a) * (a_1' + a_2')$	(3) B_3, Distributive Law
(4) $\quad = (a + a_1') * (a_1' + a_2')$	(4) B_1, Commutative Law
(5) $\quad = U * (a_1' + a_2')$	(5) Hypothesis
(6) $\quad = (a_1' + a_2') * U$	(6) B_1, Commutative Law
(7) $\quad = a_1' + a_2'$	(7) B_4, Identity
(8) $a_2' = a_2' + a_1'$	(8) Steps (1) through (7)
(9) $a_1' + a_2' = a_2' + a_1'$	(9) B_1, Commutative Law
(10) $a_1' = a_2'$	(10) Substitution

8. Prove (Absorption Law): (i) $a + (a * b) = a$, (ii) $a * (a + b) = a$.

Solution:

(i)

Statement	Reason
(1) $a + (a * b) = (a * U) + (a * b)$	(1) B_4, Identity
(2) $\quad = a * (U + b)$	(2) B_3, Distributive Law
(3) $\quad = a * (b + U)$	(3) B_1, Commutative Law
(4) $\quad = a * U$	(4) Theorem 16.1.2
(5) $\quad = a$	(5) B_4, Identity

(ii) Principle of Duality

ORDER

9. Prove Theorem 16.2: The following conditions are equivalent:

$$(1)\ a * b' = 0,\quad (2)\ a + b = b,\quad (3)\ a' + b = U,\quad (4)\ a * b = a$$

Solution:

(Only the equivalence of (1), (2) and (3) is proven here. That (4) is equivalent to the other statements is left as an exercise for the reader.) The proof is in three steps:

(i) (1) implies (2), (ii) (2) implies (3), (iii) (3) implies (1)

Proof of (i), $a * b' = 0$ implies $a + b = b$:

$$(a+b) = (a+b) * U = (a+b) * (b+b') = (b+a) * (b+b') = b + (a * b') \overset{H}{=} b + 0 = b.$$

Here $\overset{H}{=}$ means that the hypothesis is used in the step in the proof. The other steps use the axioms or previous theorems.

Proof of (ii), $a + b = b$ implies $a' + b = U$:

$$a' + b \overset{H}{=} a' + (a+b) = (a'+a) + b = (a+a') + b = U + b = U.$$

Proof of (iii), $a' + b = U$ implies $a * b' = 0$:

$$a' + b = U \text{ implies } (a'+b)' = U' \text{ implies } a'' * b' = U' \text{ implies } a * b' = 0.$$

10. Prove Theorem 16.3: For any $a, b \in B$:

$$(1)\ a \precsim a,\quad (2)\ a \precsim b \text{ and } b \precsim a \text{ implies } a = b,\quad (3)\ a \precsim b \text{ and } b \precsim c \text{ implies } a \precsim c.$$

Solution:

(1) Note that $a \precsim b$ iff $a + b = b$. Hence $a + a = a$ implies $a \precsim a$.

(2) Note that $a \precsim b$ iff $a + b = b$, and $b \precsim a$ iff $b + a = a$. Hence $a = b + a = a + b = b$.

(3) Note that $a \precsim b$ iff $a + b = b$, and $b \precsim c$ iff $b + c = c$. Hence

$$a + c = a + (b + c) = (a + b) + c = b + c = c$$

Consequently, $a \precsim c$.

11. Prove: Let $a, b \in B$, a Boolean algebra. Then $a + b$ is an upper bound for the set $\{a, b\}$.

Solution:

Note that $a + (a + b) = (a + a) + b = a + b$. Hence, by definition, $a \precsim (a + b)$. Similarly, $b \precsim (a + b)$. Hence $a + b$ is an upper bound for $\{a, b\}$.

12. Prove Theorem 16.4: Let $a, b \in B$, a Boolean algebra. Then

$$(i)\ a + b = \sup\{a, b\},\qquad (ii)\ a * b = \inf\{a, b\}$$

Solution:

(Only (i) is proven here. The proof of (ii) is left as an exercise for the reader.)

By the preceding problem, $a + b$ is an upper bound for $\{a, b\}$. In order to show that $a + b$ is the least upper bound for $\{a, b\}$, i.e. $\sup\{a, b\}$, it is only necessary to show that if c is also an upper bound for $\{a, b\}$ then $a + b$ precedes c. In other words,

$$a \precsim c \text{ and } b \precsim c \quad \text{implies} \quad (a + b) \precsim c.$$

Note that $a \precsim c$ iff $a + c = c$, and $b \precsim c$ iff $b + c = c$. Hence

$$(a + b) + c = a + (b + c) = a + c = c, \quad \text{i.e. } (a + b) \precsim c$$

13. Prove that the dual of $a \precsim b$ is $b \precsim a$, that is, the dual relation in a Boolean algebra B induces the inverse relation of the partial order in B.

Solution:

Note that $a \precsim b$ iff $a' + b = U$. The dual of $a' + b = U$ is $a' * b = 0$; hence $b * a' = 0$. But $b \precsim a$ iff $b * a' = 0$. Therefore the dual of $a \precsim b$ is $b \precsim a$.

14. Prove: The identity elements 0 and U are universal bounds, i.e. for every $a \varepsilon B$, $0 \precsim a \precsim U$.

Solution:

Note that $0+a = a+0 = a$; hence $0 \precsim a$. Also $a \precsim U$ iff $a+U = U$, which is true by Theorem 16.1.2. In other words, $0 \precsim a \precsim U$.

SWITCHING CIRCUITS

15. Determine the Boolean polynomial for each of the given three circuits.

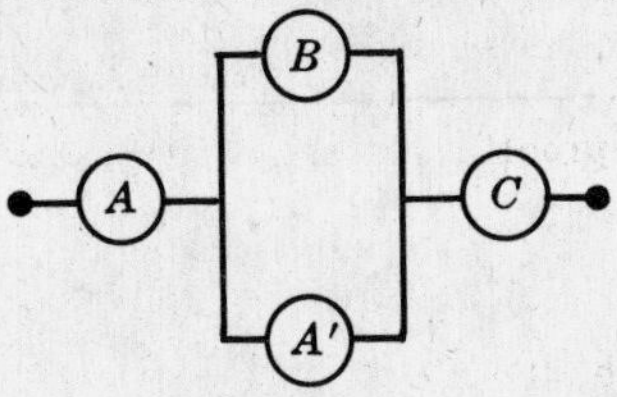

Circuit (1)

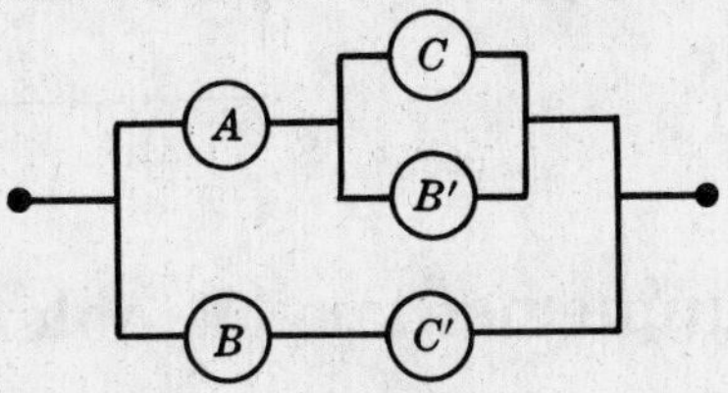

Circuit (2)

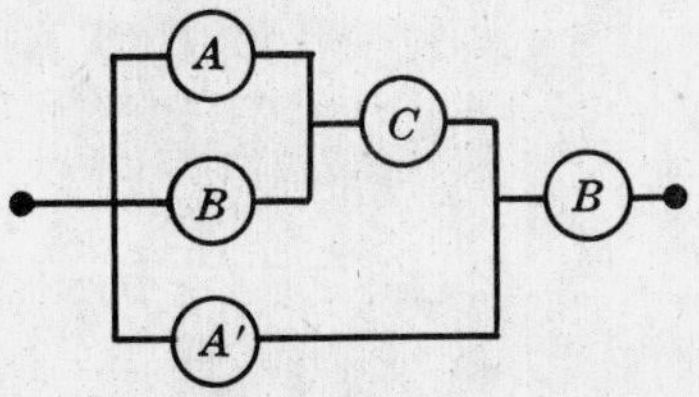

Circuit (3)

Solution:

(1) $A \wedge (B \vee A') \wedge C$ (2) $[A \wedge (C \vee B')] \vee (B \wedge C')$ (3) $\{[(A \vee B) \wedge C] \vee A'\} \wedge B$

16. Construct a circuit for each of the following Boolean polynomials:

(1) $(A \wedge B) \vee [A' \wedge (B' \vee A \vee B)]$, (2) $(A \vee B) \wedge C \wedge (A' \vee B' \vee C')$

Solution:

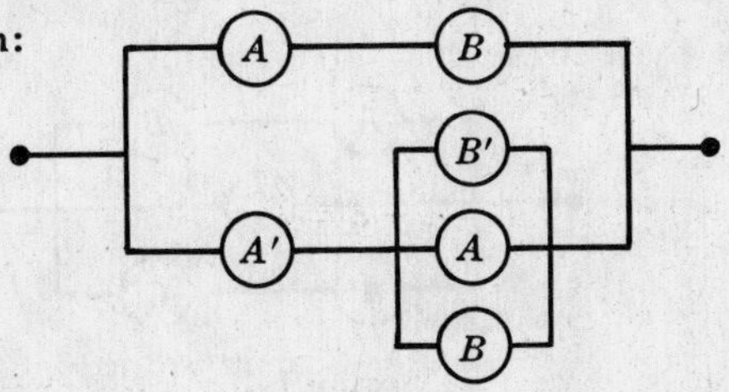

Circuit (1)

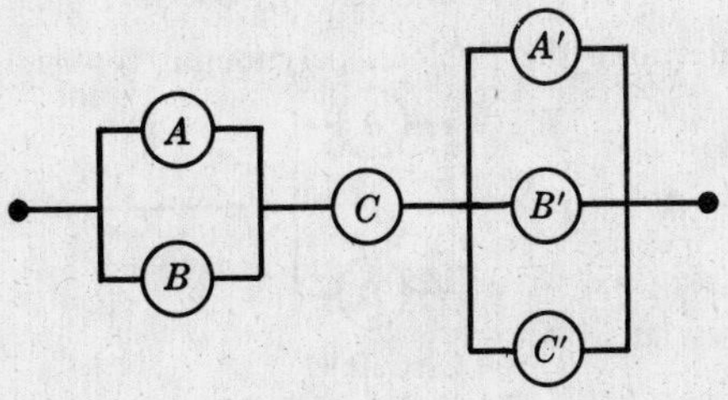

Circuit (2)

(1) Note that the series circuit $A \wedge B$ is in parallel with $A' \wedge (B' \vee A \vee B)$ which is A' in series with the parallel combination $B' \vee A \vee B$.

(2) Note that the parallel circuit $A \vee B$ is in series with C and in series with the parallel circuit $A' \vee B' \vee C'$.

17. Construct an equivalent simpler circuit of the adjacent diagram:

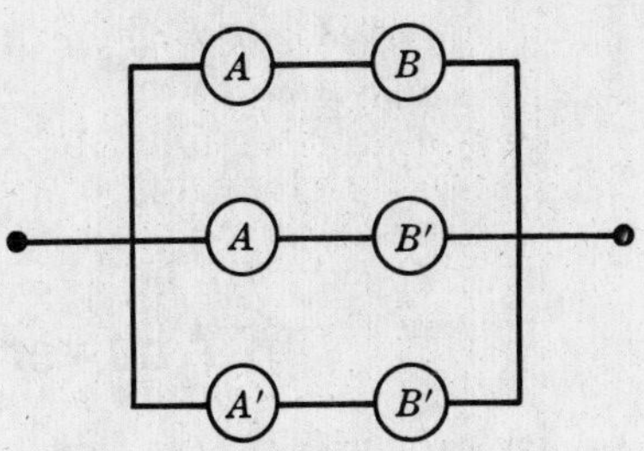

Solution:

First write a Boolean polynomial which represents the circuit:

$$(A \wedge B) \vee (A \wedge B') \vee (A' \wedge B')$$

Then simplify:

$$\begin{aligned}(A \wedge B) \vee (A \wedge B') \vee (A' \wedge B') &\equiv [A \wedge (B \vee B')] \vee (A' \wedge B') \\ &\equiv [A \wedge U] \vee (A' \wedge B') \\ &\equiv A \vee (A' \wedge B') \\ &\equiv (A \vee A') \wedge (A \vee B') \\ &\equiv U \wedge (A \vee B') \\ &\equiv A \vee B'\end{aligned}$$

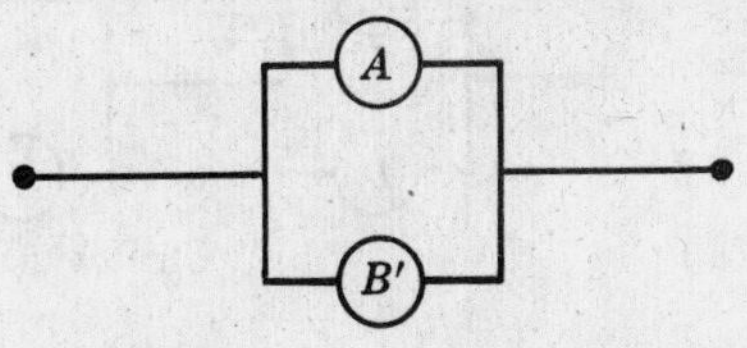

Hence the adjacent figure is an equivalent circuit.

18. Verify the solution in Problem 17 by "truth" tables:

Solution:

A	B	$(A$	$\wedge$	$B)$	$\vee$	$(A$	$\wedge$	$B')$	$\vee$	$(A'$	$\wedge$	$B')$
1	1	1	1	1	1	1	0	0	1	0	0	0
1	0	1	0	0	1	1	1	1	1	0	0	1
0	1	0	0	1	0	0	0	0	0	1	0	0
0	0	0	0	0	0	0	0	1	1	1	1	1
Step		1	2	1	3	1	2	1	4	1	2	1

A	B	B'	$A \vee B'$
1	1	0	1
1	0	1	1
0	1	0	0
0	0	1	1

Supplementary Problems

19. Prove Theorem 16.1.1 (i): $a + a = a$ (without using the principle of duality).

20. Prove Theorem 16.1.2 (ii): $a * 0 = 0$ (without using the principle of duality).

21. Prove Theorem 16.1.4 (ii): $0' = U$ (without using the principle of duality).

22. Prove Theorem 16.1.5 (ii): $(a * b)' = a' + b'$ (without using the principle of duality).

23. Prove Absorption Law (ii): $a * (a + b) = a$ (without using the principle of duality).

24. Complete the proof of Theorem 16.2: $a * b = a$ if and only if $a * b' = 0$. (Refer to Problem 9.)

25. Prove: $a \precsim b$ if and only if $b' \precsim a'$.

26. Determine the Boolean polynomial for each of the given circuits.

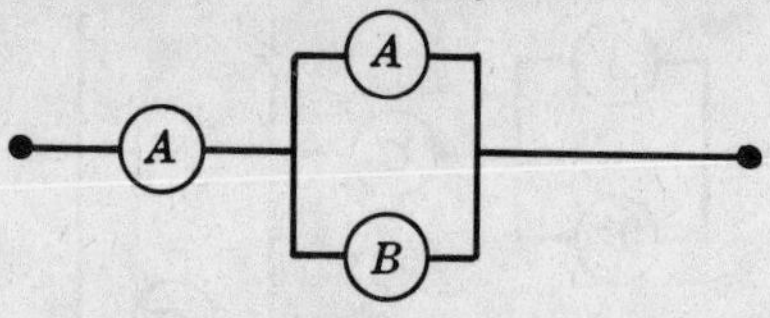

Circuit (1)

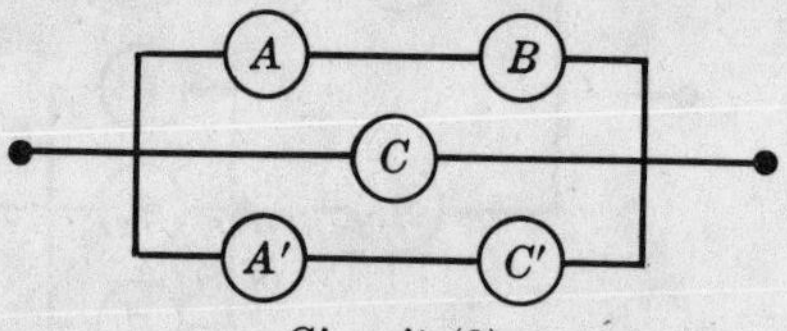

Circuit (3)

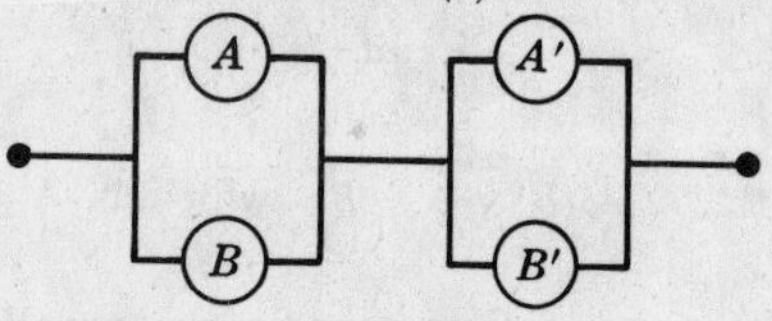

Circuit (2)

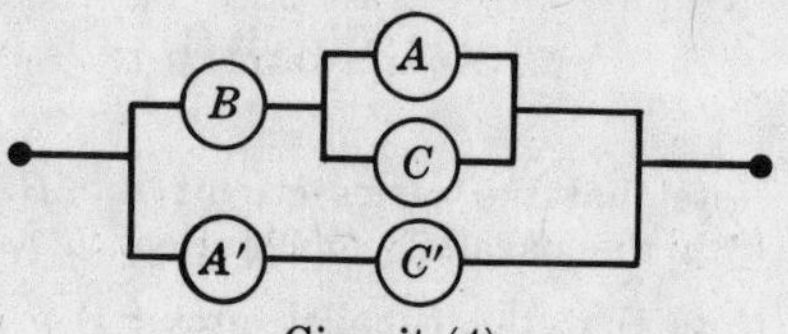

Circuit (4)

27. Construct a circuit for each Boolean polynomial:

(1) $A \vee (B \wedge C)$ (3) $(A \vee B) \wedge (C \vee D)$ (5) $(A \vee B) \wedge [A' \vee (C \wedge B')]$

(2) $A \wedge (B \vee C)$ (4) $(A \wedge B) \vee (C \wedge D)$ (6) $[(A \wedge B) \vee C] \wedge [D \vee (A' \wedge B)]$

Answers to Supplementary Problems

26. (3) $(A \wedge B) \vee C \vee (A' \wedge C')$, (4) $[B \wedge (A \vee C)] \vee (A' \wedge C')$

27.

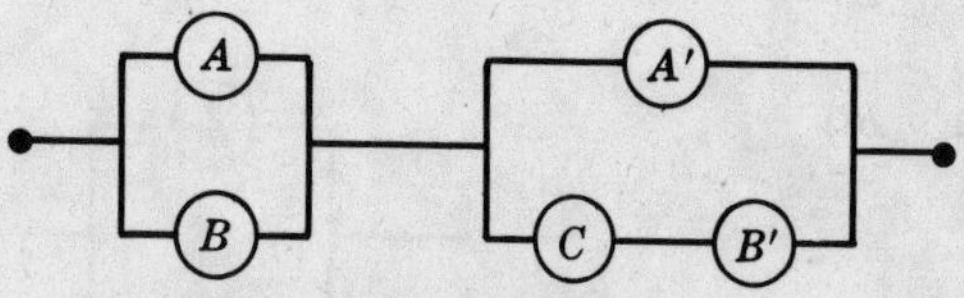

Circuit (5)

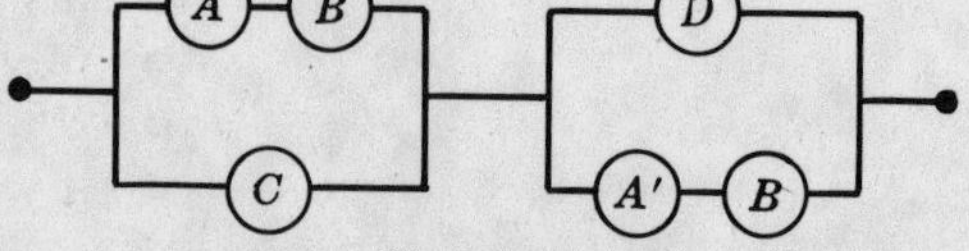

Circuit (6)

Chapter 17

Logical Reasoning

ARGUMENTS

An *argument* is an assertion that a given set of statements $S_1, \ldots, S_n$, called *premises*, yields (has as a consequence) another statement S, called the *conclusion*. Such an argument will be denoted by

$$S_1, S_2, \ldots, S_n \vdash S$$

Note that an argument is a statement and therefore has a truth value. If an argument is true, it is called a *valid* argument; if an argument is false, it is called a *fallacy*.

Example 1.1: Consider the following argument:

S_1: Some animals can reason.
S_2: Man is an animal.
....
S: Man can reason.

Here, the statement S below the line denotes the conclusion, and the statements S_1 and S_2 above the line denote the premises. Although each statement is true, it can be shown that the argument $S_1, S_2 \vdash S$ is a fallacy.

Example 1.2: Consider the following argument:

S_1: Babies are illogical.
S_2: Nobody is despised who can manage a crocodile.
S_3: Illogical people are despised.
....
S: Babies cannot manage crocodiles.

(The above argument is adapted from Lewis Carrol, *Symbolic Logic*; he is also the author of *Alice in Wonderland*.) The argument $S_1, S_2, S_3 \vdash S$ is valid.

Remark 17.1: Note that the truth value of an argument $S_1, \ldots, S_n \vdash S$ does not depend upon the particular truth value of each of the statements in the argument.

ARGUMENTS AND VENN DIAGRAMS

Many verbal statements can be translated into equivalent statements about sets, which can then be described by Venn diagrams. Hence Venn diagrams are very often used to determine the validity of an argument.

Example 2.1: Consider the argument in Example 1.2. By S_1, the set of babies is a subset of the set of illogical people, i.e.,

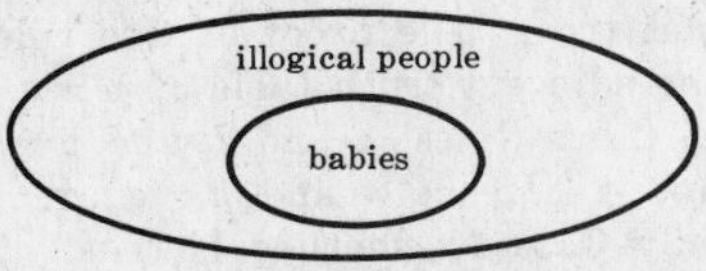

By S_3, the set of illogical people is contained in the set of despised people, i.e.,

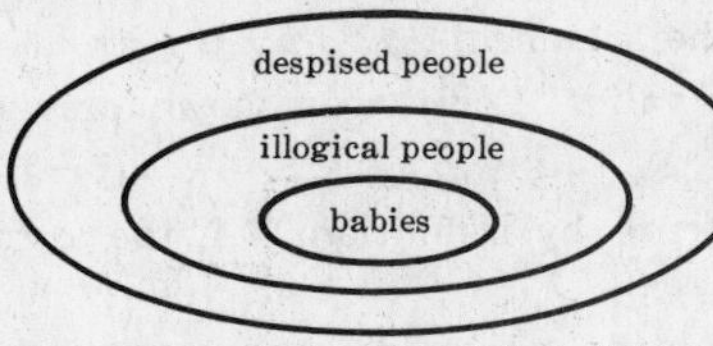

By S_2, the set of despised people and the set of people who can manage a crocodile are disjoint, i.e.,

Note that the set of babies and the set of people who can manage crocodiles are disjoint. In other words, "Babies cannot manage crocodiles" is a consequence of S_1, S_2 and S_3, that is,

$$S_1, S_2, S_3 \vdash S$$

is a valid argument.

ARGUMENTS AND PROPOSITIONS

A statement that a set of propositions $P_1, \ldots, P_n$ yields another proposition P, denoted by

$$P_1, \ldots, P_n \vdash P$$

is called an *argument on propositions* or, simply, an *argument*.

Definition 17.1: An argument on propositions $P_1, \ldots, P_n \vdash P$ is said to be *valid* if P is true whenever $P_1, \ldots, P_n$ are true or, equivalently, if

$$P_1 \wedge \cdots \wedge P_n \Rightarrow P$$

that is, if

$$(P_1 \wedge \cdots \wedge P_n) \rightarrow P$$

is a tautology.

By the Principle of Substitution, propositions can be substituted for variables in any tautology. Therefore,

Theorem 17.1: If the argument

$$P_1(p, q, \ldots), \ \ldots, \ P_n(p, q, \ldots) \ \vdash \ P(p, q, \ldots)$$

is valid, then, for any propositions $P', Q', \ldots$, the argument

$$P_1(P', Q', \ldots), \ldots, P_n(P', Q', \ldots) \ \vdash \ P(P', Q', \ldots)$$

is also valid.

Example 3.1: The argument $p, p \rightarrow q \vdash q$ (Law of Detachment) is valid. In other words, if p and $p \rightarrow q$ are true then q is true. The proof of this rule follows directly from the adjacent truth table of $p \rightarrow q$. Notice that p is true in Cases (lines) 1 and 2, and $p \rightarrow q$ is true in Cases 1, 3 and 4. Hence p and $p \rightarrow q$ are true simultaneously in only Case 1, where q is true. In other words, p and $p \rightarrow q$ yields q is a valid proposition.

p	q	$p \rightarrow q$
T	T	T
T	F	F
F	T	T
F	F	T

Example 3.2: The argument $p \rightarrow q, q \rightarrow r \vdash p \rightarrow r$ (Law of Syllogism) is valid. For it was shown previously that

$$(p \rightarrow q) \wedge (q \rightarrow r) \Rightarrow p \rightarrow r$$

Hence, by Definition 17.1, the argument is valid.

The relationship between valid arguments on propositions and valid arguments in general is as follows:

Basic Principle on Arguments: Consider the argument on propositions

$$P_1(p,q,\ldots),\ \ldots,\ P_n(p,q,\ldots)\ \vdash\ P(p,q,\ldots)$$

If statements $p_0, q_0, \ldots$ are substituted for the variables $p, q, \ldots$, then the argument

$$P_1(p_0,q_0,\ldots),\ \ldots,\ P_n(p_0,q_0,\ldots)\ \vdash\ P(p_0,q_0,\ldots)$$

is valid if and only if the given argument on propositions is valid.

Example 3.3: Consider the argument,

S_1: If a man is a bachelor, he is unhappy.
S_2: If a man is unhappy, he dies young.
....
S: Bachelors die young.

Let p be "He is a bachelor", let q be "He is unhappy", and let r be "He dies young". Then $S_1, S_2 \vdash S$ can be written

$$p \to q,\ q \to r\ \vdash\ p \to r$$

which is a valid argument on propositions (Law of Syllogism, Example 3.2). Consequently, the given argument is valid.

ARGUMENTS AND QUANTIFIERS

Let $p(x)$ be a propositional function on a set A. If

$$(\forall x \in A)\ p(x)$$

is true, then, in particular, $p(x_0)$ is also true for any specific element $x_0 \in A$. Similarly, if $p(x_0)$ is true for a specific element $x_0 \in A$, then the quantified statement

$$(\exists x \in A)\ p(x)$$

is also true. In other words,

Basic Principle on Arguments and Quantifiers: Let $p(x)$ be a propositional function on a set A. Then each of the following arguments is valid:

$$(\forall x \in A)\, p(x),\ x_0 \in A\ \ \vdash\ \ p(x_0)$$
$$x_0 \in A,\ p(x_0)\ \ \vdash\ \ (\exists x \in A)\ p(x)$$

Example 4.1: Consider the classical argument:

S_1: All men are mortal.
S_2: Socrates is a man.
....
S: Socrates is mortal.

Let M be the set of men, let $p(x)$ be "x is mortal", and let x_0 denote Socrates. Then the above argument can be written in the form

S_1: $(\forall x \in M)\ p(x)$
S_2: $x_0 \in M$
....
S: $p(x_0)$

Therefore, by the Basic Principle, the given argument is valid.

CONDITIONAL STATEMENTS AND VARIATIONS

Consider the conditional proposition $p \to q$ and other simple conditional propositions which contain p and q, i.e. $q \to p$, $\sim p \to \sim q$, and $\sim q \to \sim p$, called, respectively, the *converse, inverse,* and *contrapositive* propositions. The truth tables of these four propositions are as follows:

p	q	Conditional $p \to q$	Converse $q \to p$	Inverse $\sim p \to \sim q$	Contrapositive $\sim q \to \sim p$
T	T	T	T	T	T
T	F	F	T	T	F
F	T	T	F	F	T
F	F	T	T	T	T

Note first from the above table that a conditional statement and its converse or inverse are not logically equivalent. The following theorem, though, is a consequence of the above truth table.

Theorem 17.2: A conditional statement $p \to q$ and its contrapositive $\sim q \to \sim p$ are logically equivalent.

Example 5.1: Consider the following statements about a triangle A.

$p \to q$: If A is equilateral, then A is isosceles.
$q \to p$: If A is isosceles, then A is equilateral.

Note that $p \to q$ is true, but $q \to p$ is false.

Example 5.2: Prove: $(p \to q)$ If x^2 is odd then x is odd.

Show that the contrapositive $\sim q \to \sim p$, i.e. "If x is even then x^2 is even", is true. Let x be even; then $x = 2n$ where $n \in N$, the natural numbers. Hence $x^2 = (2n)(2n) = 2(2n^2)$ is also even. Since the contrapositive $\sim q \to \sim p$ is true, the original conditional statement $p \to q$ is also true.

Remark 17.2: Generally speaking, the converse, inverse and contrapositive of a proposition $P(p, q, \ldots) \to Q(p, q, \ldots)$ are, respectively, $Q \to P$, $\sim P \to \sim Q$ and $\sim Q \to \sim P$. Furthermore, by Theorem 17.2 and the Principle of Substitution,

$$P(p, q, \ldots) \to Q(p, q, \ldots) \equiv \sim Q(p, q, \ldots) \to \sim P(p, q, \ldots)$$

Solved Problems

ARGUMENTS AND VENN DIAGRAMS

1. Show that the following arguments are not valid by constructing a Venn diagram in which the premises hold but the conclusion does not hold.

(1) Some students are lazy.
All males are lazy.
. . . .
Some students are males.

(2) All students are lazy.
Nobody who is wealthy is a student.
. . . .
Lazy people are not wealthy.

Solution:

(1) Consider the following Venn diagram:

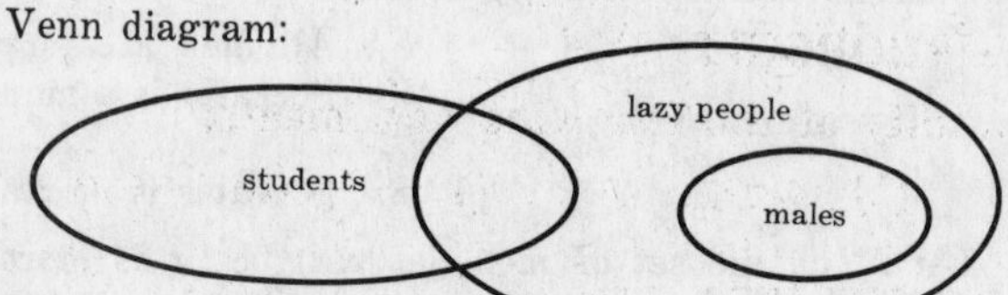

Notice that both premises hold, but the conclusion does not hold.

It is possible to construct a Venn diagram in which the premises and conclusion hold, such as

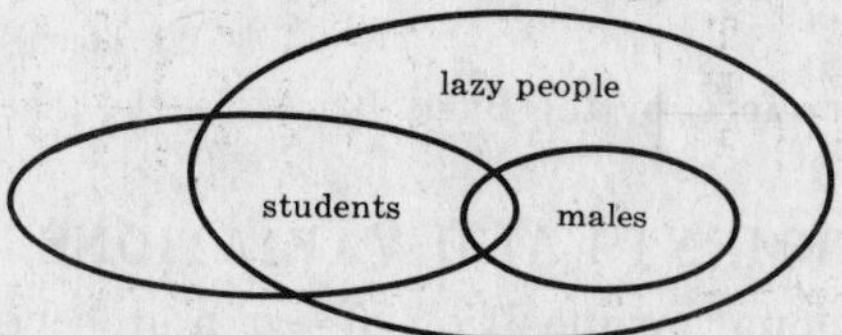

In order for an argument to be valid, the conclusion must always be true whenever the premises are true. Since the first diagram gives a case where the conclusion is not true, even though the premises are true the argument is not valid.

(2) Consider the following Venn diagram:

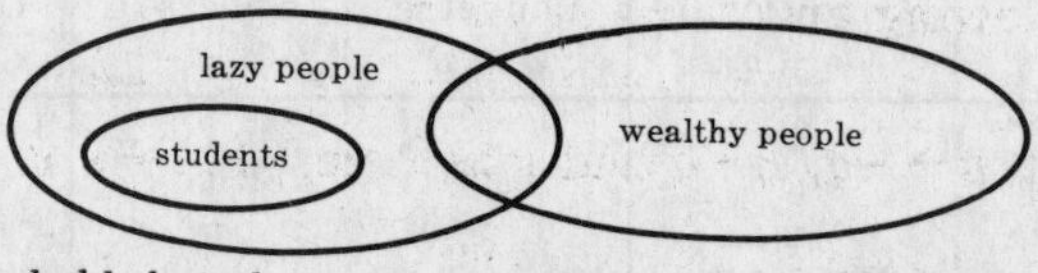

Notice that the premises hold, but the conclusion does not hold; hence the argument is not valid.

2. For each set of premises, find a conclusion such that the argument is valid and such that each premise is necessary for the conclusion.

(1) S_1: No student is lazy.
S_2: John is an artist.
S_3: All artists are lazy.
....
S: .. ..

(2) S_1: All lawyers are wealthy.
S_2: Poets are temperamental.
S_3: Marc is a lawyer.
S_4: No temperamental person is wealthy.
....
S: .. ..

Solution:

(1) By S_3, the set of artists is a subset of the set of lazy people. By S_1, the set of lazy people and the set of students are disjoint. Thus

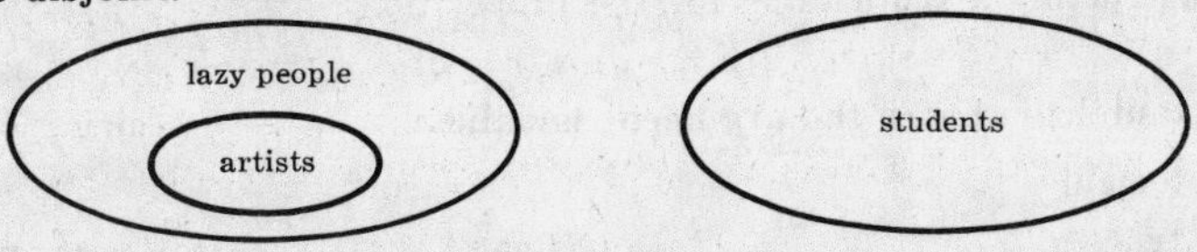

By S_2, John belongs to the set of artists; hence the correct conclusion as indicated by the Venn diagram, is "John is not a student".

(2) By S_1, the set of lawyers is a subset of the set of wealthy people. By S_4, the set of wealthy people and the set of temperamental people are disjoint. Thus

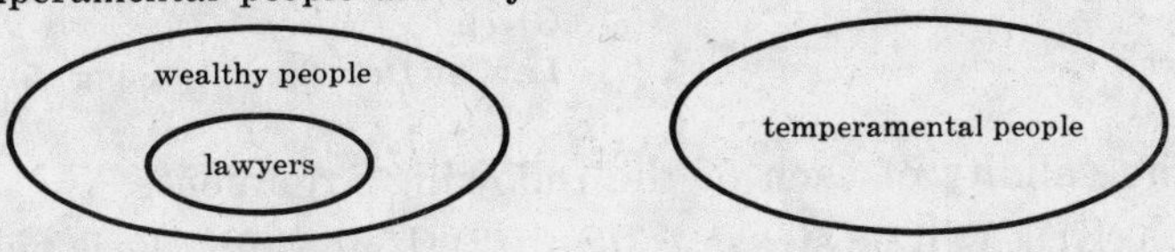

By S_2, the set of poets is a subset of the set of temperamental people, i.e.,

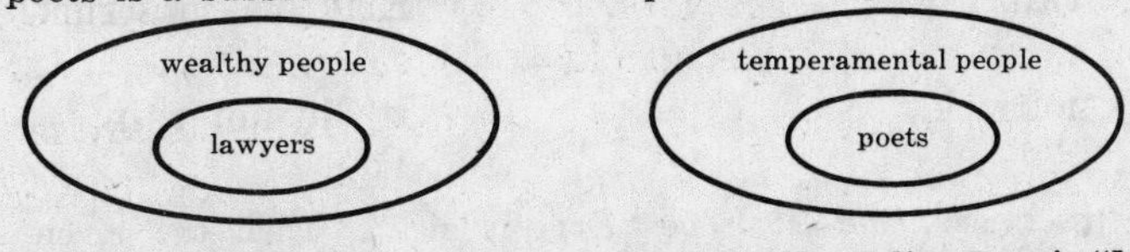

By S_3, Marc is a lawyer; hence the correct conclusion, by the Venn diagram, is "Marc is not a poet".

ARGUMENTS AND PROPOSITIONS

3. Determine the validity of the following arguments:

(1) $p \to q, \sim p \vdash \sim q$ (2) $p \leftrightarrow q, q \vdash p$

Solution:

Construct the necessary truth tables.

p	q	$[(p$	$\to$	$q)$	$\wedge$	$\sim$	$p]$	$\to$	$\sim$	q
T	T	T	T	T	F	F	T	T	F	T
T	F	T	F	F	F	F	T	T	T	F
F	T	F	T	T	T	T	F	F	F	T
F	F	F	T	F	T	T	F	T	T	F
Step		1	2	1	3	2	1	4	2	1

(1)

p	q	$p \leftrightarrow q$
T	T	T
T	F	F
F	T	F
F	F	T

(2)

(1) Since $[(p \to q) \wedge \sim p] \to \sim q$ is not a tautology, $p \to q, \sim p \vdash \sim q$ is a fallacy.

(2) Notice that $p \leftrightarrow q$ is true in Cases (lines) 1 and 4, and q is true in Cases 1 and 3; hence $p \leftrightarrow q$ and q are true simultaneously only in Case 1 where p is also true. Consequently, $p \leftrightarrow q, q \vdash p$ is a valid argument.

4. Prove that the following argument is valid:

$$p \to \sim q, r \to q, r \vdash \sim p$$

In other words, let $p \to \sim q$, $r \to q$ and r be true; then $\sim p$ is true.

Solution:

Method 1. Construct the following truth tables.

	p	q	r	$p \to \sim q$	$r \to q$	$\sim p$
1	T	T	T	F	T	F
2	T	T	F	F	T	F
3	T	F	T	T	F	F
4	T	F	F	T	T	F
5	F	T	T	T	T	T
6	F	T	F	T	T	T
7	F	F	T	T	F	T
8	F	F	F	T	T	T

Notice that $p \to \sim q$, $r \to q$ and r are true simultaneously only in Case (line) 5 where $\sim p$ is also true; hence the given argument is valid.

Method 2. Constructing a truth table for the proposition

$$[(p \to \sim q) \wedge (r \to q) \wedge r] \to \sim p$$

we find it is a tautology; hence the argument is valid.

Method 3.

Statement	Reason
(1) $p \to \sim q$ is true.	(1) Given
(2) $r \to q$ is true.	(2) Given
(3) $\sim q \to \sim r$ is true.	(3) Contrapositive of (2)
(4) $p \to \sim r$ is true.	(4) Law of Syllogism, using (1) and (3)
(5) $r \to \sim p$ is true.	(5) Contrapositive of (4)
(6) r is true.	(6) Given
(7) $\therefore \sim p$ is true.	(7) Law of Detachment, using (5) and (6)

5. Determine the validity of each of the following arguments.

(1) If it rains, Erik will be sick.
It did not rain.
....
Erik was not sick.

(2) If it rains, Erik will be sick.
Erik was not sick.
....
It did not rain.

Solution:

(1) Let p be "It rains", and let q be "Erik is sick". Then the given argument can be written $p \to q, \sim p \vdash \sim q$ which, by Problem 3, is a fallacy. Hence the given argument is a fallacy.

(2) Let p be "It rains", and let q be "Erik is sick". Then the given argument can be written $p \to q, \sim q \vdash \sim p$ which, by constructing a truth table, can be shown to be valid. Hence the given argument is valid.

CONVERSE AND VARIATIONS

6. Find and simplify: (1) Contrapositive of the contrapositive of $p \to q$. (2) Contrapositive of the converse of $p \to q$. (3) Contrapositive of the inverse of $p \to q$.

Solution:

(1) The contrapositive of $p \to q$ is $\sim q \to \sim p$. The contrapositive of $\sim q \to \sim p$ is $\sim\sim p \to \sim\sim q \equiv p \to q$, which is the original conditional proposition.

(2) The converse of $p \to q$ is $q \to p$. The contrapositive of $q \to p$ is $\sim p \to \sim q$, which is the inverse of $p \to q$.

(3) The inverse of $p \to q$ is $\sim p \to \sim q$. The contrapositive of $\sim p \to \sim q$ is $\sim\sim q \to \sim\sim p \equiv q \to p$, which is the converse of $p \to q$.

7. Determine the contrapositive of each statement.

(1) If John is a poet, then he is poor.
(2) Only if Marc studies, will he pass the test.
(3) It is necessary to snow in order for Eric to ski.
(4) If x is less than zero, then x is not positive.

Solution:

(1) Note that the contrapositive of $p \to q$ is $\sim q \to \sim p$. Hence the contrapositive of the given statement is "If John is not poor, then he is not a poet".

(2) The given statement is equivalent to "If Marc passes the test, then he studied". Therefore the contrapositive of the given statement is "If Marc does not study, then he will not pass the test".

(3) The given statement is equivalent to the statement "If Eric skis, then it snowed". Hence the contrapositive is "If it did not snow, then Eric will not ski".

(4) Note that the contrapositive of $p \to \sim q$ is $\sim\sim q \to \sim p \equiv q \to \sim p$. Hence the contrapositive of the given statement is "If x is positive, then x is not less than zero".

Supplementary Problems

8. Determine the validity of each of the following arguments:
(1) $p \to q,\ r \to \sim q \ \vdash\ r \to \sim p$ (2) $p \to \sim q,\ \sim r \to \sim q \ \vdash\ p \to \sim r$

9. For the given premises, determine a suitable conclusion so that the argument is valid.
(1) $p \to \sim q,\ q$ (2) $p \to \sim q,\ r \to q$ (3) $p \to \sim q,\ \sim p \to r$ (4) $p \to \sim q,\ r \to p,\ q$

10. Determine the validity of each of the following arguments for each proposed conclusion.

(1) No college professor is wealthy.
Some poets are wealthy.
___
(*a*) Some poets are college professors.
(*b*) Some poets are not college professors.

(2) All poets are interesting people.
Audrey is an interesting person.
___
(*a*) Audrey is a poet.
(*b*) Audrey is not a poet.

(3) All poets are poor.
In order to be a teacher, one must graduate from college.
Some mathematicians are poets.
No college graduate is poor.
___
(*a*) Some mathematicians are not teachers.
(*b*) Some teachers are not mathematicians.
(*c*) Teachers are not poor.
(*d*) Some mathematicians are not poor.
(*e*) Poets are not teachers.
(*f*) If Marc is a college graduate then he is not a poet.

(4) All mathematicians are interesting people.
Some teachers sell insurance.
Some philosophers are mathematicians.
Only uninteresting people become insurance salesmen.
___
(*a*) Some philosophers are not insurance salesmen.
(*b*) Insurance salesmen are not mathematicians.
(*c*) Some interesting people are not teachers.
(*d*) Some teachers are not philosophers.
(*e*) Some teachers are not interesting people.

11. Find: (1) Contrapositive of $p \to \sim q$.
(2) Contrapositive of $\sim p \to q$.
(3) Contrapositive of the converse of $p \to \sim q$.
(4) Converse of the contrapositive of $\sim p \to \sim q$.

12. Find the contrapositive of each of the following statements:
(1) If he has courage, he will win.
(2) It is necessary to be strong, in order to be a sailor.
(3) Only if he does not tire, will he win.
(4) It is sufficient for it to be a square, in order to be a rectangle.

Answers to Supplementary Problems

8. (1) valid (2) fallacy

9. (1) $\sim p$ (2) $p \to \sim r$ (3) $q \to r$ (4) $\sim r$

10. (1) (*a*) fallacy, (*b*) valid
(2) (*a*) fallacy, (*b*) fallacy
(3) (*a*) valid, (*b*) fallacy, (*c*) valid, (*d*) fallacy, (*e*) valid, (*f*) valid
(4) (*a*) valid, (*b*) valid, (*c*) fallacy, (*d*) fallacy, (*e*) valid

11. (1) $q \to \sim p$ (2) $\sim q \to p$ (3) $\sim p \to q$ (4) $p \to q$

12. (1) If he does not win, then he does not have courage.
(2) If he is not strong, then he is not a sailor.
(3) If he tires, then he will not win.
(4) If it is not a rectangle, it is not a square.

INDEX